SWEDEN

Sk.	Skåne	Vrm.	Värmland
Bl.	Blekinge	Dlr.	Dalarna
Hall.	Halland	Gstr.	Gästrikland
Sm.	Småland	Hls.	Hälsingland
Øl.	Øland	Med.	Medelpad
Gtl.	Gotland	Hrj.	Härjedalen
G. Sand.	Gotska Sandön	Jmt.	Jämtland
Øg.	Østergötland	Äng.	Ångermanland
Vg.	Västergötland	Vb.	Västerbotten
Boh.	Bohuslän	Nb.	Norrbotten
Dlsl.	Dalsland	Ås. Lpm.	Åsele Lappmark
Nrk.	Närke	Ly. Lpm.	Lycksele Lappmark
Sdm.	Södermanland	P. Lpm.	Pite Lappmark
Upl.	Uppland	Lu. Lpm.	Lule Lappmark
Vstm.	Västmanland	T. Lpm.	Torne Lappmark

NORWAY

Ø	Østfold	HO	Hordaland
AK	Akershus	SF	Sogn og Fjordane
HE	Hedmark	MR	Møre og Romsdal
O	Opland	ST	Sør-Trøndelag
B	Buskerud	NT	Nord-Trøndelag
VE	Vestfold	Ns	southern Nordland
TE	Telemark	Nn	northern Nordland
AA	Aust-Agder	TR	Troms
VA	Vest-Agder	F	Finnmark
R	Rogaland		

n northern   s southern   ø eastern   v western   y outer   i inner

FINLAND

Al	Alandia	Kb	Karelia borealis
Ab	Regio aboensis	Om	Ostrobottnia media
N	Nylandia	Ok	Ostrobottnia kajanensis
Ka	Karelia australis	ObS	Ostrobottnia borealis, S part
St	Satakunta	ObN	Ostrobottnia borealis, N part
Ta	Tavastia australis	Ks	Kuusamo
Sa	Savonia australis	LkW	Lapponia kemensis, W part
Oa	Ostrobottnia australis	LkE	Lapponia kemensis, E part
Tb	Tavastia borealis	Li	Lapponia inarensis
Sb	Savonia borealis	Le	Lapponia enontekiensis

USSR

Vib Regio Viburgensis   Kr Karelia rossica   Lr Lapponia rossica

FAUNA ENTOMOLOGICA SCANDINAVICA

Volume 7, part 3   1983

The Auchenorrhyncha (Homoptera) of Fennoscandia and Denmark

Part 3: The Family Cicadellidae: Deltocephalinae, Catalogue, Literature and Index

by

F. Ossiannilsson

SCANDINAVIAN SCIENCE PRESS LTD.

Copenhagen . Denmark

© *Copyright*
Scandinavian Science Press Ltd. 1983

Fauna entomologica scandinavica
is edited by »Societas entomologica scandinavica«

Editorial board
Nils M. Andersen, Karl-Johan Hedqvist, Hans Kauri,
N. P. Kristensen, Harry Krogerus, Leif Lyneborg

Managing editor
Leif Lyneborg

World list abbreviation
Fauna ent. scand.

Printed by
Vinderup Bogtrykkeri A/S
7830 Vinderup, Denmark

ISBN 87-87491-13-3
ISSN 0106-8377

Contents

SUBFAMILY DELTOCEPHALINAE

Small or medium-sized leafhoppers, often wing-polymorphous. Longitudinal veins of corium distinct even basally (in macropterous and sub-brachypterous specimens). Anterior outline of head as seen from above angular or rounded. Ocelli situated on fore border of head or immediately above or below it, usually close to eyes. Males with a distinct genital valve. Usually on grasses or other herbs, rarely on trees and bushes. In Denmark and Fennoscandia 67 genera belonging to 9 tribes (in concurrence with Emeljanov, 1962, 1964; Anufriev, 1978).

Key to tribes of Deltocephalinae

1	Aedeagus with two shafts or with two phallotremes on branches of the same shaft (Text-figs. 1951, 1960)	**Opsiini** (p. 609)
–	Aedeagus with one phallotreme	2
2 (1)	Connective forked, branches diverging or parallel, their apices not close to each other (Text-figs. 1966, 1983, 2140, 2179)	3
–	Connective with apices of branches fused or close to each other (Text-figs. 2604, 2647, 2676, 2718)	7
3 (2)	Ocelli approximately equidistant between eye and median line of head. Aedeagus asymmetrical (Text-fig. 1943)	**Grypotini** (p. 607)
–	Ocelli closer to eyes than to median line of head	4
4 (3)	Genital plates of male each with a row of flattened, apically truncate macrosetae along lateral margin (Text-fig. 1965)	**Coryphaelini** (p. 613)
–	Macrosetae of genital plates of ordinary aspect, with pointed apices	5
5 (4)	Fore wings with three subapical cells (Text-fig. 2281)	**Athysanini** (p. 668)
–	Fore wings two subapical cells (Text-fig. 1937). Genital plates apically produced into long, weakly sclerotized appendages (Text-figs. 1989, 2072, 2089)	6
6 (5)	Vertex medially approximately as long as near eyes. Sc and R in hind wings apically united into a single vein at apex (Text-fig. 1939)	**Balcluthini** (p. 615)
–	Vertex medially longer than near eyes. Sc and R in hind wings distinct at apex (Text-fig. 1938)	**Macrostelini** (p. 621)
7 (2)	Shaft of aedeagus articulated with socle, short; phallotreme large, subapical (Text-figs. 2117, 2118). Apices of branches of connective close to each other; connective long, base long (Text-fig. 2124)	**Doraturini** (p. 661)

594

–	Shaft of aedeagus firmly fused with socle, or phallotreme basal 8
8 (7)	Connective fused with socle of aedeagus (Text-figs. 2098, 2111). Pronotum laterally carinate .... **Deltocephalini** (p. 655)
–	Connective normally not fused with socle. Connective with branches apically fused (Text-figs. 2604, 2647, 2676, 2718, 2802 etc. Pronotum laterally rarely carinate .... **Paralimnini** (p. 794)

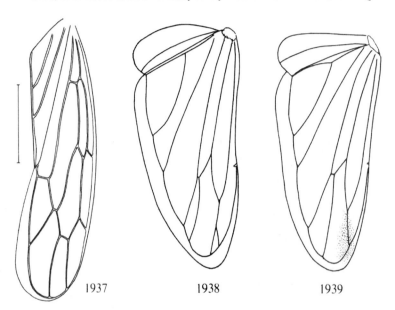

Text-figs. 1937-1939. – 1937: *Balclutha lineolata* (Horváth), right fore wing; 1938: *Macrosteles* sp., right hind wing; 1939: *Balclutha punctata* (Fabricius), right hind wing. Scale: 1 mm.

Key to genera of Deltocephalinae

1	Vertex concave 2
–	Vertex flat or convex 6
2 (1)	Vertex behind fore margin or on disk with a distinct transverse depression. Distance between eye and ocellus 2-3 × diameter of ocellus 3
–	Vertex without a distinct transverse depression behind fore margin. Each ocellus situated close to eye; distance to eye roughly equals diameter of ocellus. Transition between vertex and face abrupt .... *Platymetopius* Burmeister (p. 668)
3 (2)	Fore wings with some secondary transverse veins at least in clavus .... *Parapotes* Emeljanov (p. 796)

67 (66) Aedeagus apically with two pairs of thin appendages
(Text-figs. 2727, 2728) .... *Adarrus* Ribaut (p. 833)
– Aedeagus apically with one pair of thin appendages, toge-
ther with shaft forming an Y (Text-fig. 2509) .... *Laburrus* Ribaut (p. 767)
68 (65) Connective long, Y-shaped, base longer than branches
(Text-fig. 2423). Style with apical apophyse finger-like,
apically not curled. Aedeagus simple, without processes, phallotreme terminal. Apodemes of 2nd. abdo-
minal sternum small but distinct (Text-fig. 2428) .. *Doliotettix* Ribaut (p. 744)
– Connective not Y-shaped 69
69 (68) Apodemes of 2nd abdominal sternum long (Text-fig.
2624). Base of connective long. Styles apically not curled.
Aspect as in Plate-fig. 192 .... *Paralimnus* Matsumura (p. 798)
– Apodemes of 2nd abdominal sternum short or vestigial.
Connective with base very short, branches apically fused 70
70 (69) Styles with apical margin of apophyse curled. Apodemes
of 2nd abdominal sternum vestigial .... *Jassargus* Zachvatkin (p. 840)
– Apical margin of apophyse of style not curled. Apodemes
of 2nd abdominal sternum small but distinct .... *Sorhoanus* Ribaut (p. 864)
71 (20) Head not wider than pronotum 72
– Head wider than pronotum 81
72 (71) 7th abdominal sternum caudally with a semicircular
scale (Text-fig. 2450). Aspect as in Plate-fig. 183 *Stictocoris* Thomson (p. 750)
– 7th abdominal sternum and aspect different 73
73 (72) Caudal margin of 7th abdominal sternum medially deeply
concave (Text-fig. 2380). Anterior tibiae with one seta in
inner, 5 in outer dorsal row. Aspect as in Plate-fig. 178 ....
.... *Hesium* Ribaut (p. 730)
– Caudal margin of 7th abdominal sternum not deeply
concave medially 74
74 (73) Caudal margin of 7th abdominal sternum medially with
a distinct angular incision (Text-fig. 2173). Anteclypeus
widening towards apex. Aspect as in Plate-fig. 221 ... *Colladonus* Ball (p. 675)
– Caudal margin of 7th abdominal sternum without a dis-
tinct angular incision medially 75
75 (74) Caudal margin of 7th abdominal sternum medially with
a small angular projection (Text-fig. 2517). Fore tibiae
with 3 setae in inner, 4 in outer dorsal row. Resembling
Euscelis. (Plate-fig. 188) .... *Euscelidius* Ribaut (p. 769)
– Caudal margin of 7th abdominal sternum without an an-
gular median projection 76
76 (75) Caudal margin of 7th abdominal sternum medially mo-
derately convex (Text-figs. 2457, 2468, 2474). Upper side
normally with well-defined black markings (Plate-figs. 184, 185) ....

–   Caudal margin of 7th abdominal sternum without two
        crescent-shaped black spots near middle 86
86 (85) 7th abdominal sternum concave, caudal margin with a
        deep narrow median incision (Text-figs. 2825, 2829) ....
        .... *Arthaldeus* Ribaut (p. 859)
–   7th abdominal sternum without a median incision 87
87 (86) Caudal margin of 7th abdominal sternum with a small
        median projection (Text-fig. 2429) .... *Doliotettix* Ribaut (p. 744)
–   Shape of 7th abdominal sternum different 88
88 (87) Caudal margin of 7th abdominal sternum medially with
        two small projections or a bilobed projection surro-
        unded by a large black patch (Text-figs. 2812, 2819)   *Verdanus* Oman (p. 855)
–   Caudal margin of 7th abdominal sternum different 89
89 (88) Caudal margin of 7th abdominal sternum trilobed (Text-
        figs. 2912, 2920) .... *Mocuellus* Ribaut (p. 881)
–   Caudal margin of 7th abdominal sternum roughly
        straight without incisions or projections 90
90 (89) Larger, 4.7-5.2 mm. Fore tibiae with 3 or 4 setae in inner,
        4 in outer dorsal row .... *Laburrus* Ribaut (p. 767)
–   Smaller, 2.7-3.7 mm. Fore tibiae with one seta in inner,
        4 in outer dorsal row .... *Psammotettix cephalotes* (H.-S.) (p. 816)
91 (81) Upper side red. Caudal margin of 7th abdominal ster-
        num as in Text-fig. 2812 ... *Verdanus abdominalis* var. *rufus* (Sahlberg) (p. 855)
–   Not red, 7th abdominal sternum different 92
92 (91) 7th abdominal sternum deeply and broadly excavatex,
        not covering base of ovipositor 93
–   7th abdominal sternum covering base of ovipositor 96
93 (92) Anterior tibiae with 3 setae in inner, 4 in outer dorsal row 94
–   Anterior tibiae with one seta in inner, 3 in outer dorsal
        row .... *Limotettix* Sahlberg (p. 761)
94 (93) Vertex and pronotum with black markings (Plate-fig. 173) ....
        .... *Rhopalopyx* Ribaut (p. 695)
–   Vertex and pronotum without black markings 95
95 (94) Yellow. Vertex frontally rounded .... *Paluda* DeLong (p. 701)
–   Greyish. Vertex frontally angular .... *Rhopalopyx vitripennis* (Flor) (p. 699)
96 (92) Anterior tibiae with 1 seta in inner, 3 in outer dorsal
        row. Head much wider than pronotum, frontally broadly
        rounded (Plate-fig. 186) .... *Ophiola,* subg. *Ophiolix* Ribaut (p. 753)
–   Chaetation of anterior tibiae different 97
97 (96) Markings of vertex as in Plate-fig. 207. Anterior tibiae
        with 3 setae in inner, 4 in outer dorsal row .... *Hardya* Edwards (p. 693)
–   Markings of vertex different 98
98 (97) Anteclypeus widening towards apex 99
–   Anteclypeus not widening towards apex 100

99 (98) 7th abdominal sternum with semicircular excavation. Fore wings narrowing apically. General aspect more or less as in Plate-fig. 177 (spots in clavus often absent) .... *Mocydiopsis* Ribaut (p. 724)

– 7th abdominal sternum different. Elongate species, always macropterous with fore wings long, apically evenly rounded, antennae long .... *Cicadula* Zetterstedt (p. 708)

100 (98) 7th abdominal sternum with trapezoidal excavation, with a black spot near frontal corners of excavation (Text-fig. 2798) .... *Pinumius* Ribaut (p. 851)

– Caudal margin of 7th abdominal sternum different 101

101 (100) Caudal margin of 7th abdominal sternum practically straight 102

– Caudal margin of 7th abdominal sternum with distinct processes or incisions 104

102 (101) Anterior tibiae with 3-4 setae in inner, 4 in outer dorsal row. Broad, robust species, aspect more or less as in Plate-fig. 189 .... *Euscelis* Brullé (p. 773)

– Chaetation of anterior tibiae different 103

103 (102) Upper part of face with a broad black transverse band extending from eye to eye. General aspect as in Plate-fig. 193 .... *Paralimnus rotundiceps* (Lethierry) (p. 801)

– Pigmentation of face and aspect different .... *Psammotettix* Haupt (p. 811)

104 (101) 7th abdominal sternum with distinct median incision 105

– 7th abdominal sternum not incised medially (sometimes with two-pointed median projection) 109

105 (104) Anteclypeus widening towards apex. Large and elongate, vertex frontally rounded and with a black band along fore border (Plate-fig. 191) .... *Paramesus* Fieber (p. 794)

– Anteclypeus not widening towards apex 106

106 (105) Fore wing with a usually conspicuous black spot at apex of 3rd subapical cell (Plate-fig. 195). Incision in 7th abdominal sternum shallow, obtusely rounded (Text-fig. 2652) .... *Arocephalus punctum* (Flor) (p. 806)

– Fore wing without a conspicuous black spot at apex of 3rd subapical cell. Incision angular, well-marked 107

107 (106) 7th abdominal sternum convex caudally (Text-fig. 2905) .... *Boreotettix* Lindberg (p. 879)

– 7th abdominal sternum not convex caudally 108

108 (107) Usually sub-brachypterous, fore wings considerably shorter than abdomen, apically obliquely rounded. Head distinctly longer than pronotum, fore margin rectangular .... *Turrutus* Ribaut (p. 836)

– Macropterous, fore wings not or slightly shorter than

abdomen, apically rounded. Head not or only slightly
longer than pronotum, fore margin obtusely angular ....
.... *Cosmotettix* Ribaut, s.str. (p. 870)

109 (104) 7th abdominal sternum medially with a two-pointed pro-
cess 110

– Median process of 7th abdominal sternum, if present,
not two-pointed 114

110 (109) Process very long, almost parallel-sided (Text-fig. 2858) ....
.... *Lebradea* Remane (p. 868)

– Process shorter 111

111 (110) Lobes of process long, narrow, sharp-pointed (Text-fig.
2790) .... *Mendrausus* Ribaut (p. 850)

– Lobes of process short, blunt 112

112 (111) Median incision in projection shallow. 7th abdominal
sternum with a large black patch (Text-fig. 2805). Vertex
longer than pronotum .... *Diplocolenus* Ribaut (p. 853)

– Median incision in projection deeper 113

113 (112) Caudal margin of 7th abdominal sternum convex (Text-
fig. 2905). Elongate and small, overall length 3.7-3.9 mm ....
.... *Boreotettix* Lindberg (p. 879)

– Caudal margin of 7th abdominal sternum not convex
(Text-fig. 2419). Broad and robust, larger, overall length
4.25-5.5 mm. (Plate-fig. 180) .... *Macustus* Ribaut (p. 741)

114 (109) Caudal margin of 7th abdominal sternum with 3 projec-
tions 115

– Caudal margin of 7th abdominal sternum with one pro-
jection or divided into 3 or 4 lobes by incisions 118

115 (114) Median projection rounded, lateral projections pointed,
incisions between them deep (Text-fig. 2723). .... *Ebarrius* Ribaut (p. 830)

– Projections broadly rounded, incisions between them
shallow 116

116 (115) Elongate and long-winged. Vertex along fore border with
a black line from eye to eye. Aspect as in Plate-fig. 192 ....
.... *Paralimnus phragmitis* (Boheman) (p. 800)

– Small, oblong-ovate. No black line present along fore
border of vertex 117

117 (116) Fore wings usually black or brown with well-defined light
spots, veins entirely dark (Plate-fig. 158) .... *Neoaliturus* Distant (p. 611)

– Fore wings with diffuse markings or unicolorous light.
Veins light. Vertex with black or brown markings (Plate-
figs. 165, 166) .... *Deltocephalus* Burmeister (p. 655)

118 (114) Caudal margin of 7th abdominal sternum divided into 3
or 4 lobes by incisions 119

– Caudal margin of 7th abdominal sternum with one me-

Tribe Grypotini
Genus *Grypotes* Fieber, 1866

Grypotes Fieber, 1866a: 503.
Type-species: *Jassus puncticollis* Herrich-Schäffer, 1834, by monotypy.

Body elongate. Head conspicuously wider than pronotum, medially not longer than near eyes. Transition between vertex and face evenly rounded. Frontoclypeus strongly narrowing towards apex. Lower part of face flattened. Anteclypeus narrow, apically bent caudad. Distance from an ocellus to nearest eye equals approximately distance from ocellus to median line of vertex. Aedeagus asymmetrical. Comparatively small leafhoppers. In Fennoscandia and Denmark one species.

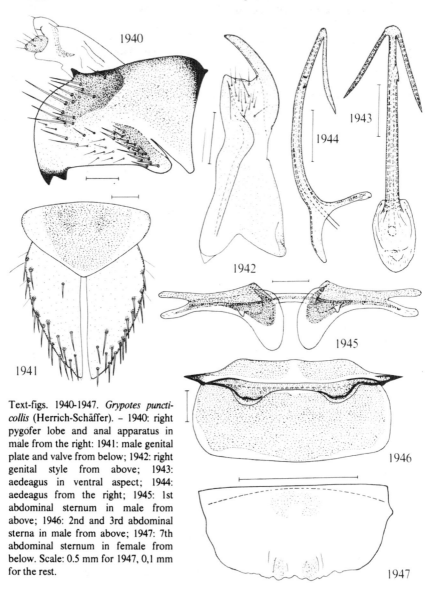

Text-figs. 1940-1947. *Grypotes puncticollis* (Herrich-Schäffer). – 1940: right pygofer lobe and anal apparatus in male from the right: 1941: male genital plate and valve from below; 1942: right genital style from above; 1943: aedeagus in ventral aspect; 1944: aedeagus from the right; 1945: 1st abdominal sternum in male from above; 1946: 2nd and 3rd abdominal sterna in male from above; 1947: 7th abdominal sternum in female from below. Scale: 0.5 mm for 1947, 0,1 mm for the rest.

273. **Grypotes puncticollis** (Herrich-Schäffer, 1834)
Plate-fig. 157, text-figs. 1940-1947.

Jassus puncticollis Herrich-Schäffer, 1834d: 7.
Jassus pinetellus Zetterstedt, 1840: 1077.

Brownish yellow with a greyish green tinge, shining. Head anteriorly with two parallel, medially arched, fuscous transverse bands, one above and one below ocelli. Laterally these bands reach eyes. The lower band usually interrupted on each side near frontal suture; also the upper band sometimes decomposing in a similar way. On vertex a fuscous spot near each eye. Below each antenna a small spot. Frontoclypeus with fuscous transverse streaks, the uppermost pair usually broader than the rest. Also genae and lora may be spotted. Pronotum with fine transverse striations, anteriorly with a row of small fuscous spots along fore border. Scutellum with 2 small dark spots in front of its transverse suture. Fore wings longer than abdomen, greenish grey with concolorous veins. Veins of hind wings dark. Thoracic venter largely light, dark spotted, with longitudinal dark streaks, hind tibiae with spines arising from dark points. Abdomen in male black with lateral and caudal margins light, pygofer lobes and genital plates yellowish. Female abdomen largely light. Male pygofer lobes and anal apparatus as in Text-fig. 1940, genital plate and valve as in Text-fig. 1941, genital style as in Text-fig. 1942, aedeagus as in Text-figs. 1943, 1944, basal abdominal sterna in male as in Text-figs. 1945, 1946, 7th abdominal sternum in female as in Text-fig. 1947. Overall length of males 4.0-5.0 mm, of females 4.6-5.0 mm. – Last instar larva uniformly brownish, head considerably wider than pronotum, abdomen with longer setae only on tergites VII and VIII.

Distribution. Common and widespread in Denmark, found in SJ, EJ, WJ, LFM, SZ, NEZ, and B. – Common in southern and central Sweden, Sk.-Dlr. – Norway: found in AK, Bø, VAy. – Moderately common in East Fennoscandia, found in Al, Ab, N, Sa, Oa; Kr. – Widespread in Europe, also recorded from Algeria and Tunisia.

Biology. On *Pinus silvestris* and *nigricans* (Melichar, 1896). Especially on large solitary trees (Linnavuori, 1959). Hibernation takes place in the egg stage (Müller, 1957). Adults in July-October.

Tribe Opsiini
Genus *Opsius* Fieber, 1866

Opsius Fieber, 1866a: 505.
Type-species: *Opsius stactogalus* Fieber, 1866, by monotypy.

Fairly elongate, in front broad, behind tapering. Fore border of head arched, head in dorsal aspect medially not distinctly longer than near eyes. Face short and broad, frontoclypeus only slightly longer than maximal width. Anteclypeus approximately parallel-sided. Transition between vertex and face smoothly rounded. Aedeagus with 2 shafts and 2 phallotremes. In Denmark and Fennoscandia one species.

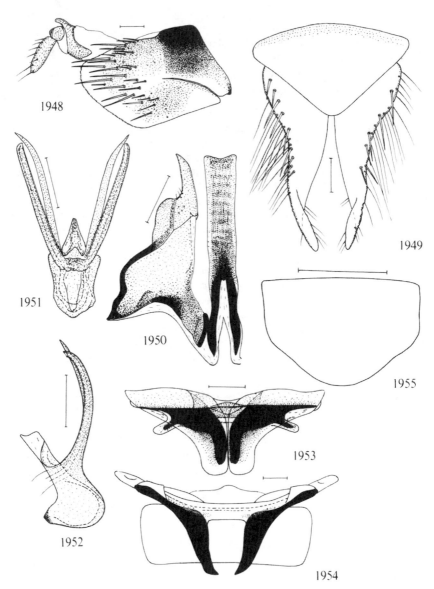

Text-figs. 1948-1955. *Opsius stactogalus* Fieber. – 1948: right pygofer lobe and anal apparatus in male from the right; 1949: male genital plates and valve from below; 1950: connective and right genital style from above; 1951: aedeagus in ventral aspect; 1952: aedeagus from the left; 1953: 1st abdominal sternum in male from above; 1954: 2 nd and 3rd abdominal sterna in male from above; 1955: 7th abdominal sternum in female from below (depressed under coverglass). Scale: 0.5 mm for 1955, 0,1 mm for the rest.

274. **Opsius stactogalus** Fieber, 1866
Plate-fig. 212, text-figs. 1948-1955.

Opsius stactogalus Fieber, 1866a: 505.

Green, shining. In male, head and anterior part of pronotum as well as venter yellowish green, caudal part of pronotum and fore wings except apical part being intensely green, almost verdigris green, with concolorous veins and more or less densely sprinkled with small whitish dots; apical part of fore wing hyaline or brownish. Legs green, abdominal dorsum black with green segmental borders. Female usually lighter and more uniformly green, sometimes almost as male. Tarsi in both sexes apically darker. Pygofer lobes and anal complex in male as in Text-fig. 1948, male genital plates and valve as in Text-fig. 1949, connective and genital style as in Text-fig. 1950, aedeagus as in Text-figs. 1951, 1952, 1st-3rd abdominal sterna in male as in Text-figs. 1953, 1954, 7th abdominal sternum in female as in Text-fig. 1955. Overall length of males 4.0-4.4 mm, of females 4.3-4.7 mm. – Larva light, not or weakly pigmented, head wider than pronotum, setae on abdominal dorsum stump, very short, inconspicuous, fore body bald.

Distribution. So far not found in Denmark, nor in East Fennoscandia. – Sweden: Sk.: Lund and Åkarp (Ossiannilsson). – Norway: AK: Oslo, Botanical Garden 25.VII.1960 (Ossiannilsson). – England, Central and southern Europe including s. Russia, n. Africa, Iran, Armenia, Georgia, Tadzhikistan; Nearctic region.

Biology. On *Tamarix*, adults in July-October. Hibernation takes place in the egg stage (Müller, 1956, 1957).

Economic importance. "In Europa und (eingeschleppt) Nordamerika an Tamarisken weit verbreitet (besonders an *Tamarix gallica*) und bisweilen beträchtlich schädlich" (Müller, 1956).

Genus *Neoaliturus* Distant, 1918

Neoaliturus Distant, 1918: 63.
Type-species: *Aliturus gardineri* Distant, 1908, by original designation.
Circulifer Zachvatkin, 1935: 111.
Type-species: *Jassus haematoceps* Mulsant & Rey, 1855, by original designation.
Distomotettix Ribaut, 1938a: 97.
Type-species: *Jassus fenestratus* Herrich-Schäffer, 1834, by original designation.

Small leafhoppers. Head as broad as pronotum or a little broader. Vertex short. Anteclypeus proximally somewhat compressed. frontoclypeus elongate. Transition between vertex and face rounded. Socle of aedeagus absent or indistinctly delimited. Shaft of aedeagus forked with two phallotremes. In Fennoscandia one species.

275. **Neoaliturus fenestratus** (Herrich-Schäffer, 1834)
Plate-fig. 158, text-figs. 1956-1963.

Jassus fenestratus Herrich-Schäffer, 1834a: 5.
Jassus (Deltocephalus) guttulatus Kirschbaum, 1868b: 128.

Vertex shorter than pronotum, head as seen from above obtusely angular. Face, vertex

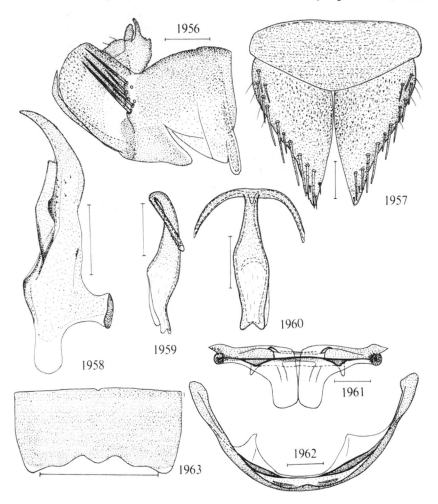

Text-figs. 1956-1963. *Neoaliturus fenestratus* (Herrich-Schäffer). – 1956: right pygofer lobe and anal complex in male from the right; 1957: male genital plates and valve from below; 1958: right genital style from above; 1959: aedeagus from the left; 1960: aedeagus in ventral aspect; 1961: 1st abdominal sternum in male from above; 1962: 2nd abdominal sternum in male in frontal aspect; 1963: 7th abdominal sternum in female from below (depressed under coverglass). Scale: 0.5 mm for 1963, 0.1 mm for the rest.

and scutellum appearing dull by very fine surface sculpture, pronotum and fore wings shining. Pronotum transversely striate. In typical *fenestratus* (Plate-fig. 158), the general colour is black, face with brownish yellow transverse streaks, fore wings with white hyaline spots in clavus and in their apical part as illustrated in the plate-figure. Hind legs black. Fore and middle legs entirely yellow (♀), or tibiae and tarsi yellowish (♂). There exists a considerable individual variation in colour (= *guttulatus* Kbm.), the general colour being paler, light spots larger and more numerous, or general colour light, spots dark. Male pygofer lobes and anal complex as in Text-fig. 1956 (the shape of the pygofer process may be somewhat varying), male genital valve and plates as in Text-fig. 1957, genital styles as in Text-fig. 1958, aedeagus as in Text-figs. 1959, 1960, 1st and 2nd abdominal sterna in male as in Text-figs. 1961, 1962, 7th abdominal sternum in female as in Text-fig. 1963. Overall length of males 2.8-3.3 mm, of females 3.0-3.4 mm. – Nymphs, see Walter (1978).

Distribution. Not found in Denmark, Sweden and Norway. – In East Fennoscandia only found in Kr: Kuujärvi, Sept. 10, 1941 (Tiensuu). – Widespread in Europe (not in Great Britain), North Africa, West, Central and East Asia.

Biology. "An *Helichrysum arenarium*" (Kuntze, 1937). "Auf Trockenrasen" (Wagner & Franz, 1961). "Stenotope Art der Trockenrasen"; "Imaginal-Überwinterer". Two generations (Schiemenz, 1969b). In dry sandy places 7. VII-23.VIII (Vilbaste, 1974).

Tribe Coryphaelini
Genus *Coryphaelus* Puton, 1886

Coryphaeus Fieber, 1866a: 503 (nec Gistl, 1847).
    Type-species: *Jassus gyllenhalii* Fallén, 1826, by monotypy.
Coryphaelus Puton, 1886: 81 (n.n.).
    Type-species: *Jassus gyllenhalii* Fallén, 1826, by monotypy.

Body broad, robust. Head with eyes as broad as pronotum, fore border faintly curved, parallel to hind border. Transition between face and vertex smoothly rounded. Eyes large, protruding. Fore wings approximately as long as abdomen, surface of cells transversely wrinkled. Male genital plates with short and broad truncate macrochaetae. – Monotypic.

276. *Coryphaelus gyllenhalii* (Fallén, 1826)
    Plate-fig. 159, text-figs. 1964-1973

Jassus Gyllenhalii Fallén, 1826: 61.

Upper side yellow with extended black markings. Fore body largely dull, fore wings shining. Head above with a pair of large black spots usually covering major part of vertex and each more or less completely enclosing an ocellus. Frontoclypeus above with a pair of black spots consisting of confluent transverse streaks. Below these there are

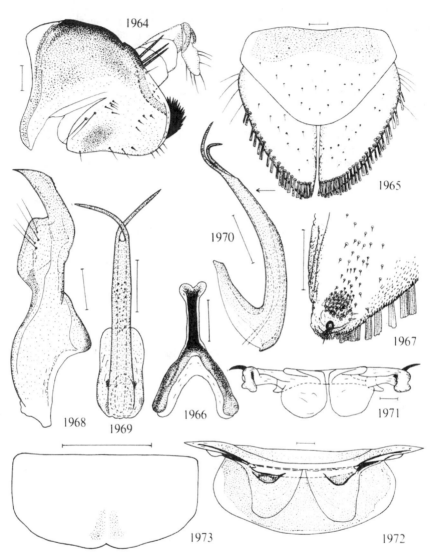

Text-figs. 1964-1973. *Coryphaelus gyllenhalii* (Fallén). – 1964: left pygofer lobe and anal apparatus in male from the left; 1965: male genital plates and valve from below; 1966: connective from above; 1967: apex of right genital plate from above; 1968: right genital style from above; 1969: aedeagus in ventral aspect (in direction of arrow in Text-fig. 1970); 1970: aedeagus from the left, 1971: 1st abdominal sternum in male from above; 1972: 2nd and 3rd abdominal sterna in male from above; 1973: 7th abdominal sternum in female from below (depressed under coverglass). Scale: 0.5 mm for 1973, 0.1 mm for the rest.

three longitudinal black streaks, the lateral ones consisting of confluent rests of transverse streaks, the median one continuing on anteclypeus. Also genae and lora partly black. Pronotum with two black transverse bands, one along fore border and one caudally of middle. In very light specimens, these bands may be reduced in size or disintegrated into spots. Anterior part of scutellum black. Fore wings between veins with broad black longitudinal bands usually not completely filling the cell space. Apical part of fore wing fumose. Legs often partly orange-coloured, with black longitudinal streaks, setae of hind tibiae arising from black spots. Apices of hind tibiae and of hind tarsal segments black, also tarsi of fore and middle legs partly black. Abdomen black with yellow lateral margins, segment borders yellow or whitish; abdominal venter in females often largely yellow. Male genital valve entirely or partly yellow, female genital segment yellow. Male pygofer and anal apparatus as in Text-fig. 1964, genital plates and valve as in Text-figs. 1965, 1967, connective as in Text-fig. 1966, genital style as in Text-fig. 1968, aedeagus as in Text-figs. 1969, 1970, 1st and 2nd abdominal sterna in male as in Text-figs. 1971, 1972, 7th abdominal sternum in female as in Text-fig. 1973. Overall length (♂♀) 4.5-6 mm. – Larvae yellow, fore body black spotted, abdomen transversely black-banded. "Abdomen with hairs on all segments, hairs not obviously arranged in rows; occasionally, longitudinal rows of hairs may be present on last three tergites, amongst irregularly distributed hairs; additional sclerite paired" (Vilbaste, 1982).

Distribution. So far not found in Denmark, nor in Norway. – Sweden: not uncommon, found in Sk., Hall., Sm., Öl., Ög., Vg., Dlr., Gstr., Hls. – Fairly common in southern and central East Fennoscandia, found in Al, Ab, N, St, Ta, Oa, Tb, Kb, Om; Kr. – Austria, German D. R. and F. R., Poland, Estonia, Latvia, Lithuania, n. and m. Russia, Ukraine, Azerbaijan, Kazakhstan.

Biology. On *Scirpus* species at the shores of lakes and rivers, July-September (Sahlberg, 1871). On *Scirpus lacustris* (Ossiannilsson, 1947b). On *Scirpus lacustris* and *S. Tabernaemontani* (Lindberg, 1947). On *Sparganium* (Wagner & Franz, 1961). On *Scirpus* and *Hippuris vulgaris* (Linnavuori, 1969b). On *Scirpus* (Vilbaste, 1974).

Tribe Balcluthini
Genus *Balclutha* Kirkaldy, 1900

Gnathodus Fieber, 1866a: 505 (nec Pander, 1856).
  Type-species: *Cicada punctata* Thunberg, 1784, by monotypy.
Balclutha Kirkaldy, 1900: 243.
  Type-species: *Cicada punctata* Thunberg, 1784, by subsequent designation.

Elongate leafhoppers, always macropterous. Head anteriorly rounded angular, transition between face and vertex rounded. Vertex medially not longer than near eyes, 1/4-1/3 of length of pronotum. Fore wings with only two subapical cells (Text-fig. 1937). Sc and R in hind wing distally united, reaching peripheric vein as one vein (Text-fig. 1939).

Male genital plates apically abruptly narrowing. Macrosetae of legs, male pygofer, and genital plates finely pubescent. Aedeagus simple, without appendages, phallotreme dorso-terminal. Apodemes of 2nd abdominal sternum in male vestigial. Apex of female ovipositor not extending behind pygofer lobes. Nymphs strongly dorsoventrally flattened, abdomen hairless. In Denmark and Fennoscandia four species.

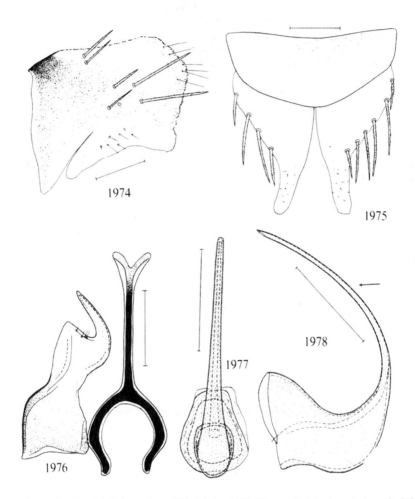

Text-figs. 1974-1978. *Balclutha punctata* (Fabricius). – 1974: left pygofer lobe in male from the left; 1975: male genital plate and valve from below; 1976: right genital style and connective from above; 1977: aedeagus in ventral aspect (in direction of arrow in Text-fig. 1978); 1978: aedeagus from the left. Scale: 0.1 mm.

Key to species of *Balclutha*

1   Pronotum distinctly wider than head, sides strongly diverging caudad. Fore wings usually opaque. Hind wings apically fumose (around vein Sc + R, Text-fig. 1939) 2

–   Pronotum not or only slightly wider than head, sides less strongly diverging caudad (Text-fig. 1986). Fore wings hyaline. Hind wings not pigmented 3

2 (1) Smaller, 3.6-4.1 mm. Shaft of aedeagus very slender, apically almost straight (Text-fig. 1978). Second apical cell in fore wing 2.5-3 times as long as wide .... 277. *punctata* (Fabricius)

–   Larger, 4.0-4.9 mm. Shaft of aedeagus strongly curved (Text-fig. 1978). Second apical cell in fore wing 3.5-4.3 times as long as wide (Text-fig. 1937) .... 280. *lineolata* (Horváth)

3 (1) Larger, 3.7-3.9 mm. Shaft of aedeagus strongly and evenly curved (Text-figs. 1980, 1982) .... 278. *rhenana* Wagner

–   Smaller, 3.3-3.65 mm. Shaft of aedeagus less strongly curved, apical part almost straight (Text-fig. 1984) .... 279. *calamagrostis* Ossiannilsson

277. *Balclutha punctata* (Fabricius, 1775)
Plate-fig. 160, text-figs. 1939, 1974-1978.

Cicada punctata Fabricius, 1775: 687.
Cicada punctata Thunberg, 1784: 21.
Cicadula spreta Zetterstedt, 1838: 298.

Elongate, shining. Head medially not or only slightly longer than near eyes. Upper side sordid yellowish white, light green or orange-coloured with or without black markings. Vertex, pronotum and scutellum with or without some indistinctly delimited orange-coloured or brownish spots. Fore wings in moderately pigmented specimens with the following black or fuscous markings: an oblique row of spots proceeding from scutellar corner to a point on claval suture somewhat proximally of its middle and continuing on the adjacent part of corium; a spot in distal corner of clavus, another in the M-fork on level with the clavo-distal spot, and one in distal end of the median cell. Apical part of fore wing partly fumose. In strongly pigmented individuals, these markings coalesce forming a broad zigzag-shaped longitudinal band. In such specimens the longitudinal streaks are often dark brownish, the venter being largely brown. On the other hand, one often finds specimens without any markings on upper side. Thoracic venter black-spotted, dorsum of abdomen black with light lateral margins and segment borders, venter of abdomen in males coloured as dorsum, in females largely light. Pygofer lobes in male as in Text-fig. 1974, genital plates and valve as in Text-fig. 1975, genital style and connective as in Text-fig. 1976, aedeagus as in Text-figs. 1977, 1978. Caudal border of 7th abdominal sternum in female almost straight. Overall length of males 3.6-3.9 mm, of females 3.7-4.1 mm. – In nymphs, pattern of anterior part of body indistinct; longitudinal bands on abdomen, if present, indistinctly delimited.

Distribution. Very common in Denmark, Sweden, and East Fennoscandia. In Norway found in most districts up to NTi. – Widespread in the Palaearctic, also found in the Australian, Nearctic, and Oriental regions.

Biology. "Während des ganzen Jahres, überwintert an Koniferen" (Kuntze, 1937). Larvae can often been found densely clustered in the panicles of e.g. *Phalaris arundinacea* and other grasses (Ossiannilsson, 1947b). "An Gräsern in Wäldern, aber auch auf Wiesen und Trockenrasen. Steigt in den Alpen bis in subalpine Lagen empor" (Wagner & Franz, 1961). "Hibernation als Imagines" (Schiemenz, 1964). "Imaginal-Überwinterer, 1 Generation" (Schiemenz, 1969b). "Very frequent throughout Finland in meadows on moderately damp soil, but also in wet or even dry meadows. . . . In fields *B. punctata* chiefly occurs in leys and cereals, and in natural meadows on wasteland. . . . In winter it lives on *Picea abies, Pinus silvestris* and *Juniperus communis*" (Raatikainen & Vasarainen, 1976).

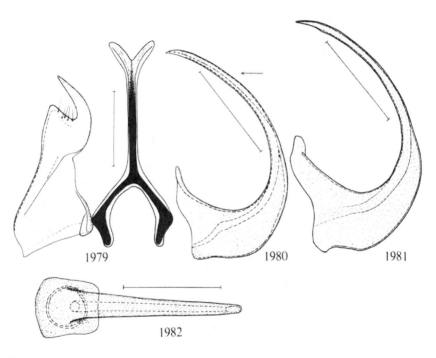

Text-figs. 1979-1982. *Balclutha rhenana* Wagner. – 1979: connective and right genital style from above (specimen from vicinity of Mainz, W. Wagner); 1980: aedeagus from the left (same specimen); 1981: same in ventral aspect (in direction of arrow in 1980); 1982: aedeagus from the left (specimen from Denmark, Trolle). Scale: 0.1 mm.

278. **Balclutha rhenana** Wagner, 1939
   Text-figs. 1979-1982.

Balclutha rhenana W. Wagner, 1939: 155.

Pronotum only slightly broadening caudad. Fore wings hyaline, hind wings unspotted. Fore wings and dorsum of fore body spotted as in normal specimens of *punctata*. Genital styles and connective as in Text-fig. 1979, aedeagus as in Text-figs. 1980-1982. Length (according to Wagner, 1939) 4-4.25 mm; length of body (sec. Wagner, 1950b) 3.7-3.9 mm.

   Distribution. Denmark: B: Ypnasted strand 21.VIII.1976 (Trolle). – Sweden: Gtl.: Hörsne, Gothem 20.VI.1934 (Lohmander); Vg.: Alingsås, Mjörn, Nolhagaviken 4.VII.-

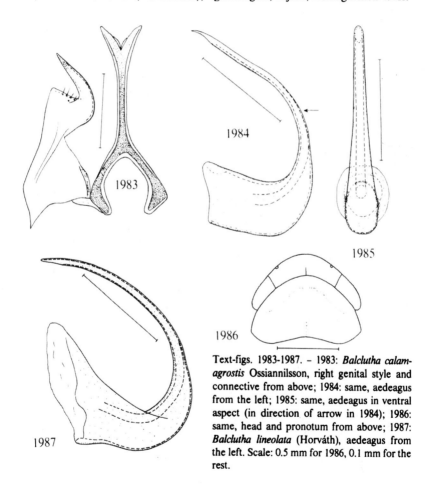

Text-figs. 1983-1987. – 1983: *Balclutha calamagrostis* Ossiannilsson, right genital style and connective from above; 1984: same, aedeagus from the left; 1985: same, aedeagus in ventral aspect (in direction of arrow in 1984); 1986: same, head and pronotum from above; 1987: *Balclutha lineolata* (Horváth), aedeagus from the left. Scale: 0.5 mm for 1986, 0.1 mm for the rest.

1969 (K.-H. Larsson); Upl.: Ekerö 1.X.1959 (Ossiannilsson). – Not yet recorded from Norway. – East Fennoscandia: Ab: Åbo (Linnavuori). – Austria, Bulgaria, Bohemia, Moravia, Slovakia, German D.R. and F. R., Greece, Italy, Netherlands, Yugoslavia; Afghanistan, Iran, Kazakhstan, Uzbekistan, Kirghizia.

Biology. On *Calamagrostis epigeios* (Linnavuori, 1953b). "An Gräsern, besonders in der Nähe von Flussufern" (Wagner & Franz, 1961). On *Phalaris arundinacea* (Trolle, in litt.).

279. *Balclutha calamagrostis* Ossiannilsson, 1961
   Text-figs. 1983-1986.

Balclutha calamagrostis Ossiannilsson, 1961a: 59.

Body small, elongate. Head as seen from above with fore margin evenly curved, medially only slightly longer than near eyes. Vertex distinctly convex. Pronotum not or indistinctly wider than head (Text-fig. 1986). Sordid yellow or light yellowish green. Dark (fuscous or orange) markings on head indistinct or diffuse, consisting of the usual transverse lateral streaks on frontoclypeus, broad vertical band on genae below eyes, and sometimes 2-4 very diffuse spots on vertex. Also anteclypeus may be fuscous, and sometimes the face is entirely dark. Fore margins of pronotum often fuscous. Behind fore border of pronotum there is a transverse arched dark line parallel with the fore border (Text-fig. 1986). This is the most constant dark marking of the dorsal aspect of this species. Caudally of this five longitudinal dark stripes are sometimes present. Scutellum with the scuto-scutellar transverse furrow black, for the rest spotless or with very diffuse lateral spots. Fore wings hyaline with light yellow veins, immaculate or with 6-8 small dark spots arranged as in *punctata*. Hind wings white, spotless. Thoracic venter and abdomen usually black with yellow segmental margins, male genital plates light. Legs dirty yellow. Male pygofer lobes each with 8 macrosetae, genital plates each with 5 macrosetae. Genital style and connective as in Text-fig. 1983, aedeagus as in Text-figs. 1984, 1985. Overall length: males 3.3-3.6 mm, females 3.4-3.65 mm.

Distribution. So far not found in Denmark and Norway. – Sweden: Upl.: Ekerö 15.IX and 1.X 1959 (Ossiannilsson). – East Fennoscandia: Ta: Tammela 1973 (Huldén); several localities in Ab and N (Albrecht, 1977). – German D.R., Estonia, Latvia, Lithuania, Bohemia.

Biology. On *Calamagrostis epigeios* in dry and sandy localities (Ossiannilsson; Albrecht, l.c.; Vilbaste, 1974).

280. *Balclutha lineolata* (Horváth, 1904)
   Text-figs. 1937, 1987.

Gnathodus punctatus var. *lineolatus* Horváth, 1904: 586.
Balclutha boica W. Wagner, 1950b: 97.

620

The following is essentially a translation of Wagner's original description of *B. boica*, with some modifications. – Head distinctly narrower than pronotum. Second apical cell in fore wing 3.5-4.3 times as long as wide (Text-fig. 1937). Ground-colour light brownish with pitch-brown markings. Dark areas: on pronotum behind fore border an arched line shaped as a sine curve, behind that a longitudinal streak reaching hind border, and on each side two spots, the more lateral spot extending further caudad than the other; lateral corners of scutellum with an irregularly defined spot, the transverse furrow and a caudal spot touching the latter; on the fore wings two rows of spots, one extending from scutellar apex to middle of costal border, the other reaching from apex of clavus to first and second apical cells. The second apical cell is wholly covered by brownish pigment, the third apical cell is transparent, while the fourth is apically faintly obscure. (Hind wing apically pigmented as in *punctata*). Shaft of aedeagus (Text-fig. 1987) very strongly curved, its ventral border strongly bent backwards. Measurements in mm: lengths: vertex medially 0.09-0.12; pronotum 0.52; fore wing 3.5-3.8; aedeagus (greatest distance from apex to ventral border) 0.22-0.23; total length 4.1-4.3. (Overall length of the Swedish specimen 4.03). Widths: head 0.91-0.93; pronotum 1.2-1.3; fore wing in apical part max. 0.91-0.93. (Wagner described the species on 3 males only). Total length in material from Czechoslovakia and Bulgaria (♂♀) 4.34-4.9 (Dlabola, 1959a).

Distribution. Not recorded from Denmark, Norway, and East Fennoscandia. – Sweden: one male found in Upl.: Solna, Bergshamra 22.V.1946 (Ossiannilsson). – Latvia, Lithuania, Estonia, Bulgaria, Slovakia, Moravia, German F.R., Poland, Kazakhstan, Kamchatka.

Biology. In forests and forest clearings, 14.-29.VIII. (15.V.) (Vilbaste, 1974). The Swedish specimen was found on *Picea abies*, indicating that hibernation does take place in the adult stage.

Tribe Macrostelini

Genus *Macrosteles* Fieber, 1866

Macrosteles Fieber, 1866a: 504.
   Type-species: *Cicada sexnotata* Fallén, 1806, by subsequent designation.
Acrostigmus Thomson, 1869: 76.
   Type-species: *Cicada sexnotata* Fallén, 1806, by subsequent designation.
Erotettix Haupt, 1929: 255.
   Type-species: *Thamnotettix cyane* Boheman, 1845, by original designation.

Body elongate. Head anteriorly rounded obtusely angular or almost evenly arched. Anteclypeus tapering towards apex or parallel-sided. Fore wings longer than abdomen, with two subapical cells. 1st abdominal sternum in male usually with distinct apodemes, apodemes of 2nd abdominal sternum in males well developed in many species, vestigial in others. Macrosetae on legs, pygofer and male genital plates pubescent. Male pygofer caudally with a fringe of stout macrosetae, usually also with a caudo-ventral tubercle.

Genital plates in male elongate, apical part narrow, lobiform. Shaft of aedeagus with a pair of apical appendages. 7th abdominal sternum in females with caudal border roughly straight or evenly curved without distinct incisions and processes. Nymphs with 2 hairs on 7th, 4 on 8th abdominal tergum. In Fennoscandia and Denmark 19 species.

Key to species of *Macrosteles*

1	Head distinctly narrower than pronotum (Subgenus *Erotettix* Haupt) .... 299. *cyane* (Boheman)
–	Head as wide as pronotum or wider (Subgenus *Macrosteles* s.str.) 2
2 (1)	Black markings on junction of vertex and frontoclypeus absent, vertex with or without a pair of black spots near caudal margin. Aedeagus as in Text-figs. 1995, 1996 282. *oshanini* Razvyazkina
–	Head with black markings on junction of vertex with frontoclypeus, often also on disk of vertex 3
3 (2)	Black markings on head as seen from above more or less as in Plate-fig. 162, three pairs of markings being present (sometimes fused): a pair of roundish spots near caudal margin, a pair of transversely oblong spots on junction of frontoclypeus and vertex, and between these a pair of transverse clavate "intermediate" spots 7
–	Black markings on vertex not as above, never fused 4
4 (3)	"Intermediate" spots on vertex present, spots near caudal margin absent or punctiform .... 288. *fascifrons* (Stål)
–	"Intermediate" spots absent 5
5 (4)	Apical appendages of aedeagus branched (Text-figs. 2043-45) .... 293. *quadripunctulatus* (Kirschbaum)
–	Apical appendages simple 6
6 (5)	Maximum overall length 3.6 mm (♂) or 4 mm (♀). Ground-colour lemon or orange yellow. Thoracic venter unspotted. Aedeagus as in Text-figs. 1990-1992 .... 281. *septemnotatus* (Fallén)
–	Overall length not less than 3.8 mm (♂) or 4.1 mm (♀). Ground colour light yellow to sordid greenish yellow. Thoracic venter with black markings behind coxae. Aedeagus as in Text-figs. 2000, 2001 .... 283. *variatus* (Fallén)
7 (3)	Shaft of aedeagus distad of middle with a pair of small triangular lobes distinctly visible in lateral aspect 18
–	Shaft of aedeagus without such lobes 8
8 (7)	Shaft of aedeagus with 3 longitudinal crests, one dorsal and two lateral (Text-figs. 2019, 2020) .... 287. *cristatus* (Ribaut)
–	Shaft of aedeagus without a dorsal longitudinal crest (la-

direction of shaft (Text-figs. 2051-2054) .... 295. *empetri* (Ossiannilsson)
– Lobes on shaft of aedeagus ventrolateral. Appendages
either recurrent or strongly bent near base 19
19 (18) Appendages long, approximately half as long as shaft, re-
current, more or less parallel with shaft (Text-figs. 2060,
2061). Lobes on shaft larger .... 297. *horvathi* (W. Wagner)
– Appendages shorter, in side view forming a varying
angle with shaft, lobes on shaft smaller (Text-figs. 2064-
2066) .... 298. *nubilus* (Ossiannilsson)

281. *Macrosteles (Macrosteles) septemnotatus* (Fallén, 1806)
Plate-fig. 213, text-figs. 1988-1994.

Cicada septemnotata Fallén, 1806: 35.

Light yellow, shining. Head usually with a pair of rounded black spots below and between ocelli. In addition the following black spots may be present: a pair of varying size near posterior border of vertex, one medially to each antenna, and one on lower part of frontoclypeus. Scutellum with a black spot near anterior corners. Fore wings often with a pale longitudinal streak along claval suture; apical part fumose. Also the costal border may be paler. Macrosetae of hind tibiae arising from usually indistinctly darker points, claws fuscous. Male pygofer as in Text-fig. 1988, genital valve, plates, connective and styles as in Text-fig. 1989, aedeagus as in Text-figs. 1990-1992, 1st abdominal sternum in male as in Text-fig. 1993, 2nd and 3rd abdominal sterna as in Text-fig. 1994. Overall length of males 3.1-3.6 mm, of females 3.7-4.0 mm. – "Nymph either uniformly pale brownish yellow or with 2 black spots on last tergite, or with darkened last tergite and anal segment, or, rarely, black spots on head and crown; if spots on crown, then the anterior body bears a brown pattern, but the wing pads are always pale" (Vilbaste, 1982).

Distribution. Common in Denmark, found in SJ, EJ, NEJ, F, LFM, NEZ, B. – Common in Sweden, Sk. – Vb. – Norway: found in AK, Os, Bø, VE, TEy, Ry. – Common in East Fennoscandia, Al, Ab, N – ObN and Ks; Vib, Kr. – Widespread in n., w., c., and e. Europe, also recorded from Italy, Morocco, Altai Mts., Kazakhstan, and Japan.

Biology. On *Filipendula ulmaria* (Kuntze, 1937; Ossiannilsson, 1947b; Vilbaste, 1982). Adults in June-September.

Text-figs. 1988-1994. *Macrosteles septemnotatus* (Fallén). – 1988: left pygofer lobe in male from the left; 1989: genital plates, styles, and connective from above; 1990: aedeagus in ventral aspect; 1991: aedeagus from the left; 1992: another specimen, aedeagus in ventral aspect; 1993: 1st abdominal sternum in male in antero-dorsal aspect; 1994: 2nd and 3rd abdominal sterna in male from above. Scale: 0.1 mm.

624

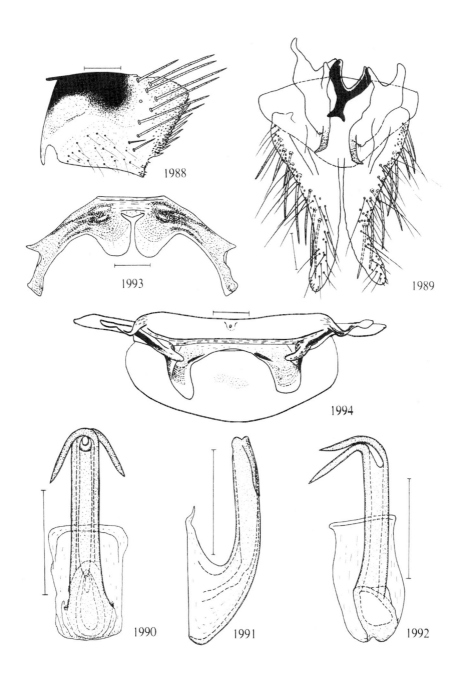

1988

1989

1993

1994

1990

1991

1992

625

282. *Macrosteles (Macrosteles) oshanini* Razvyazkina, 1957
Text-figs. 1995-1998.

Cicadula opacipennis Edwards, 1891: 30 (nec Lethierry, 1876).
Macrosteles oshanini Razvyazkina, 1957: 525 (n.n.).

Elongate, yellowish. Vertex sometimes with a pair of dark spots near caudal margin.
Frontoclypeus usually with some dark transverse streaks. Fore wings greyish green.
Dorsum og abdomen black with light lateral margins. Male pygofer lobes without a
caudo-ventral tubercle. Aedeagus as in Text-figs. 1995, 1996, 1st abdominal sternum in
male as in Text-fig. 1997, 2nd and 3rd abdominal sterna in male as in Text-fig. 1998.
Overall length of male 3.0-3.4 mm, of female 3.4-3.6 mm.

Distribution. Not found in Denmark, Norway and East Fennoscandia. – Sweden:
rare, found in Sm.: Ö. Torsås, Ingelstad 19.VIII.1974 (Gyllensvärd); Öl.: Halltorps hage
19.VII.-31.VIII.1976 (H. Andersson and R. Danielsson); Gtl.: Östergarn, Vike 18.VIII.-
1969 (Gyllensvärd); Upl.: Djursholm 6.IX.1947 (Ossiannilsson). – England, German
D.R. and F. R., Romania, Moravia, Slovakia, Poland, Lithuania, m.Russia.

Biology. "Scheint spezifisch zu sein für schattige Waldwiesen" (Kuntze, 1937).
"Restricted to the graminaea vegetation of the shady albuetum" (Razvyazkina, 1957).
In a fen (Vilbaste, 1974).

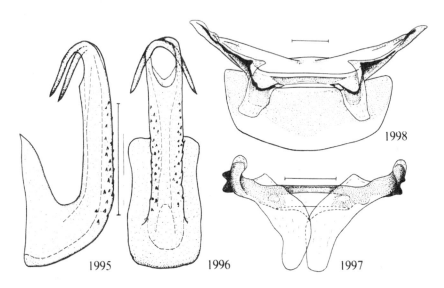

Text-figs. 1995-1998. *Macrosteles oshanini* Razvyazkina. – 1995: aedeagus from the left; 1996:
aedeagus in ventral aspect; 1997: 1st abdominal sternum in male from above; 1998: 2nd and 3rd
abdominal sterna in male from above. Scale: 0.1 mm.

283. *Macrosteles (Macrosteles) variatus* (Fallén, 1806)
Plate-fig. 161, text-figs. 1999-2003.

Cicada variata Fallén, 1806: 34.

Resembling *M. septemnotatus* but a little larger, ground-colour paler with a greenish tinge, dark markings more extended. The pattern of dark markings on the head is more constant than in *septemnotatus*, consisting of a pair of fairly large spots on fore border and another pair near hind border; spot on lower end of frontoclypeus always missing, the spot medially of antenna continuing downwards as a black streak in the frontal suture. Frontoclypeus often with traces of dark transverse lines. Pronotum light or with extended brownish patches. Fore wings usually each with a broad S-shaped brownish band sometimes breaking up into spots. Abdominal dorsum black with yellow side and

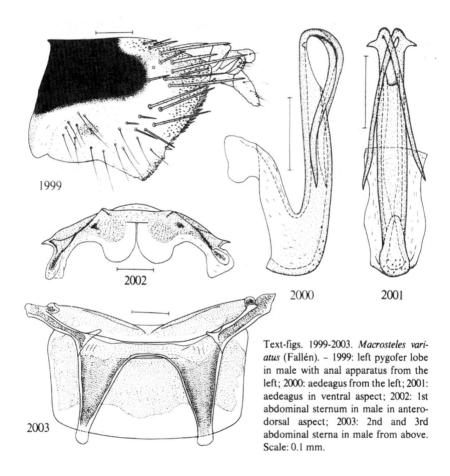

1999

2002

2000   2001

2003

Text-figs. 1999-2003. *Macrosteles variatus* (Fallén). – 1999: left pygofer lobe in male with anal apparatus from the left; 2000: aedeagus from the left; 2001: aedeagus in ventral aspect; 2002: 1st abdominal sternum in male in antero-dorsal aspect; 2003: 2nd and 3rd abdominal sterna in male from above. Scale: 0.1 mm.

segmental margins, venter as dorsum or largely light. Legs with more or less distinct longitudinal streaks, macrosetae on hind tibiae arising from distinct black points. Male pygofer and anal apparatus as in Text-fig. 1999, aedeagus as in Text-figs. 2000, 2001, 1st-3rd abdominal sterna in male as in Text-figs. 2002, 2003. Overall length of males 3.8-4.0 mm, of females 4.1-4.7 mm. – "Nymph uniformly brown or with both head and tergites IV-VI pale; 2 pairs of black spots always present on head" (Vilbaste, 1982).

Distribution. Widespread but uncommon in Denmark, found in EJ, NEJ, LFM, SZ, NEZ, and B. – Comparatively scarce in Sweden, Sk. – Nb., also in G. Sand. – Norway: found in Ø, Bø, VE, Nsy, Nsi, TRi. – Scarce in East Fennoscandia but found in most provinces up to ObN; Kr. – Widespread in Europe, also recorded from Kazakhstan, Tuva, Kirghizia, Kamchatka, Korean Peninsula, Maritime Territory, Japan; Nearctic region.

Biology. On *Urtica dioica* (Kuntze, 1937; Wagner & Franz, 1961). Among *Vaccinium myrtillus* (Linnavuori, 1969a). Inhabits undergrowth of forests (Vilbaste, 1982). Adults in June-August.

284. *Macrosteles (Macrosteles) sexnotatus* (Fallén, 1806)
Plate-fig. 162, text-figs. 2004-2009.

Cicada sexnotata Fallén, 1806: 34.
Macrosteles sexnotatus Le Quesne, 1968: 190.

Greyish green-yellow, shining, markings much varying. Vertex in light specimens marked as in Plate-fig. 162. Sutures of face and transverse streaks on frontoclypeus black; frontoclypeus often with a dark median line In dark specimens these markings are confluent, leaving but little of the pale ground-colour: on vertex a narrow median line and a pair of spots surrounding ocelli are usually light. Pronotum entirely light or with two brown-grey patches sometimes covering almost the whole surface except for a narrow median line and a spot on each side; near fore border some small black spots are often present. Scutellum usually with black basal triangles. Fore wings more or less uniformly greyish green-yellow, or with dark streaks in the cells between the light veins; apical part fumose. Commissural border proximally and distally dark with a light section between these dark parts. Venter in male largely black, in female usually largely light; dorsum of abdomen black or bluish black with yellow lateral margins and segment borders. Legs with dark longitudinal streaks and spots, black points on hind tibiae strongly marked. Male pygofer lobes as in Text-fig. 2004, genital plates, valve, connective, and styles as in Text-fig. 2005, aedeagus as in Text-figs. 2006, 2007, 1st abdominal sternum in male as in Text-fig. 2008, 2nd and 3rd abdominal sterna in male as in Text-fig. 2009. Overall length of males 3.2-3.7 mm, of females 3.3-3.8 mm. – Nymphs with "3 pairs of spots usually present on vertex; lateral margin of wing pads usually pale (sometimes with dark "incisions"). Abdominal tergites form central white patch which broadens posteriorly; a double pale patch usually present on the tip of each wing pad" (Vilbaste, 1982).

Distribution. Denmark: so far known from several places in EJ and NEZ. – Sweden: found in Sk. (Fallén), Sm., Öl., Gtl., Ög., Vg., Upl., Nrk., Dlr., Ång. – Norway: found in AK, Ø, VAy. – Common in East Fennoscandia, found in most provinces up to Ks; Vib, Kr, Lr. – According to available records widespread in the Palaearctic region but confusion with the following species cannot be excluded.

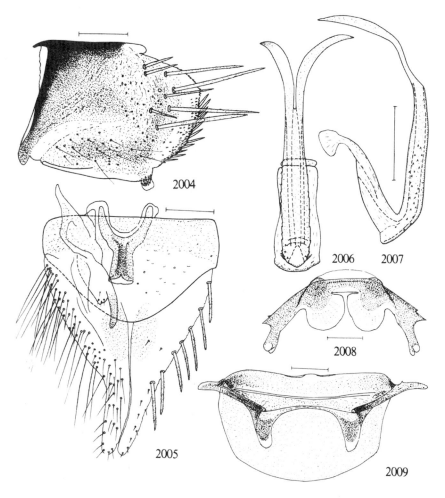

Text-figs. 2004-2009. *Macrosteles sexnotatus* (Fallén). – 2004: left pygofer lobe of male from the left; 2005: genital plates, valve, connective and left style; 2006: aedeagus from the left; 2007: aedeagus in ventral aspect; 2008: 1st abdominal sternum in male in antero-dorsal aspect; 2009: 2nd and 3rd abdominal sterna in male from above. Scale: 0.1 mm.

Biology. On grasses and in clover fields (Le Quesne, 1969). "Frequent almost everywhere in Finland in moist seashore and lakeshore meadows and treeless fens. Nowadays it also inhabits pastures and fields. Feeds on grasses" (Raatikainen & Vasarainen, 1976). "Bivoltiner Eiüberwinterer" (Müller, 1978). Adults in June-September.

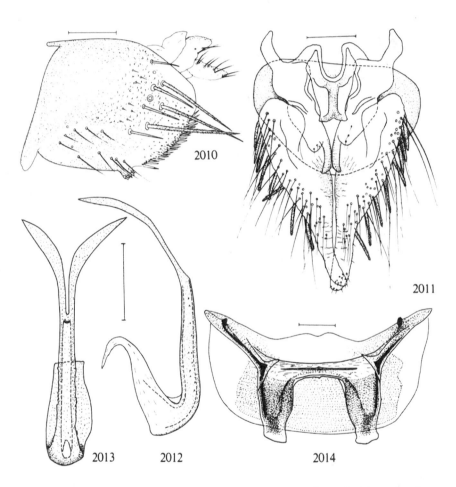

Text-figs. 2010-2014. *Macrosteles ossiannilssoni* Lindberg. – 2010: left pygofer lobe and anal apparatus in male from the left; 2011: genital plates, style and connective from above; 2012: aedeagus from the left; 2013: aedeagus in ventral aspect; 2014: 2nd and 3rd abdominal sterna in male from above. Scale: 0.1 mm.

285. *Macrosteles (Macrosteles) ossiannilssoni* Lindberg, 1953
Text-figs. 2010-2014.

Macrosteles sexnotatus Ossiannilsson, 1951c: 110 (nec Fallén, 1806).
Macrosteles ossiannilssoni Lindberg, 1953: 236.
Macrosteles ossiannilssoni Le Quesne, 1968: 190.

Adult resembling *M. sexnotatus*, males also in structure of genitalia, differing by shape and size of the apodemes of 2nd abdominal sternum (Text-fig. 2014). Male pygofer and anal apparatus as in Text-fig. 2010, genital plates, valve, styles and connective as in Text-fig. 2011, aedeagus as in Text-figs. 2012, 2013. Overall length of males 2.9-3.6 mm, of females 3.2-3.7 mm. – Nymphs differing from those of *sexnotatus* by the colouration of abdomen: rows of tiny spots present, sometimes together with brown spots, within pale patches on abdominal tergites (Vilbaste, 1982).

Distribution. In Denmark known from NEJ: Skagen, EJ: Strandkær, and NEZ: Bøllemosen. – Widespread in Sweden, found in Ög. (many localities), Dlsl., Dlr., Hls., Ång., P.Lpm. – Norway: widespread, found in HEn, AAi, STi, TRi (Holgersen). – East Fennoscandia: so far only in Sb: Virtasalmi (Linnavuori, 1969b). – England, Scotland, Germany, Estonia, Latvia, Lithuania, Canary Is., Madeira, Greece, Crete, Iran.

Biology. "Sometimes in montane areas associated with *Juncus squarrosus* L. or *Sphagnum* patches" (Le Quesne, 1969). "In damp and dry meadows, fens, on lake shores" (Vilbaste, 1974). Adults in June-October.

286. *Macrosteles (Macrosteles) alpinus* (Zetterstedt, 1828)
Text-figs. 2015-2018.

Cicada alpina Zetterstedt, 1828: 533.

Resembling *sexnotatus*. Black markings often strongly extended, confluent. Ground-colour often brownish yellow or sordid yellow. In dark specimens also the fore wings may be partly fuscous, especially basally. Aedeagus as in Text-figs. 2015, 2016, 1st abdominal sternum in male as in Text-fig. 2017, 2nd and 3rd abdominal sterna in male as in Text-fig. 2018. Overall length of males 3.0-3.4 mm, of females 3.2-3.5 mm.

Distribution. Not found in Denmark. – Sweden: found in Jmt., Ång., Ås.Lpm., P.Lpm, Lu.Lpm., T.Lpm. – Widespread in Norway, found in AK, On, TEi, Ri, HOy, HOi, Nsi, Nnv, TRy, TRi, Fi, Fn, Fø. – East Fennoscandia: rare in the south-west (Ab: Raisio; Sammatti), common in the north: Kb, ObN, Ks, Le, Li; Vib, Lr. – Austria, Bohemia, France, German F.R., England, Scotland, Italy, Switzerland, Estonia, Latvia, n. Russia, Altai Mts., n. Siberia, Kirghizia, Kamchatka, Mongolia, Tuva.

Biology. "On *Menyanthes trifoliata* growing in quagmires" (Linnavuori, 1952a). "In damp meadows and wet treeless fens, and nowadays also in fields" (Raatikainen & Vasarainen, 1976). Adults in July-August.

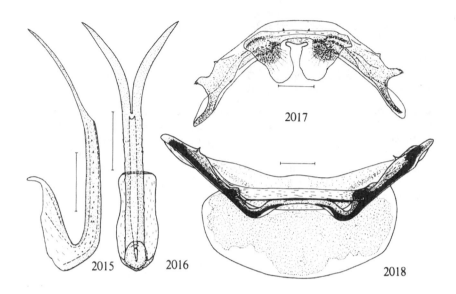

Text-figs. 2015-2018. *Macrosteles alpinus* (Zetterstedt). – 2015: aedeagus from the left; 2016: aedeagus in ventral aspect; 2017: 1st abdominal sternum in male in antero-dorsal aspect; 2018: 2nd and 3rd abdominal sterna in male from above. Scale: 0.1 mm.

287. *Macrosteles (Macrosteles) cristatus* (Ribaut, 1927)
Text-figs. 2019-2022.

Cicadula cristata Ribaut, 1927: 164.

Resembling *sexnotatus,* on an average a little larger. Usually largely light-coloured. Aedeagus as in Text-figs. 2019, 2020, 1st abdominal sternum in male as in Text-fig. 2021, 2nd and 3rd abdominal sterna in male as in Text-fig. 2022. Overall length of males 3.3-3.9 mm, of females 3.7-4.2 mm. In nymphs, markings on vertex as in *sexnotatus;* "pale patches on abdominal tergites each contain 2 brown spots, between which there are no rows of tiny spots" (Vilbaste, 1982).

Distribution. Denmark: so far known from NEZ: Tisvilde, and B: Boderne. – Widespread and common in Sweden, Sk. – Lu.Lpm. – Widespread in Norway, found in AK, Os, Bv, VAy, Ri, HOi, Nsi, TRy, TRi, and Fi. – East Fennoscandia: found in Ab, N, Oa, Sb, Om, Ok, ObN. – Widespread in Europe, also in Altai Mts., Kazakhstan, Tuva, Mongolia, Manchuria, Kirghizia, Kamchatka, Maritime Territory, Kurile Is.; Nearctic region.

Biology. In potato fields (Ossiannilsson, 1943a). Often on *Polygonum* and *Linum* (Ossiannilsson, 1947b). "Bewohner mesophiler Wiesen" (Wagner & Franz, 1961). In lucerne fields (Obrtel, 1969). In oatfields. Hibernates in egg-stage (Raatikainen, 1971).

A dominant species in cereals (Raatikainen & Vasarainen, 1971). Adults in June-August.

Economic importance. Vector of the European aster yellows to cereals (Murtomaa, 1969). But "of very minor importance, for very few specimens of this species carry aster yellows in Finland" (Raatikainen & Vasarainen, 1971).

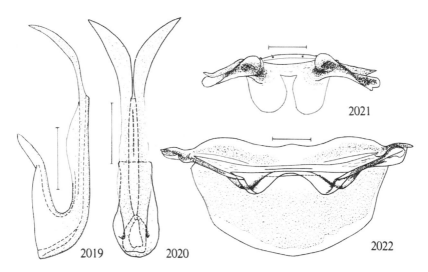

Text-figs. 2019-2022. *Macrosteles cristatus* (Ribaut). – 2019: aedeagus from the left; 2020: aedeagus in ventral aspect; 2021: 1st abdominal sternum in male in antero-dorsal aspect; 2022: 2nd and 3rd abdominal sterna in male from above. Scale: 0.1 mm.

288. *Macrosteles (Macrosteles) fascifrons* (Stål, 1858)
Text-figs. 2023-2028.

Thamnotettix fascifrons Stål, 1858a: 194.
Macrosteles fasciifrons lindbergi Dlabola, 1963: 322.
Amer.: six-spotted leafhopper, Aster leafhopper.

Resembling *sexnotatus* but much varying in extension of markings (cf. Beirne, 1952). In Swedish specimens the markings of the head are usually much reduced: the caudal spots (cf. Plate-fig. 162) represented by a small point or absent, the spots on junction of vertex and frontoclypeus not or just visible from above, the intermediate transverse lines reduced to a pair of short streaks and a small spot medially of each eye. Pronotum and scutellum yellowish without markings, those on fore wings obsolete. In f. *lindbergi* Dlabola, markings on vertex are absent. Aedeagus as in Text-figs. 2023-2026, 1st abdominal sternum in male as in Text-fig. 2027, 2nd and 3rd abdominal sterna in male as in Text-fig. 2028. Overall length of one Swedish male 3.9 mm, of one Swedish female 4.1 mm; length of f. *lindbergi* according to Dlabola (1963) ♂ 4.2 mm, ♀ 3.9-4.3 mm, of specimens from Canada and Alaska according to Beirne (1956) (♂♀) 3.5-4.2 mm.

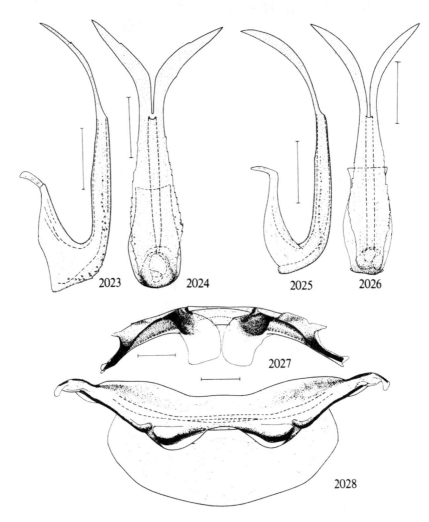

Text-figs. 2023-2028. *Macrosteles fascifrons* (Stål). – 2023: aedeagus from the left; 2024: aedeagus in ventral aspect; 2025: (another specimen) aedeagus from the left; 2026: (same) aedeagus in ventral aspect; 2027: 1st abdominal sternum in male in antero-dorsal aspect; 2028: 2nd and 3rd abdominal sterna in male from above. Scale: 0.1 mm.

Distribution. So far not found in Denmark, nor in Norway. – Sweden: Hls.: Edsbyn 22.VII.1978, 1.VIII.1980, 3.IX.1980 (Bo Henriksson); Ång.: Ådalsliden, Krångesjön 20.VII.1966 (Ossiannilsson). – East Fennoscandia: Ab: Tenala 16.VII.1920 (Håkan Lindberg). – Nearctic region.

Biology. "The nymphs feed on grasses and cereals in the spring, and may cause appreciable damage when abundant. In late spring the adults migrate to other herbaceous plants, where three or more overlapping broods develop during the summer. In late autumn the adults migrate back to grasses and cereals for the winter" (Beirne, 1952). Müller (1956) enumerates a great number of herbaceous hostplants. "Eiablagen subepidermal in Blattbasen; ein ♀ bis 127. Eizeit 7-17 Tage. Larvalzeit 20 Tage. Alle Stadien können überwintern" (Müller, l.c.).

Economic importance. In North America this species acts as a vector of several plant diseases caused by virus or mycoplasma, in the first place the "East American aster yellows" on many plants, and the "Californian aster yellows" (Müller, 1956; Heinze, 1959).

Note. Vilbaste (1980, p. 386) thinks that *fascifrons* is conspecific with *alpinus* (Zett.), but "the problem requires further examination".

289. *Macrosteles (Macrosteles) laevis* (Ribaut, 1927)
Text-figs. 2029-2032.

Cicadula sexnotata Tullgren, 1925: 5 (nec Fallén, 1806).
Cicadula laevis Ribaut, 1927: 162.
Danish: dværgcikade. Swedish: dvärgstrit. Norwegian: dvergsikade. Finnish: kääpiökaskas. German: Zwergzikade.

Resembling *M. sexnotatus*. Aedeagus as in Text-figs. 2029, 2030, 1st abdominal sternum in male as in Text-fig. 2031, 2 nd and 3rd abdominal sterna in male as in Text-fig. 2032. Overall length of males 3.2-3.5 mm, of females 3.3-4.0 mm. – Nymph (according to Vilbaste, 1982): "middle spots tend to be absent from vertex; 3 separate pale spots usually present on lateral margin of wing pads, or wing pads almost entirely brownish grey. Colour of abdomen brown or dark brown".

Distribution. Denmark: so far found only in Bornholm, where it is common and abundant. – Sweden: common and abundant in the entire country, Sk. – T.Lpm. – Widespread in Norway, found in AK, HEn, Bv, Ry, Ri, HOi, Nsy, TRy, TRi, Fi. – Common in the eastern and northern parts of East Fennoscandia, elsewhere rare; found in Ab, N, Oa, Sb, Ok, ObS, ObN, Ks, Li; Kr, Lr. – Widespread in Europe, including Iceland, also in Iran, Afghanistan, Anatolia, Georgia, Altai Mts., Kazakhstan, Uzbekistan, Tuva, Mongolia, Manchuria, Kirghizia, Kamchatka, Maritime Territory, Kurile Isl.

Biology. In Sweden normally 2, sometimes 3 generations (Tullgren, 1925). In "Stranddünen, Sandfeldern" (Kuntze, 1937). "Bevorzugt trockene Wiesen" (Marchand, 1953). Hibernation takes place in the egg stage (Müller, 1956, 1957). "An Gräsern, auch an Getreide und sogar Kartoffeln" (Wagner & Franz, 1961). "Ei-Überwinterer, 2 Generationen" (Schiemenz, 1969b). In lucerne fields (Obrtel, 1969). "Very frequent in leys and migrated to some extent to cereal crops. . . . The first adults were captured from leys on 27.6." (Raatikainen & Vasarainen, 1973). "Originally, this species probably occurred in dry meadows, but its major habitats nowadays are cultivated fields, especially leys and cereals. In cages it fed and reproduced on oats, barley and *Lolium perenne,* and fed on other cereals, e.g. wheat and rye" (Raatikainen & Vasarainen, 1976).

Economic importance. A serious outbreak of this species in Sweden in 1918 was described by Tullgren (1925). Cereals (wheat, rye, barley and oats) were attacked resulting in withering, discolouration, stunting, and death of plants. In this case at least part of the damage was the direct result of sap losses caused by the feeding of the insects. *Macrosteles laevis* has also been shown to be a vector of European aster yellows on cereals and other plants (Murtomaa, 1967), and of oat blue dwarf (Lindsten, Vacke & Gerhardson, 1970), and of the agent causing stolbur disease of tomato and asters (Heinze, 1959).

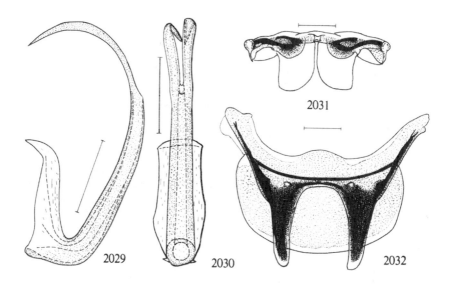

Text-figs. 2029-2032. *Macrosteles laevis* (Ribaut). – 2029: aedeagus from the left; 2030: aedeagus in ventral aspect; 2031: 1st abdominal sternum in male in antero-dorsal aspect; 2032: 2nd and 3rd abdominal sterna in male from above. Scale: 0.1 mm.

290. *Macrosteles (Macrosteles) fieberi* (Edwards, 1889)
   Text-figs. 2033-2035.

Cicadula frontalis Fieber, 1885: 45 (nec Scott, 1875).
Cicadula fieberi Edwards, 1889: 703 (n.n.).

Resembling *sexnotatus*, black markings on head often more or less fused. Aedeagus as in Text-figs. 2033, 2034, 2nd and 3rd abdominal sterna in male as in Text-fig. 2035. Overall length of males 3.3-4.2 mm, of females 3.6-4.6 mm. – Nymph: abdomen either uniformly pale brownish yellow or with darker hind margins of tergites; black spots on front of head smaller [than in *viridigriseus*]; between these and the spots on ocellocular area are distinct interruptions; pattern of head more distinct (Vilbaste, 1982).

Distribution. So far not found in Denmark. – Widespread but scarce in Sweden, found in Bl., Ög., Sdm., Nrk., Dlr., Hls., Ång., Vb., Nb., P.Lpm., Lu.Lpm. – Appears to be scarce in Norway, recorded from AK, VAy, VAi, and Ry. – Rare and sporadic in East Fennoscandia, found in Ab, Ta, Oa, Kb. – Austria, Bulgaria, Moravia, Slovakia, France, German D.R. and F.R., England, Scotland, Ireland, Netherlands, Poland, Romania, Yugoslavia, Ukraine, Anatolia, Armenia, Moldavia, Altai Mts., Iran, Kazakhstan, Tuva, Mongolia, Kirghizia, Kamchatka; Nearctic region.

Biology. On *Scirpus* (Kuntze, 1937). "Auf seggenreichen Weissmooren in Juli und August" (Kontkanen, 1938). "Eiüberwinterer mit 2 Generationen" (Remane, 1958). "Ausgesprochen hygrophil" (Schiemenz, 1965).

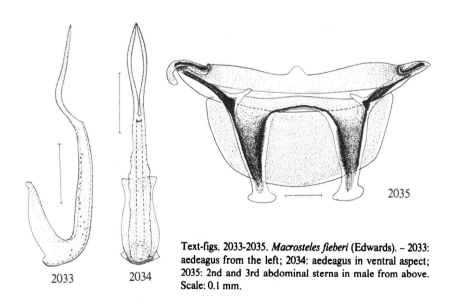

Text-figs. 2033-2035. *Macrosteles fieberi* (Edwards). – 2033: aedeagus from the left; 2034: aedeagus in ventral aspect; 2035: 2nd and 3rd abdominal sterna in male from above. Scale: 0.1 mm.

2033   2034

2035

291. *Macrosteles (Macrosteles) lividus* (Edwards, 1894)
     Text-figs. 2036-2038.

Cicadula livida Edwards, 1894: 104.

Resembling *sexnotatus*, dark markings often confluent, fore wings usually largely dark, in females often blackish or brownish with pale veins. Aedeagus as in Text-figs. 2036, 2037, 2nd and 3rd abdominal sterna in male as in Text-fig. 2038. Overall length of males 3.5-4.0 mm, of females 4.0-4.5 mm.

Distribution. Denmark: known from NEZ: Damhussøen 18.IX. and 23.IX.1917 (C. C. R. Larsen), and B: Svaneke 27.VIII.1972 (Trolle). – Rare in Sweden, found in Bl., Öl., Vg., Upl. Hls., Med. – Norway: one (pipunculised) male found in Nnø: Bonnå 16.VIII.1961 (Holgersen). – East Fennoscandia: found in Al, Ab, N, Ka. – England, Netherlands, German D. R. and F. R., Lithuania, Latvia, Estonia, Poland, Slovakia, Altai Mts., Kazakhstan, Uzbekistan, Tuva, Mongolia, Kirghizia, Maritime Territory.

Biology. On *Scirpus Tabernaemontani* and *Heleocharis* spp. (Linnavuori, 1952a). Among *Trichophorum pumilum, Carex distans, Puccinellia distans, Juncus Gerardi, Camphorosoma ovatum, Plantago maritima,* and *Suaeda maritima* (Lauterer, 1980).

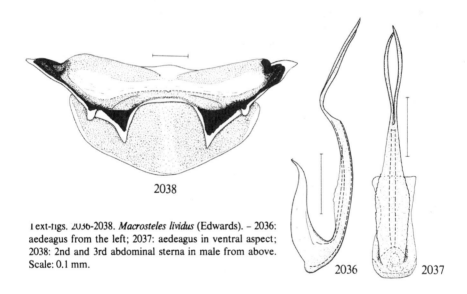

2038

Text-figs. 2036-2038. *Macrosteles lividus* (Edwards). – 2036: aedeagus from the left; 2037: aedeagus in ventral aspect; 2038: 2nd and 3rd abdominal sterna in male from above. Scale: 0.1 mm.

2036   2037

292. *Macrosteles (Macrosteles) viridigriseus* (Edwards, 1924)
     Text-figs. 2039-2042.

Cicadula viridigrisea Edwards, 1924: 54.

Resembling *sexnotatus*, black markings of head often reduced. Aedeagus as in Text-figs. 2039, 2040, 1st abdominal sternum in male as in Text-fig. 2041, 2nd and 3rd abdominal sterna in male as in Text-fig. 2042. Overall length of males 2.9-3.3 mm, of females 3.5-3.8 mm. – Nymphs: abdomen as in *fieberi;* "large black spots on front of head; these are continuous with brownish spot on ocellocular area with at most a very narrow interruption; pattern of head rather unclear with exception of small hind spots which may be rather dark" (Vilbaste, 1982).

Distribution. Denmark: so far found in EJ: Kasted Mose (Trolle), and LFM: Rødbyhavn (Trolle). – Not uncommon in Sweden, found in Sk., Bl., Sm., Öl., Ög., Boh., Sdm., Upl., Nrk., Dlr., Med., Vb. – Norway: found in AK, VAy, and HOi. – East Fennoscandia: common in the southern part (Al, Ab-Ta; Vib), for the rest rare (Oa). – England, Scotland, Wales, Netherlands, France, Switzerland, German D. R. and F. R., Estonia, Latvia, Lithuania, Poland, Austria, Bulgaria, Bohemia, Moravia, Slovakia, Hungary, Italy, n. and m. Russia, Ukraine, Georgia, Moldavia. – Nearctic region.

Biology. Belongs to the "Flachmoor" (Kuntze, 1937). To a certain degree halophilous (Linnavuori, 1952a). In the "Molinietum hydrocotyletosum" (Marchand, 1953). "Ein Feuchtwiesentier mit Ausstrahlungen bis hinein in die trockneren Glatthaferwiesen" (Strübing, 1955). In damp meadows, in fens, on shores of water bodies (Vilbaste, 1974). "Particularly in wet or moist seashore meadows and other moist meadowland" (Raatikainen & Vasarainen, 1976). – Adults in June-October.

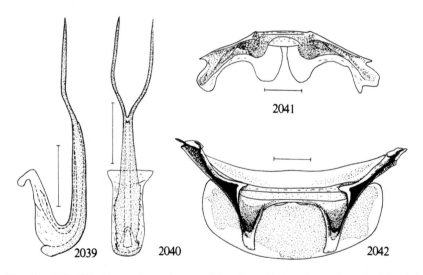

Text-figs. 2039-2042. *Macrosteles viridigriseus* (Edwards). – 2039: aedeagus from the left; 2040: aedeagus in ventral aspect; 2041: 1st abdominal sternum in male in antero-dorsal aspect; 2042: 2nd and 3rd abdominal sterna in male from above. Scale: 0.1 mm.

293. *Macrosteles (Macrosteles) quadripunctulatus* (Kirschbaum, 1868)
Text-figs. 2043-2047.

Jassus (Thamnotettix) quadripunctulatus Kirschbaum, 1868b: 99.

Resembling *sexnotatus* but black markings on vertex never confluent, the intermediate transverse pair of streaks missing. Aedeagus as in Text-figs. 2043-2045, 1st abdominal sternum in male as in Text-fig. 2046, 2nd and 3rd abdominal sterna in male as in Text-fig. 2047. Overall length of males 2.9-3.1 mm, of females 3.2-3.5 mm.

Distribution. Very rare in Denmark, one male found 27-30.VIII.1918 in NEZ: Tisvilde (Oluf Jacobsen). – Very rare in Sweden, found in Sk.: Veberöd, Hasslamöllan (Kemner) and in Sm.: Ö Torsås, Ingelstad 8.VIII. and 14.VIII.1974 (Gyllensvärd). – Not recorded from Norway. – East Fennoscandia: found in N: Helsingfors; Sa: Joutseno; Kb: Hammaslahti. – England, Netherlands, German D. R. and F. R., Latvia, Lithuania, Poland, Romania, Bulgaria, Hungary, Bohemia, Moravia, Slovakia, Italy, Greece, Yugoslavia, m. and s. Russia, Ukraine, Anatolia, Afghanistan, Iran, Iraq, Azerbaijan, Kazakhstan, Tadzhikistan, Uzbekistan, Tuva, Kirghizia.

Biology. Wagner (1935a) found *M. quadripunctulatus* on *Corispermum hissopifolium* L. *(Chenopodiaceae)*. This plant is not indigenous in Fennoscandia. Kuntze (1937) "streifte die Art von *Setaria* und *Panicum*-Arten" on "Binnendünen, Sandfeldern, besonnten

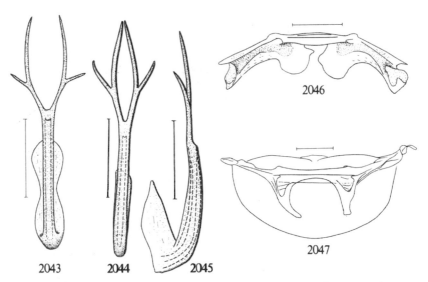

Text-figs. 2043-2047. *Macrosteles quadripunctulatus* (Kirschbaum). – 2043: aedeagus in ventral aspect (Swedish specimen); 2044: aedeagus in ventral aspect (specimen from Saloniki); 2045: aedeagus from the left (Saloniki); 2046: 1st abdominal sternum in male from above (Saloniki); 2047: 2nd and 3rd abdominal sterna in male from above (teneral, Saloniki). Scale: 0.1 mm.

Hängen". "In xerophilen bis hygrophilen Biotopen, wobei erstere bevorzugt werden. . .
. Ei-Überwinterer, 2 Generationen" (Schiemenz, 1969b).

294. *Macrosteles (Macrosteles) sordidipennis* (Stål, 1858)
Text-figs. 2048-2050.

Thamnotettix sordidipennis Stål, 1858a: 194.
Cicadula sexnotata salina Reuter, 1886: 211.

Resembling *sexnotatus* but frontoclypeus in lateral aspect more convex, head of female as seen from above conspicuously angularly produced, almost as long as pronotum. Frontoclypeus often largely black. Aedeagus as in Text-figs. 2048, 2049, 2nd and 3rd abdominal sterna in male as in Text-fig. 2050. Overall length of males 2.6-3.4 mm, of females 3.1-4.1 mm. – Nymph (one specimen): abdominal terga dark brown, frontoclypeus swollen, dark brown with narrow yellow transverse streaks; pale stripe below black spot on transition to vertex narrow, spots on vertex well-defined.

Distribution. Rare in Denmark, found only once: EJ: Kalø Vig 31.VII.1966 (Trolle). – Sweden: Sk.: Landskrona 22.VII.1937 (Ossiannilsson); Öl.: Borgholm 10.VII.1940 (Wieslander), Gärdslösa, seashore 13.VIII.1964 (Ossiannilsson); Boh.: Kosterö (Reuter), Stenungsund 9.VII.1953 (Ossiannilsson), Herrestad, vicinity of Kärranäs 29.VII.1967 (Ossiannilsson). – So far not found in Norway. – East Fennoscandia: rare and sporadic, found in Al: Eckerö (Reuter); Ab: Raisio (Linnavuori); N: Hangö (Håkan Lindberg); Oa. – England, Netherlands, German D. R. and F. R., Poland, Austria, Bohemia, Moravia, Hungary, n. Russia, Altai Mts., Kazakhstan, Kirghizia, m. Siberia, Tuva, Mongolia.

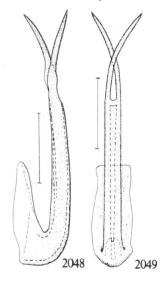
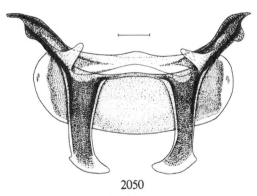

2050

2048   2049

Text-figs. 2048-2050. *Macrosteles sordidipennis* (Stål). 2048: aedeagus from the left; 2049: aedeagus in ventral aspect; 2050: 2nd and 3rd abdominal sterna in male from above. Scale: 0.1 mm.

Biology. On *Festuca distans* Griseb. (Wagner, 1937a). "Common at the end of June and in the beginning of July on *Juncus Gerardi* on seashores" (Linnavuori, 1952a). Halobiont (Wagner & Franz, 1961). Swedish adults were captured 9.VII-22.VIII; two generations.

295. *Macrosteles (Macrosteles) empetri* (Ossiannilsson, 1935)
Text-figs. 2051-2055.

Limotettix 6-notata J. Sahlberg, 1871: 250 (Obs.).
Cicadula empetri Ossiannilsson, 1935a: 127.

Resembling *sexnotatus,* smaller, wings comparatively shorter, little longer than abdomen, fore wings with a more or less distinct colour pattern consisting of two dark patches in clavus, one at middle and one at apex, two dark patches in corium, viz. one

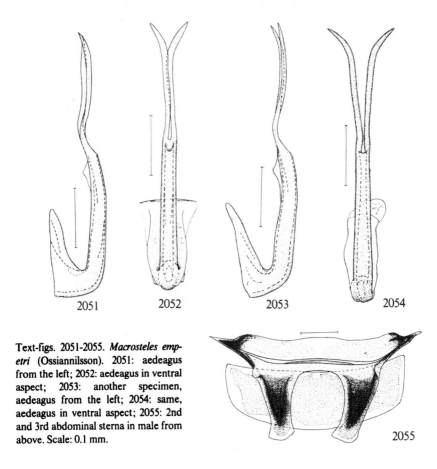

2051   2052   2053   2054

Text-figs. 2051-2055. *Macrosteles empetri* (Ossiannilsson). 2051: aedeagus from the left; 2052: aedeagus in ventral aspect; 2053: another specimen, aedeagus from the left; 2054: same, aedeagus in ventral aspect; 2055: 2nd and 3rd abdominal sterna in male from above. Scale: 0.1 mm.

2055

642

at level of the interspace between claval patches, the other in 5th apical cell and the distal end of median cell, and also some indistinctly delimited patches in 2nd and 3rd apical cells. Eyes of the live insect reddish. Aedeagus as in Text-figs. 2051-2054, 2nd and 3rd abdominal sterna in male as in Text-fig. 2055. Overall length (♂♀) 2.75-3.2 mm.

Distribution. So far not found in Denmark. – Sweden: found in Dlr., Ång., Vb., Ås. Lpm., P. Lpm. (several localities). – Norway: HOi: Kinsarvik, Stavali south of Sysenvann 12.VIII.1969; TRy: Tromsdal (Ardö). – East Fennoscandia: found in Om, Ok, Ks. – N. Russia.

Biology. Always with *Empetrum nigrum* Sahlberg, 1871; Ossiannilsson, 1935a; Linnavuori, 1969a). However, these is no evidence that *Empetrum* is the true hostplant of this insect; the distribution of *empetri* so far established is much smaller than that of *Empetrum*. Adults in July and August.

296. *Macrosteles (Macrosteles) frontalis* (Scott, 1875)
Text-figs. 2056-2059.

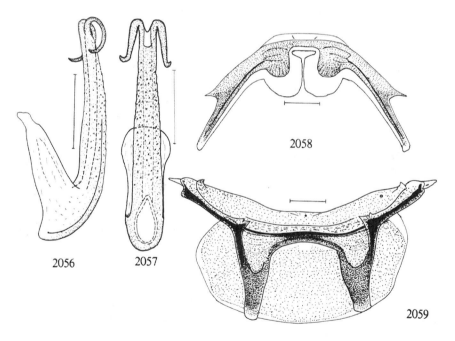

Text-figs. 2056-2059. *Macrosteles frontalis* Scott). 2056: aedeagus from the left; 2057: aedeagus in ventral aspect; 2058: 1st abdominal sternum in male in antero-dorsal aspect; 2859: 2nd and 3rd abdominal sterna in male in dorsal aspect.

Cicadula frontalis Scott, 1875a: 231.
Cicadula hamata Ossiannilsson, 1936b: 10.

Resembling *sexnotatus*, larger, black markings usually much extended. Anteclypeus comparatively long and narrow, almost parallel-sided, ratio length: maximal width about 1.5. Black spots on junction of vertex and frontoclypeus large, often confluent with caudal spots on vertex. Frontoclypeus light with black transverse streaks, or largely brownish or black. Pronotum with or without a pair of large black patches; in light specimens there are some small black spots along fore border. Fore wings in dark specimens largely brownish, the subcostal cell remaining light even in such individuals. Aedeagus as in Text-figs. 2056, 2057, 1st abdominal sternum in male as in Text-fig. 2058, 2nd and 3rd abdominal sterna in male as in Text-fig. 2059. Overall length of males 3.4–4.0 mm, of females 4.0–4.7 mm.

Distribution. Not found in Denmark. – Sweden: only in the central and northern parts of the country, scarce, found in Upl., Dlr., Hls., Jmt., Ång., Vb., Nb., P. Lpm., T. Lpm. – Scarce in Norway, found in HEn: Folldals Verk, and Tolga (Holgersen), and in MRi: Myrebo, Romsdal (Holgersen). – Scarce and sporadic in East Fennoscandia, found in N, Sa, Sb, Kb, Om, Ks; Vib, Kr. – England, Scotland, Ireland, Netherlands, France, Belgium, Switzerland, German D. R. and F. R., Austria, Italy, Bohemia, Moravia, Slovakia, Hungary, Romania, Yugoslavia, Poland, Estonia, Latvia, Lithuania, Ukraine, Tuva, Kamchatka.

Biology. On *Equisetum palustre* (Kuntze, 1937; Wagner & Franz, 1961). On *Equisetum* species (Vilbaste, 1974). "On *Equisetum* species both in swamps and in dry places" (Raatikainen & Vasarainen, 1976). Adults in June-September.

297. *Macrosteles (Macrosteles) horvathi* (W. Wagner, 1935)
Plate-fig. 163, text-figs. 2060-2063.

Cicadula fasciifrons Edwards, 1891: 29 (nec Stål, 1858).
Cicadula horvathi W. Wagner, 1935a: 18 (n.n.).

Resembling *sexnotatus* but with a strong tendency to melanism. The light ground-colour of the head is often reduced to small spots or narrow streaks. Also pronotum and scutellum are often largely dark, the former with a light median line. Fore wings often largely brownish with several light patches, viz. one in clavus around the distal end of the longer vein, another in corium along the proximal third of claval suture. In very dark specimens also the subcostal cell is black or brownish. Anteclypeus tapering towards apex, ratio length:maximal width 1.27-1.35. Aedeagus as in Text-figs. 2060, 2061, 1st abdominal sternum in male as in Text-fig. 2062, 2nd and 3rd abdominal sterna in male as in Text-fig. 2063. Overall length of males 3.0-3.7 mm, of females 3.8-4.2 mm. – Nymph: abdomen dark, anterior margins of tergites usually more darkened. Pale stripe below black spots on transition to frons is very narrow (about as wide as uppermost arch-line; hind pairs of spots on vertex are usually connected with brownish "shadows" (Vilbaste, 1982).

Distribution. Denmark: known from several places in F: Turø, Vindeby; LFM: the wood around Nykøbing Falster, and B: Dueodde and Ypnasted. – Widespread and fairly common in Sweden, found in most provinces, Sk. – T. Lpm. – Widespread in Norway, found in AK: Valle and Hovind (Siebke); TEi: Vigdesjå, Kviteseid (Holgersen); Ri: Stakstøl, Hjelmeland (Holgersen); Nnø: Bonnå (Holgersen); TRy: Tromsdal (Ardö). – Common in southern and central East Fennoscandia, found in Al, Ab, N, Ka, Sa, Oa, Kb, Ok; Vib, Kr. – England, Scotland, Wales, Ireland, France, Netherlands, German D. R. and F. R., Switzerland, Austria, Bohemia, Moravia, Slovakia, Hungary, Romania, Albania, Yugoslavia, Poland, Estonia, Latvia, Lithuania, Ukraine,. Canary Is., Iran. Altai Mts., Tuva, Manchuria.

Biology. On *Juncus lamprocarpus* and *Juncus Gerardi* (Wagner, 1935a). Halophilous (Kuntze, 1937). In moist meadows (Vilbaste, 1974). "Originally chiefly on seashore and lakeshore meadowland and on treeless fens" (Raatikainen & Vasarainen, 1976). "A coastal species" (Trolle, in litt.). Adults in June-October.

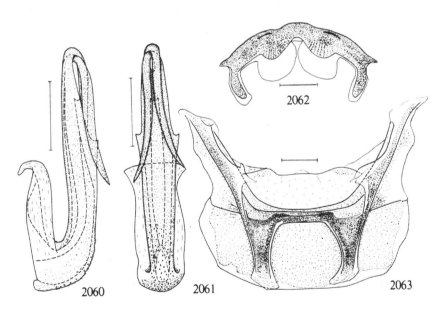

Text-figs. 2060-2063. *Macrosteles horvathi* (Wagner). 2060: aedeagus from the left; 2061: aedeagus in ventral aspect; 2062: 1st abdominal sternum in male in antero-dorsal aspect; 2063: 2nd and 3rd abdominal sterna in male from above. Scale: 0.1 mm.

298. *Macrosteles (Macrosteles) nubilus* (Ossiannilsson, 1936)
   Text-figs. 2064-2067.

Cicadula nubila Ossiannilsson, 1936b: 10.
Cicadula mannerheimi Kontkanen, 1937: 146.

Resembling *horvathi,* on an average a little larger. Anteclypeus as in *horvathi.* Subcostal cell in fore wing of dark specimens concolorous. Angle of appendages of aedeagus with shaft varying (Text-figs. 2064-2066), 2nd and 3rd abdominal sterna in male as in Text-fig. 2067. Overall length of males 3.3-4.1 mm, of females 3.9-4.4 mm.

Distribution. Not found in Denmark, nor in Norway. – Not uncommon in northern Sweden, found in Med., Ång., Vb., Nb., P. Lpm., T. Lpm. – East Fennoscandia: found in N: Tvärminne (Håkan Lindberg); Ta: Tuulos (Linnavuori); Kb: Pyhäleskä, Ham-

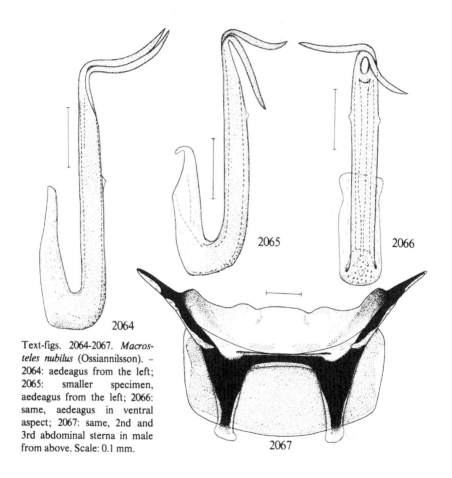

Text-figs. 2064-2067. *Macrosteles nubilus* (Ossiannilsson). – 2064: aedeagus from the left; 2065: smaller specimen, aedeagus from the left; 2066: same, aedeagus in ventral aspect; 2067: same, 2nd and 3rd abdominal sterna in male from above. Scale: 0.1 mm.

maslahti (Kontkanen); Ks: Kuusamo (Kontkanen); Kr: Prääshä (Kontkanen). – Estonia (Ossiannilsson, 1951b); N. Germany (Wagner, 1939).

Biology. In "*Carex-vesicaria*-Weissmoor" (Kontkanen, 1938). In a "*Saxifraga hirculus*-Braunmoor" (Kontkanen, 1948). On *Heleocharis palustris* (Linnavuori, 1952a). Adults in June-August.

299. *Macrosteles (Erotettix) cyane* (Boheman, 1845).
Plate-fig. 218, text-figs. 2068-2070.

Thamnotettix cyane Boheman, 1845b: 158.

Body elongate. Head rounded obtuse angular. Anteclypeus parallel-sided, frontal sutures above S-shaped. Wax-secretion abundant. – Blackish brown, in life usually covered by a bluish white wax-layer giving the impression of the ground-colour being bluish black. Head above usually black with a small elongate yellowish spot near apex, hind border of vertex light except for medially. Sometimes there are additional light spots; ocelli light. Pronotum black with a more or less complete light transverse band consisting of light patches. Fore wings black or brownish yellow with black veins. In some specimens the brownish yellow colour dominates. Venter dirty yellow with more or less extended black patches. Aedeagus as in Text-figs. 2068, 2069, 2nd and 3rd abdominal sterna in male as in Text-fig. 2070. Overall length of males 3.8-4.2 mm, of females 4.8-5.4 mm. – "Nymph entirely black, with pale narrow longitudinal line and narrow margins of tergites; bluish sheen present when alive" (Vilbaste, 1982).

Distribution. Very rare in Denmark, found only once in F: Kirkeby skov outside Svendborg 22.VIII.1916 (Oluf Jacobsen). – Scarce in southern and central Sweden, found in Sk., Bl., Hall., Ög., Sdm., Upl. – Norway: AK: Elle near Drøbak, Aug. 17, 1917 (Warloe). – Rare and sporadic in East Fennoscandia, found in Ab: Pargas (Lundström, Reuter, J. Sahlberg), Tenala (Håkan Lindberg), Lojo (Krogerus and Lindberg), Vichtis

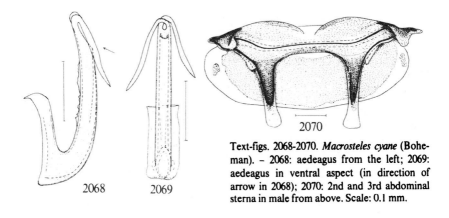

Text-figs. 2068-2070. *Macrosteles cyane* (Boheman). – 2068: aedeagus from the left; 2069: aedeagus in ventral aspect (in direction of arrow in 2068); 2070: 2nd and 3rd abdominal sterna in male from above. Scale: 0.1 mm.

(Lindberg); N: Hangö, Täcktom (Lindberg); Sa: Kerimäki (Saalas). – England, Belgium, Netherlands, France, German D. R. and F. R., Poland, Estonia, Latvia, Bohemia, Slovakia, Romania, Yugoslavia, Maritime Territory, Japan.

Biology. On the floating leaves of *Nymphaea* spp., *Nuphar luteum, Potamogeton natans*. Adults in July-September.

Genus *Sonronius* Dorst, 1937

Sonronius Dorst, 1937: 9.
Type-species: *Cicadula dahlbomi* Zetterstedt, 1838, by original designation.

As *Macrosteles* but pygofer lobes in male without a caudal fringe of macrosetae and without a ventro-caudal tubercle. Macrosetae on legs, pygofer and male genital plates not pubescent. Phallotreme situated on dorsal side of aedeagus near apex. Base of connective 1.5 times as long as length of branches. Vertex much shorter than pronotum. In Fennoscandia three species.

Key to species of *Sonronius*

1　Veins of fore wings black .... 300. *dahlbomi* (Zetterstedt)
–　Veins of fore wings light 2
2 (1) Body robust. Width of head 1.3 mm or more .... 300. *dahlbomi* (Zetterstedt)
–　Body elongate. Width of head 1.21 mm or less 3
3 (2) Larger, overall length of males 3.6-4.4 mm, of females 4.1-4.8
　　mm. Head often with an orange tinge, especially in females ....
　　.. 301. *binotatus* (J. Sahlberg)
–　Smaller, overall length of males 3.1-3.3 mm, of females 3.7-
　　3.9 mm .... 302. *anderi* (Ossiannilsson)

300. *Sonronius dahlbomi* (Zetterstedt, 1838)
Plate-fig. 164, text-figs. 2071-2076.

Cicada quadripunctata Fallén, 1806: 32 (nec De Villers, 1789).
Cicadula maculipes Zetterstedt, 1838: 297.
Cicadula dahlbomi Zetterstedt, 1838: 297.

Pale greyish green with black markings, shining. Head distinctly wider than pronotum. Male usually with the following black markings on vertex: a large rounded spot caudally of each ocellus, a small point behind that spot, a longitudinal streak or a pair of small dots near each eye, and a pair of transverse crescent-shaped spots on junction of vertex and frontoclypeus. Face with black transverse streaks and black sutures. Sometimes these markings are confluent, the black colour dominating. Pronotum light with fore border partially black, or black with a more or less narrow light median line

648

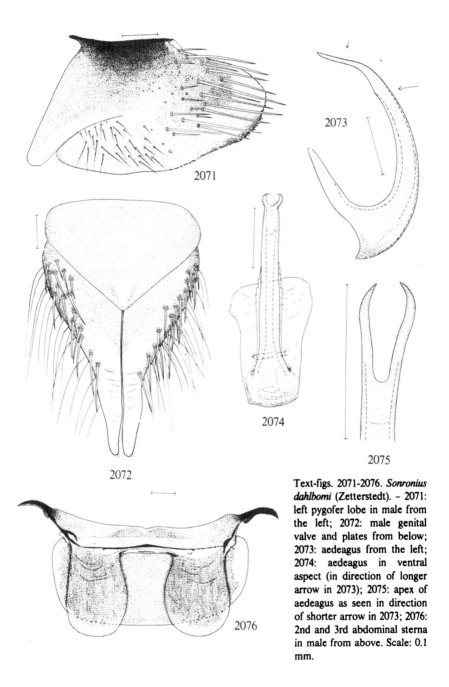

Text-figs. 2071-2076. *Sonronius dahlbomi* (Zetterstedt). – 2071: left pygofer lobe in male from the left; 2072: male genital valve and plates from below; 2073: aedeagus from the left; 2074: aedeagus in ventral aspect (in direction of longer arrow in 2073); 2075: apex of aedeagus as seen in direction of shorter arrow in 2073; 2076: 2nd and 3rd abdominal sterna in male from above. Scale: 0.1 mm.

649

and a couple of light lateral patches. Scutellum light with black basal triangles, or largely black. Veins of fore wings in male usually black, black-bordered. Females usually largely light with black markings more or less reduced, fore wings sometimes as in males but more often entirely light. Venter largely dark. Legs with black spots and streaks, setae of hind tibiae arising from black points. Male pygofer lobes as in Text-fig. 2071, aedeagus as in Text-figs. 2073-2075, 2nd and 3rd abdominal sterna in male as in Text-fig. 2076. Overall length of males 4.25-4.8 mm, of females 4.7-5.4 mm. – Nymphs with "only 2 hairs present on tergites VII and VIII; dorsal body surface almost uniformly black" (Vilbaste, 1982).

Distribution. Not found in Denmark. – Sweden: rare in the south, fairly common in the central and northern provinces, found in Sk., Vg., Sdm., Upl., Vrm., Dlr., Hls., Med., Hrj., Jmt., Nb., Ås. Lpm., Ly. Lpm., P. Lpm., T. Lpm. – Norway: found in HEn: Dallvang (Holgersen); On: Dovre (Boheman); STi: Stören (Holgersen); TRi: Rundhaug (Holgersen), Nordreisa n. f. Nedvefors (A. Löken); Fi: Sarggejak (Huldén). – Scarce and sporadic in East Fennoscandia, found in Ab, N, Ka, Ta, Kb, ObN; Vib, Kr, Lr. – England, Ireland, France, Belgium, German F. R., Switzerland, Italy, Slovakia, Poland, Latvia, n. and m. Russia, Ukraine, Altai Mts., Kazakhstan, Tuva, Mongolia, Kamchatka, Sakhalin; Nearctic region.

Biology. Among *Epilobium angustifolium* and *Rubus idaeus* (Sahlberg, 1971). I found *S. dahlbomi* on *Alchemilla* sp., *Chamaenerion angustifolium*, and *Filipendula ulmaria*. Adults in June-September.

301. *Sonronius binotatus* (J. Sahlberg, 1871)
Text-figs. 2077-2081.

Limotettix binotata J. Sahlberg, 1871: 242.

Elongate, greenish yellow, frontoclypeus and vertex often orange-coloured. Vertex with a pair of roundish black spots behind ocelli. In males, additional black or fuscous markings may be present, viz. 1-3 small spots or streaks near each eye, and a pair of short longitudinal streaks on apex. Frontoclypeus with more or less distinct fuscous transverse streaks. Usually with a black spot between antenna and eye. Pronotum light, often with a dark transverse arch-line behind fore border. Scutellum entirely light or with more or less distinct dark basal triangles. Fore wings hyaline, veins yellow, apical part of fore wing and apex of clavus more or less distinctly fumose. Fore and middle femora with dark spots and streaks, setae of hind tibiae arising from black points. Male genital plates and valve as in Text-fig. 2077, aedeagus as in Text-figs. 2078-2080, 2nd and 3rd abdominal sterna in male as in Text-fig. 2081. Overall length of males 3.6-4.4 mm, of females 4.1-4.8 mm. – Nymphs as in *dahlbomi* (Vilbaste, 1982).

Distribution. So far not found in Denmark. – Rare in Sweden, found in Sk. (Zetterstedt); Sm.: Barnarp, Flahult (Ossiannilsson); Dlr.: Rättvik, Ångtjärnen and Råberget (Tord Tjeder); Hls.: Edsbyn (several localities) and Voxna, Rävsaxtjärn (Bo Henriksson); Med.: Liden, Järkvissle, and Hässjö (Gyllensvärd). – Norway: only found in AK: near Oslo (Siebke); HEs: Löten 29.VIII.1961 (Holgersen). – Rare and sporadic in East Fennoscandia, found in Al (sec. Linnavuori, 1969a); Ab: Tenala (Håkan Lindberg), Karislojo (J. Sahlberg), Sammatti (J. Sahlberg, Saalas), Helsingfors (Lindberg, Rahm); St: Yläne (J. Sahlberg, Saalas); Tb: Viitasaari (P. H. Lindberg); Vib, Kr. – German F. R., Austria, Moravia, Slovakia, Poland, Latvia, Lithuania, n. Russia, Ukraine, m. Siberia, Altai Mts., Tuva, Kazakhstan, Kamchatka, Maritime Territory.

Biology. On dry meadows in August and September (Sahlberg, 1871). In dry sandy places, with e.g. *Calamagrostis epigeios* (Linnavuori, 1969a). On sand dunes, on *Chamaenerium angustifolium;* on willows (Vilbaste, 1974). Adults in July-September.

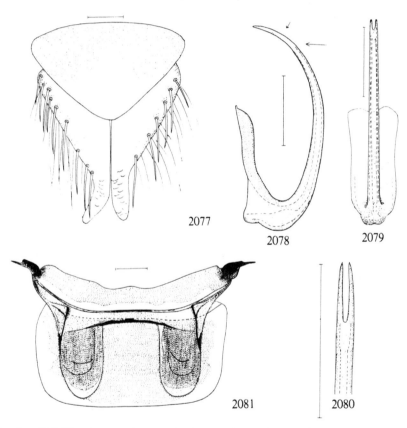

Text-figs. 2077-2081. *Sonronius binotatus* (Sahlberg). – 2077: male genital valve and plates from below; 2078: aedeagus from the left; 2079: aedeagus in ventral aspect (in direction of longer arrow in 2078); 2080: apex of aedeagus in ventro-terminal aspect (in direction of shorter arrow in 2078); 2081: 2nd and 3rd abdominal sterna in male from above. Scale: 0.1 mm.

302. **Sonronius anderi** (Ossiannilsson, 1948)
    Text-figs. 2082-2087.

Macrosteles anderi Ossiannilsson, 1948b: 26.

Male: head anteriorly roundish (Text-fig. 2082), medially a little longer than near eyes, wider than pronotum, above sordid greenish yellow with a large roundish spot on each side half-way between middle and eye, and with a pair of minute brownish spots close together at middle of fore border. Interrupted black side-lines present along eyes. Face yellowish with indistinct darker markings, frontoclypeus partly or entirely brownish.

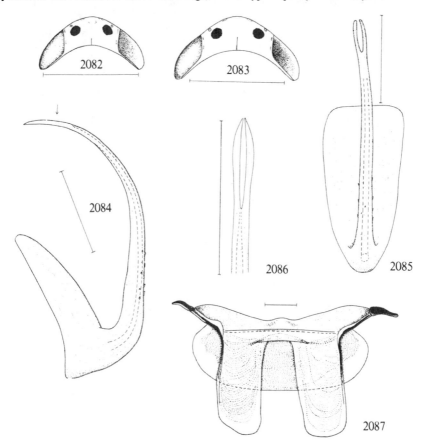

Text-figs. 2082-2087. *Sonronius anderi* (Ossiannilsson). – 2082: head of male from above; 2083: head of female from above; 2084: aedeagus from the left; 2085: aedeagus in ventral aspect; 2086: apex of aedeagus in ventro-terminal aspect (in direction of arrow in 2084); 2087: 2nd and 3rd abdominal sterna in male from above. Scale: 1 mm for 2082 and 2083, 0.1 mm for the rest.

Pronotum and scutellum greenish yellow. Fore wings longer than abdomen, semi-transparent, apex fumose, veins yellow. Hind wings with dark veins. Fore and middle femora with fuscous transverse bands and spots as in *dahlbomi;* hind tibiae dark spotted. Thoracic sternum and abdomen largely black. Female with head more angularly produced (Text-fig. 2083), venter of abdomen largely yellow. Aedeagus as in Text-figs. 2084-2086, 2nd and 3rd abdominal sterna as in Text-fig. 2087. Overall length of male 3.1-3.3 mm, of female 3.7-3.9 mm.

Distribution. So far not found in Denmark and Norway. – Very rare in Sweden, only found in Vg.: Kinnekulle, Österplana hed 24.VII.1947, 2 ♂♂, 10 ♀♀ (K. Ander). – Very rare in East Fennoscandia, only one male found in Ab: Rymättylä 30.VI.1949 (Linnavuori, 1950). – Not recorded outside Fennoscandia.

Biology unknown.

Genus *Sagatus* Ribaut, 1948

Sagatus Ribaut, 1948: 57.
Type-species: *Cicada punctifrons* Fallén, 1826, by original designation.

Related to *Macrosteles* and *Sonronius*. Macrosetae of legs, pygofer, and male genital plates not pubescent, those on genital plates disordered. Uppermost part of frontoclypeus visible from above. Male pygofer without a caudal fringe of macrosetae and without a ventro-caudal tubercle. Phallotreme situated on the ventral side of aedeagus. One species.

303. *Sagatus punctifrons* (Fallén, 1826)
Plate-fig. 214, text-figs. 2088-2094.

Cicada punctifrons Fallén, 1826: 42.

Greenish white to greenish yellow. Head short and broad, in male only slightly longer medially than near eyes, in female with frontoclypeus somewhat swollen, the median line being considerably longer than border-line on eyes as seen from above. Frontoclypeus often yellowish brown, especially in females. Vertex with two roundish black spots anteriorly touching ocelli, situated closer to eyes than to hind border. Between these spots and the hind border of vertex a pair of minute black spots may be present. In addition there may be some less distinct dark spots. Pronotum often with fumose patches, scutellum at most with fumose spots. Fore wings greenish with whitish veins, especially in females often with light brownish shades in the cells, particularly in clavus. Femora often with black longitudinal streaks on in- and outside. Dorsum of abdomen largely black, venter light or partly black. Male pygofer lobes as in Text-fig. 2088, genital plates, valve, style and connective as in Text-fig. 2089, style as in Text-fig. 2090, aedeagus as in Text-figs. 2091, 2092, 1st abdominal sternum as in Text-fig. 2093, 2nd and 3rd abdominal sterna as in Text-fig. 2094. Overall length of males 4-5 mm, of females 4.75-6.3 mm. – Nymphs with "2 longitudinal rows of hairs present laterally on abdominal tergites, with 2 median hairs on tergite VIII; hairs on hind angles of tergites V-VIII" (Vilbaste, 1982).

Distribution. Fairly common in Denmark, found in EJ, NWJ, NEJ, NEZ, B. – Common in southern and central Sweden, found in Sk., Hall., Gtl., Upl., Vrm., Dlr., Gstr. – Norway: found in AK, Bø, VE. – Common in southern and central East Fennoscandia,

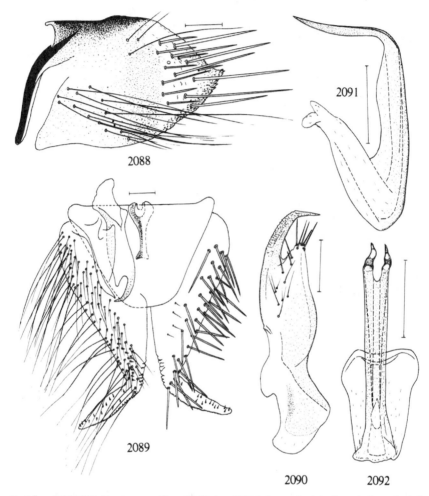

Text-figs. 2088-2092. *Sagatus punctifrons* (Fallén). – 2088: left pygofer lobe in male from the left; 2089: left genital plate with genital valve, connective and left genital style from above with dorsal pilosity (left) and from below with ventral pilosity (right); 2090: left genital style from above; 2091: aedeagus from the left; 2092: aedeagus in ventral aspect. Scale: 0.1 mm.

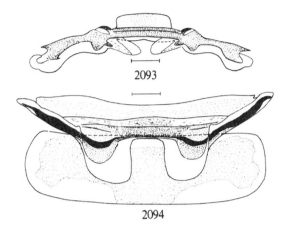

2093

2094

Text-figs. 2093-2094. *Sagatus punctifrons* (Fallén). – 2093: 1st abdominal sternum in male in antero-dorsal aspect; 2094: 2nd and 3rd abdominal sterna in male from above. Scale: 0.1 mm.

found in Ab, N, Ka, St, Ta, Sb; Vib, Kr. – England, France, Belgium, Netherlands, Switzerland, German D. R. and F. R., Austria, Italy, Hungary, Bohemia, Moravia, Slovakia, Romania, Albania, Poland, Estonia, Latvia, Lithuania, n. and m. Russia, Ukraine, Moldavia, Azerbaijan, Kazakhstan, Tuva, Altai Mts., m. Siberia: Nearctic region.

Biology. On *Salix repens* especially in sandy places, July-September (Ossiannilsson, 1947b). "Auf schmalblättrigen *Salix*-Arten *(S. repens, triandra, fragilis, incana, purpurea)* (Wagner & Franz, 1961).

Tribe Deltocephalini
Genus *Deltocephalus* Burmeister, 1838

Deltocephalus Burmeister, 1838b: [15].
Type-species: *Cicada pulicaris* Fallén, 1806, by subsequent designation.

Small leafhoppers. Body broad, robust. Head above comparatively convex. Frontoclypeus broad, convex, anteclypeus tapering towards apex or almost parallel-sided. Wing-dimorphous species, fore wings in f. sub-brachyptera a little shorter than abdomen. Anal tube in male not sclerotized. Connective fused with socle of aedeagus. Two species.

Key to species of *Deltocephalus*

1   Upper side straw-coloured, vertex with two sharply defined
     black spots, pronotum and fore wings entirely light .. 305. *maculiceps* (Boheman)
–   Upper side brownish yellow with indistinctly delimited fus-
     cous or blackish brown markings on head, pronotum, and
     fore wing .... 304. *pulicaris* (Fallén)

655

304. *Deltocephalus pulicaris* (Fallén, 1806)
Plate-fig. 165, text-figs. 2095-2101.

Cicada pulicaris Fallén, 1806: 21.

Upper side shining brownish with darker markings varying in extension, venter largely black. Vertex usually with the following black-brown markings, sometimes indistinct,

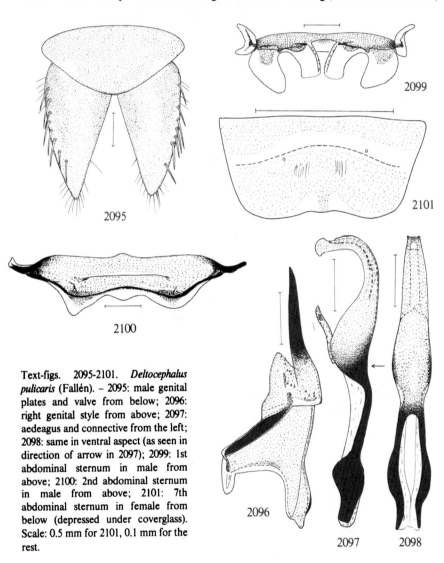

Text-figs. 2095-2101. *Deltocephalus pulicaris* (Fallén). – 2095: male genital plates and valve from below; 2096: right genital style from above; 2097: aedeagus and connective from the left; 2098: same in ventral aspect (as seen in direction of arrow in 2097); 2099: 1st abdominal sternum in male from above; 2100: 2nd abdominal sternum in male from above; 2101: 7th abdominal sternum in female from below (depressed under coverglass). Scale: 0.5 mm for 2101, 0.1 mm for the rest.

sometimes extending and confluent: near fore border two pairs of small spots, one near ocelli, another near apex of head; behind these a pair of fairly large roughly triangular spots, often confluent and thus forming a transverse band. Immediately in front of hind border of vertex one or two pairs of spots may be present. Black transverse streaks on frontoclypeus usually confluent; face often almost entirely black. Pronotum unicolorous brownish yellow or with a few dark spots along fore border; the dark colour behind these often forming longitudinal bands. Scutellum often with a pair of small dark spots on lateral angles and a cross-shaped marking between these. Veins of fore wings light, longitudinal veins usually dark-bordered, transverse veins broadly light-bordered; dark streaks in cells often confluent forming a large dark patch in each cell. Wing-dimorphous: male macropterous with fore wings a little longer than abdomen, female usually sub-brachypterous, fore wings being shorter than abdomen. Thoracic venter and abdomen black, the latter with quite narrow light segmental margins; coxae, apices of femora, and tibiae largely brownish yellow. Male genital plates and valve as in Text-fig. 2095, style as in Text-fig. 2096, aedeagus and connective firmly attached to each other as in Text-figs. 2097, 2098, 1st abdominal sternum in male as in Text-fig. 2099, 2nd abdominal sternum in male as in Text-fig. 2100, 7th abdominal sternum in female as in Text-fig. 2101. Overall length of males 2.15-2.8 mm, of females 2.4-3.4 mm. – "Whole nymph uniformly pale brown to black-brown, except for anal segment, which is clearly darker" (Vilbaste, 1982).

Distribution. Common in Denmark, found in almost all provinces. – Common and widespread in Sweden, Sk.-T. Lpm. – Norway: common and widespread, Ø, AK-Fi. – Common and widespread in East Fennoscandia, found in almost all provinces, Al, AK, N-ObN, Ks, Li; Vib, Kr, Lr. – Widespread in Europe, also found in Algeria, Armenia, Moldavia, Georgia, Altai Mts., Kazakhstan, Tuva, m. and w. Siberia, Kirghizia, Kamchatka, Kurile Isl.; Nearctic region.

Biology. In open fields and meadows, both dryish fields, moist sloping meadows, peaty meadows, and cultivated fields (Linnavuori, 1952a). In the "Molinio-Arrhenatheretea" (Marchand, 1953). Hibernation takes place in the egg-stage, in central Europe 2 generations (Remane, 1958; Schiemenz, 1969b). "Frequent and abundant in leys and pastures, and migrated to cereal crops" (Raatikainen & Vasarainen, 1973). "Its natural food-plants are grasses" (Raatikainen & Vasarainen, 1976). Adults in June-September.

305. *Deltocephalus maculiceps* Boheman, 1847
   Plate-fig. 166, Text-figs. 2102-2106.

Deltocephalus maculiceps Boheman, 1847b: 264.

Upper side shining straw-coloured or light yellow, unspotted except for a pair of jet-black spots on vertex (Plate-fig. 166). Transverse streaks on frontoclypeus in female usually not or little darker than interspaces but in male the lower part of frontoclypeus is largely black. Thoracic venter in female largely light, in male partly or largely black.

legs entirely light, or trochanters and femora with broad black transverse bands, tibiae with black points, hind tibiae and hind tarsi often largely dark. Abdomen in male black with narrow yellow segmental margins and lateral borders; in female the light colour is more extended. Male genital plates and valve as in Text-fig. 2102, genital style as in Text-fig. 2103, aedeagus with attached connective as in Text-figs. 2104, 2105. 7th abdominal sternum in female as in Text-fig. 2106. Overall length of male 2.6-3.1 mm, of female 2.2-3.6 mm.

Distribution. Widespread but rather scarce in Denmark, found in EJ, LFM, NEZ, and B. – Rare in Sweden, only found in Gtl.: Läderbro, Lummelund (Boheman), Myrvälder (Boheman, Ossiannilsson), Tingstäde (Ossiannilsson), Sundre, Muskmyr (Danielsson). Also Stål found *maculiceps* in Gotland. – Not found in Norway, nor in East Fennoscandia. – England, Ireland, France, German D. R. and F. R., Netherlands, Switzerland, Bohemia, Poland.

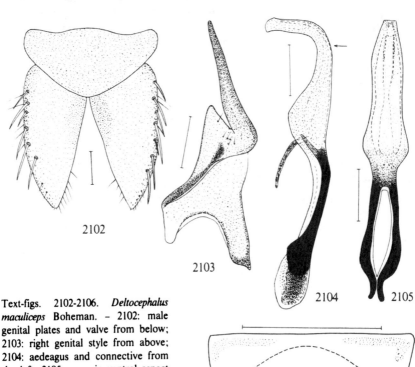

2102

2103

2104   2105

Text-figs. 2102-2106. *Deltocephalus maculiceps* Boheman. – 2102: male genital plates and valve from below; 2103: right genital style from above; 2104: aedeagus and connective from the left; 2105: same in ventral aspect (as seen in direction of arrow in 2104); 2106: 7th abdominal sternum in female from below (depressed under coverglass). Scale: 0.5 mm for 2106, 0.1 mm for the rest.

2106

Biology. In "Hochmooren und Flachmooren. An *Eriophorum?*" (Kuntze, 1937). "Eiüberwinterer mit einer Generation" (Remane, 1958).

Genus *Endria* Oman, 1949

Endria Oman, 1949: 175.
    Type-species: *Jassus inimicus* Say, 1830, by original designation.

As *Deltocephalus*. Fore wings with some secondary transverse veins in clavus, sometimes also in corium. In the Palaearctic region one species.

306. *Endria nebulosa* (Ball, 1900)
    Text-figs. 2107-2114.

Lonatura nebulosa Ball, 1900: 341.
Endria nebulosa Remane, 1961b: 73.

Body robust. Fore border of head as seen from above obtusely angular, less rounded than in *Deltocephalus pulicaris*, transition of vertex and frontoclypeus rounded. Head a little shorter than pronotum. Setae on legs and pygofer (♂♀) conspicuously long. Wing-dimorphous, with a macropterous and a brachypterous form in both sexes. Fore wings of macropters a little longer or (in egg-swollen females) a little shorter than abdomen; fore wings of brachypters covering roughly 2/3 of abdomen. Macropters in general aspect suggestive of male *Deltocephalus pulicaris*, differing by greater size and less confluent markings. Head above sordid yellow, usually with 5 pairs of black markings, viz. 2 pairs of large spots immediately behind fore border, behind these a pair of transverse sickle-shaped spots, and 2 pairs of smaller more or less ovoid spots near caudal border. Face largely brownish with light patches, or largely light with brown transverse streaks. Pronotum whitish with a few black spots near fore border and three pairs of indistinctly limited brown longitudinal bands. Scutellum whitish or sordid yellow with caudal apex and basal triangles fuscous. Fore wings whitish, veins partly dark-bordered, the dark pigment more extended in clavus. Abdomen and venter of thorax largely dark brown or black, segmental borders light. Femora light, transversely black-banded, tibiae with large black spots and streaks. Hind tarsi light with 1st and 2nd segments largely black, the former with basal 1/3 or 1/4 light. – Brachypters largely light yellow, black markings much reduced. On the head, only the two frontal pairs of black spots are quite distinct, remaining markings obsolete or absent. Face, pronotum and scutellum entirely or largely light yellow, fore wings transparent, veins white, partly white-bordered. Thorax and abdomen largely light yellow, the latter with more or less distinct dark transverse streaks and lateral dots. Legs largely light, tibiae and hind tarsi more or less as in macropters. Male pygofer lobes as in Text-fig. 2107 (pigmentation not always as shown in figure!), male genital plates and valve as in Text-fig. 2108, genital style as in Text-fig. 2109, aedeagus as in Text-figs. 2110-2112, 1st abdominal sternum in male as in text-fig. 2113, 7th abdominal sternum in female as in Text-fig. 2114. Overall length of

659

macropters 3.8-4.1 mm, of brachypterous males 3.5-3.8 mm, of brachypterous females 4.1-4.3 mm (measurements according to Remane, 1961b).

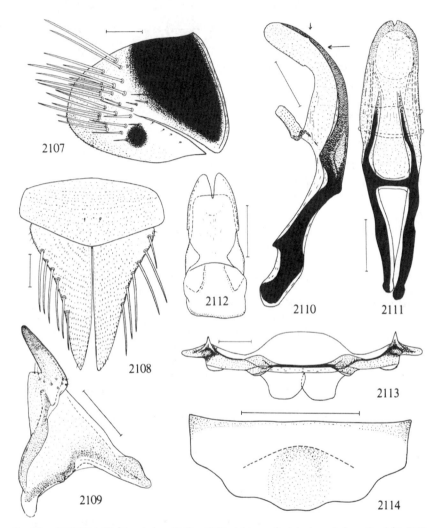

Text-figs. 2107-2114. *Endria nebulosa* (Ball). – 2107: right pygofer lobe in male from the right; 2108: male genital plates and valve from below; 2109: right genital style in male from above; 2110: aedeagus and connective from the left; 2111: same in ventral aspect (as seen in direction of longer arrow in 2110); 2112: apex of aedeagus in ventro-terminal aspect (in direction of shorter arrow in 2110); 2113: 1st abdominal sternum in male from above; 2114: 7th abdominal sternum in female from below (depressed under coverglass). Scale: 0.5 mm for 2114, 0.1 mm for the rest.

Distribution. Rare in Denmark, found in several places in B: Arnager, Ølene and Poulsker plantage (Trolle). – Very rare in Sweden, only found in Gtl. in 1974 by R. Remane (in litt.). – Not recorded from Norway. – East Fennoscandia: found in N: Tusby, and Vanda, Sjöskog, 1974, and in Tvärminne and Bromarv, Täcktom in 1976 by A. Albrecht. – German F. R. and D. R., Bohemia, Moravia, m. Russia, Mongolia, Korean Peninsula; Nearctic region.

Biology. On *Calamagrostis epigeios* (Remane, 1961b; Albrecht, 1977; Trolle, in litt.). On *Carex limosa* and *Carex* spp. in peat-bogs (Lauterer, 1980). Hibernation in the egg-stage; one generation (Remane, 1961b). In vole tunnels (Trolle, 1982). Adults in July-October.

Tribe Doraturini
Genus *Doratura* J. Sahlberg, 1871

Doratura J. Sahlberg, 1871: 291.
Type-species: *Athysanus stylatus* Boheman, 1847, by subsequent designation.
Doraturina Emeljanov, 1964: 403.
Type-species: *Jassus homophylus* Flor, 1861, by original designation.

Fairly small, robust species. Head rounded angular, longer than pronotum, vertex concave, frontoclypeus above transversely impressed. Pronotum short, sides not carinate. Wing-dimorphous species, usually brachypterous. Fore wings in brachypters very short, leaving most of abdomen uncovered (Plate-fig. 167), in macropters about as long as abdomen. Macrosetae of male genital plates placed in disorder. Shape of male genital style characteristic, see Text-figs. 2116 etc. Aedeagus simple, shaft without appendages. Phallotreme subapical, on dorsal side of aedeagus. Female saw-case extending far beyond caudal end of pygofer (Text-figs. 2119, 2135). In Denmark and Fennoscandia four species.

Key to species of *Doratura* (adults)

1 Males 2
– Females 5
2 (1) Genital plates apically angular as in Text-fig. 2130. Shaft of
aedeagus on apical half laterally serrate (Text-fig. 2134). Apical part of genital style (distally of medially directed tooth)
comparatively short (Text-figs. 2131, 2132) (Subgenus *Dora-
turina* Emeljanov) .... 310. *homophyla* (Flor)
– Shape of genital plates different. Shaft of aedeagus not ser-
rate. Apical part of genital style long (Text-figs. 2116, 2121,
2126) (Subgenus *Doratura* s.str.) 3
3 (2) Shaft of aedeagus on ventral side with minute spinules (Text-
fig. 2117). .... 307. *stylata* (Boheman)

Shaft of aedeagus ventrally quite smooth. 4

♪ (3) Large species, length at least 4 mm. Fore wings in brachypters usually without conspicuous dark markings. Apical part
of genital style comparatively longer (Text-fig. 2126) .... 309. *impudica* Horváth

- Small species, length at most 3.3 mm. Fore wings in brachypters usually with a dark streak along claval suture, often
also with additional dark markings. Apical part of genital
style comparatively shorter .... 308. *exilis* Horváth

5 (1) Caudal border of 7th abdominal sternum strongly convex
(Text-fig. 2135) .... 310. *homophyla* (Flor)

- Caudal border of 7th abdominal sternum almost straight or
with a small median projection 6

6 (5) Large species, length at least 5.3 mm .... 309. *impudica* Horváth

- Smaller species, length at most 4.5 mm 7

7 (6) Fore wing in brachypters longer, not or little shorter than
visible part of saw-case .... 307. *stylata* (Boheman)

- Fore wing in brachypters considerably shorter than visible
part of saw-case (ratio 0.8 or less) .... 308. *exilis* Horváth

Key to species of *Doratura* (nymphs)
(After Vilbaste, 1982)

1 Tergum of abdomen pale, speckled dark; large spots pre-
sent around bases of spines .... 309. *impudica* Horváth

- Tergum of abdomen mostly dark, either uniformly so, or
with dark longitudinal bands 2

2 (1) Abdominal tergum with 2 wide, sharply delimited dark longitudinal bands separated by a broad (1/4-1/5 width of dark
bands) whitish longitudinal band; lateral margins of body
narrowly whitish .... 308. *exilis* Horváth

- Abdominal tergum without dark brown longitudinal bands;
pale median longitudinal band narrower 3

3 (2) Abdomen usually wholly unicoloured, sometimes with a narrow pale median longitudinal band; lateral margins much
paler than central areas, the limits of these parts being con-
tinuous .... 310. *homophyla* (Flor)

- Abdomen often with whitish patches on tergites IV and V;
if unicoloured, then dark pigment extends almost to the la-
teral margin, or margin only slightly paler .... 307. *stylata* (Boheman).

307. ***Doratura (Doratura) stylata*** (Boheman, 1847)
Plate-fig. 167, text-figs. 2115-2119

Athyanus (sic) *stylatus* Boheman, 1847a: 31.

662

Brownish yellow, shining. Head above on fore border with 3 black spots extending also on upper part of face, the middle one being the largest. On vertex caudally of these spots additional black or fuscous spots arranged in two transverse rows may be present, the anterior one often missing. Face below marginal spots with two black transverse bands, one between eyes, the other below eyes, the latter continuing caudad on thoracic pleura. Anteclypeus and lora with black markings. Pronotum entirely light or with some indistinctly delimited fuscous spots. Scutellum entirely light or with a blackish brown longitudinal band. Fore wings of brachypters covering abdominal terga I-IV, caudally rounded, leathery, unicolorous light or with more or less distinct dark streaks or spots especially in apical cells. Fore wings of macropters hyaline, in females extending to apex of pygofer. Abdominal tergum with a more or less distinct whitish, sometimes black-edged, median longitudinal line; each segment with a transverse row

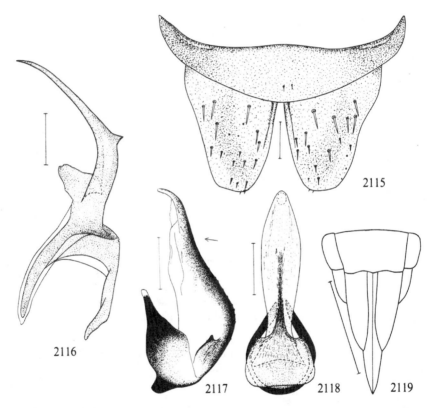

Text-figs. 2115-2119. *Doratura stylata* (Boheman). – 2115: male genital plates and valve from below; 2116: right genital style from above; 2117: aedeagus from the left; 2118: aedeagus in ventral aspect; 2119: apical part of female abdomen from below. Scale: 1 mm for 2119, 0.1 mm for the rest.

of small black spots, two of which tend to extend on each side longitudinally resulting in the formation of two pairs of longitudinal bands. Male pygofer with 3 black spots. Female with a roundish or polygonal median black spot on 8th abdominal tergum and with a median longitudinal streak on 9th tergum. Venter black-spotted to largely black. Male genital plates and valve as in Text-fig. 2115; in intact specimens the anterior part of the valve is covered by the caudal border of 8th abdominal sternum, thereby appearing to be much shorter. Male genital style as in Text-fig. 2116, aedeagus as in Text-figs. 2117, 2118. Apical part of female abdominal sternum as in Text-fig. 2119. Overall length of male 3.0-3.7 mm, of female 3.5-4.5 mm.

Distribution. Common in Denmark, found in almost all provinces. – Common in Sweden, Sk.-Vb., Lu.Lpm. – Norway: fairly common in the south, found in Ø, AK, Os, On, Bø, Bv, TEy, TEi, AAy, AAi, VAy. – Very common in southern and central East Fennoscandia, found in Al, Ab-Om; Vib, Kr, Lr. – Widespread in Europe, also recorded from Algeria, Tunisia, Moldavia, Georgia, Kazakhstan, Tuva, Altai Mts., m. and w. Siberia, Kirghizia, Maritime Territory; Nearctic region.

Biology. In "Binnendünen, Sandfeldern, Wiesen" (Kuntze, 1937). In dryish fields and moist sloping meadows (Linnavuori, 1952a). In the "Molinietum typic. Var. v. *Nardus stricta*" (Marchand, 1953). Hibernation in the egg stage, one generation (Remane, 1958; Schiemenz, 1969b). Adults in June-October.

308. *Doratura (Doratura) exilis* Horváth, 1903
Text-figs. 2120-2124.

Doratura exilis Horváth, 1903a: 454.

Resembling *stylata,* somewhat smaller. Fore wings in brachypters shorter, caudally more square, more or less hyaline, with a dark streak along claval suture, usually also with additional dark markings. Light longitudinal line on abdominal tergum distinct, sharply limited, usually dark-bordered; two pairs of lateral longitudinal bands present on abdominal tergum as in strongly pigmented specimens of *stylata.* Male genital plates and valve as in text-fig. 2120, genital style as in Text-fig. 2121, aedeagus as in Text-figs. 2122, 2123, connective as in Text-fig. 2124. 7th abdominal sternum in female much as in *stylata,* caudal margin medially often obtusely produced. Saw-case in female comparatively longer than in *stylata,* its ventral margin almost straight. Overall length of male 2.5-3.0 mm, of female 3.5-4.0 mm.

Distribution. So far not recorded from Denmark, Norway and East Fennoscandia. – Sweden: scarce but often abundant in suitable localities, found in Sk.: Rinkaby (D. Gaunitz); Öl.: Borgholm; Vickleby; Resmo; Karlevi; Kastlösa; Kalkstad (D. Gaunitz, Bornfeldt, Ossiannilsson); Vg.: Kinnekulle, Österplana hed; Karlevi; Valtorp, Öja hed; Vilske-Kleva-heden (Ander). – France, Portugal, German D.R. and F.R., Austria, Italy, Albania, Andorra, Bohemia, Moravia, Slovakia, Hungary, Yugoslavia, Bulgaria, Romania, Poland, Estonia, Latvia, Lithuania, Ukraine, Armenia, Moldavia, Iran, Kazakhstan, Mongolia, Altai Mts., w. Siberia.

Biology. In "besonnten Hängen" (Kuntze, 1937). Xerophilous, belonging to the "Corynephoretum cladonietosum" (Marchand, 1953). Hibernation in egg-stage, 1 generation (Remane, 1958; Schiemenz, 1969b). In dry meadows on sandy soils (Vilbaste, 1974). Adults in June-August.

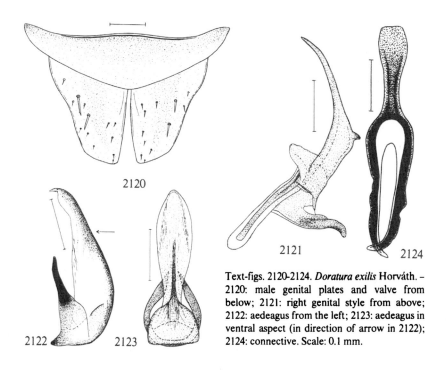

2120

2121

2124

2122

2123

Text-figs. 2120-2124. *Doratura exilis* Horváth. – 2120: male genital plates and valve from below; 2121: right genital style from above; 2122: aedeagus from the left; 2123: aedeagus in ventral aspect (in direction of arrow in 2122); 2124: connective. Scale: 0.1 mm.

309. *Doratura (Doratura) impudica* Horváth, 1897
Text-figs. 2125-2129.

Doratura impudica Horváth, 1897: 629.

Resembling *D. stylata* but larger. Fore wings in brachypters caudally fairly square, covering the four basal abdominal terga. Male genital plates and valve as in Text-fig. 2125, genital style as in Text-fig. 2126, aedeagus as in Text-figs. 2127-2129. Caudal border of 7th abdominal sternum in female almost straight. Overall length of males 4.0-4.3 mm, of females 5.0-5.75 mm.

Distribution. Scarce in Denmark, found in EJ, WJ, NEZ, and B. – Very rare in Sweden, only found in Sk.: Maglehem, St. Juleboda 13.VIII.1930 (D. Gaunitz), and in Sk.: Ravlunda, middle of August, 1933 (K. Ander). – Not recorded from Norway and East Fennoscandia. – England, France, Netherlands, German D.R. and F.R., Austria,

Italy, Bohemia, Moravia, Slovakia, Bulgaria, Albania, Romania, Yugoslavia, Hungary, Poland, Estonia, Latvia, Lithuania, Ukraine, Anatolia, Moldavia, Kazakhstan, Uzbekistan, Kirghizia.

Biology. Xerothermophilous (Schiemenz, 1965). "Stenotope Art der Trockenrasen" (Schiemenz, 1969b). "On sand dunes, in dry meadows, in young pine forests, etc." (Vilbaste, 1974). Adults in July-August.

310. *Doratura (Doraturina) homophyla* (Flor, 1861)
Text-figs. 2130-2135.

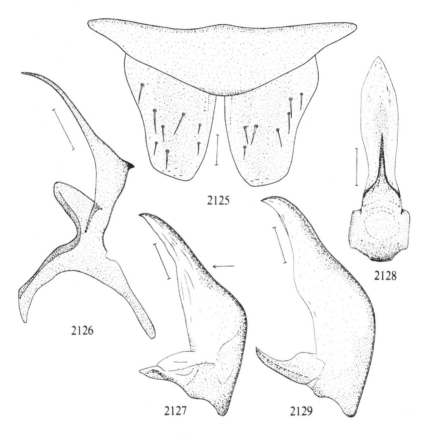

Text-figs. 2125-2129. *Doratura impudica* Horváth. – 2125: male genital valve and plates from below; 2126: right style from above; 2127: aedeagus of Swedish specimen from the left; 2128: same in ventral aspect (in direction of arrow in 2127); 2129: aedeagus of Polish specimen from the left. Scale: 0.1 mm.

Jassus (Athysanus) homophylus Flor, 1861a: 276.

Resembling *D. stylata,* upper side mostly more strongly dark-marked, also fore wings often with dark streaks and dots in cells. Fore wings in brachypters covering the 4 basal abdominal terga. Male genital segment large, visible part of genital valve long, genital plates (Text-fig. 2130) as long as pygofer lobes, apically obliquely cut off, latero-apical corner rectangular; style as in Text-figs. 2131, 2132, aedeagus as in Text-figs. 2133, 2134, apex of female abdomen in ventral aspect as in Text-fig. 2135. Overall length of males 3.0-3.5 mm, of females 4.0-5.0 mm.

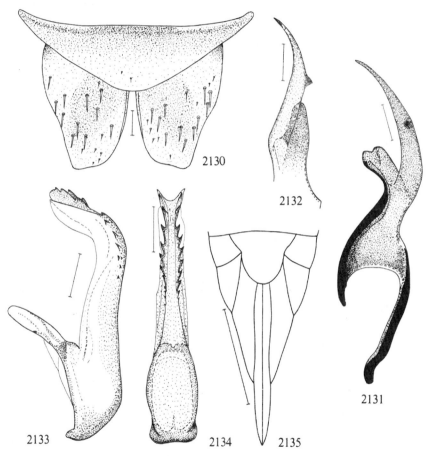

Text-figs. 2130-2135. *Doratura homophyla* (Flor). – 2130: male genital valve and plates from below; 2131: right genital style from above; 2132: apical part of genital style from inside; 2133: aedeagus from the left; 2134: aedeagus in ventral aspect; 2135: apical part of female abdomen from below. Scale: 1 mm for 2135, 0.1 mm for the rest.

Distribution. Widespread but not very common in Denmark, found in SJ, EJ, NWJ, NEJ, LFM, NEZ. – Scarce, locally abundant in Sweden, found in Sk., Bl., Hall. Öl. – So far not recorded from Norway. – Scarce in East Fennoscandia, somewhat commoner in the eastern and south-eastern parts, found in Al, Ta, Kb; Vib, Kr. – France, Belgium, Netherlands, German D.R. and F.R., Austria, Italy, Albania, Bulgaria, Bohemia, Moravia, Slovakia, Hungary, Yugoslavia, Poland, Estonia, Latvia, Lithuania, n., m., and s. Russia, Ukraine, Armenia, Moldavia, Anatolia, Iran, Iraq, Israel, Altai Mts., Azerbaijan, Kazakhstan, Uzbekistan, Tuva, Mongolia, w. Siberia, Kirghizia.

Biology. In "Binnendünen, Sandfeldern, besonnten Hängen, Heiden, Wiesen" (Kuntze, 1937). In the "Corynephoretum agrostidetosum aridae" (Marchand, 1953). "Ein Tier steppenartiger Sandfelder und besonnter Hänge" (Strübing, 1955). Hibernation in the egg-stage, one generation (Remane, 1958; Schiemenz, 1969b). In dry meadows, mainly on sandy soils (Vilbaste, 1974). Adults in June-September.

Tribe Athysanini
Genus *Platymetopius* Burmeister, 1838

Platymetopius Burmeister, 1838b: 16.

Type-species: *Cicada vittata* Fabricius, 1775 (nec Linné, 1758), by subsequent designation.

Head medially more or less angularly produced. Transition between vertex and face abrupt, angular. Ocelli situated on fore border of head close to eyes. Anteclypeus expanding towards apex. Pronotum laterally carinate. Male pygofer with a ventro-caudal process. Stem of connective long. Elongate species with wings longer than body and with characteristic elegant markings. In Denmark and Fennoscandia three species.

Key to species of *Platymetopius*

1   Fore wings brownish yellow with many milk-white spots varying in size (Plate-fig. 168) .... 313. *guttatus* Fieber
–   Fore wings yellow with a broad sinuous longitudinal brown band along commissural border and in apical part (Plate-fig. 219) 2
2 (1) Appendages of male pygofer lobes as in Text-fig. 2137; 7th abdominal sternum in female as in Text-fig. 2144 .... 311. *undatus* (De Geer)
–   Appendages of male pygofer lobes as in Text-fig. 2145. 7th abdominal sternum in female as in Text-fig. 2150 .... 312. *major* (Kirschbaum).

311. *Platymetopius undatus* (De Geer, 1773)
     Plate-fig. 219, text-figs. 2136-2144.

Cicada undata De Geer, 1773: 185.
Cicada vittata Fabricius, 1775: 684 (nec Linné, 1758).

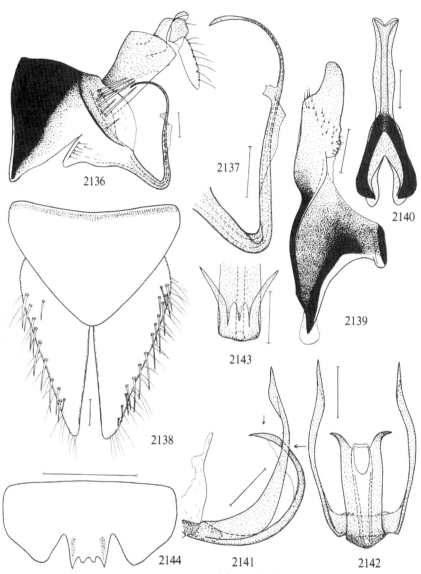

Text-figs. 2136-2144. *Platymetopius undatus* (De Geer). – 2136: left pygofer lobe and anal apparatus in male from the left; 2137: appendage of left pygofer lobe in male from the left; 2138: male genital plates and valve from below; 2139: right genital style from above; 2140: connective from above; 2141: aedeagus from the left; 2142: aedeagus in ventral aspect (as seen in direction of longer arrow in 2141); 2143: apex of canaliferous part of aedeagus as seen in direction of shorter arrow in 2141; 2144: 7th abdominal sternum in female from below. Scale: 0.5 mm for 2144, 0.1 mm for the rest.

Elongate, shining, upper side brownish, venter whitish yellow to sordid yellow. Vertex frontally fairly acute-angled, apically with a short light median streak, lateral borders light, the rest of vertex as well as pronotum and scutellum whitish mottled, pronotum with lateral borders light. Fore wings light yellow, along posterior margin with a broad brown longitudinal band with two arched indentations (Plate-fig. 219), surface of the proximal indentation (in clavus) being milk-white. In addition there are some whitish dots in the brown band. Subcostal cell with some indistinct secondary transverse veins. Male pygofer and anal apparatus as in Text-fig. 2136, appendage of pygofer as in Text-fig. 2137, genital plates and valve as in Text-fig. 2138, genital style as in Text-fig. 2139, connective as in Text-fig. 2140, aedeagus as in Text-figs. 2141-2143, 7th abdominal sternum in female as in Text-fig. 2144. Overall length of males 4.8-5.3 mm, of females 5.0-6.2 mm. - Nymphs pale yellow, lateral margins of head concave; abdomen broad, dorsoventrally flattened.

Distribution. Uncommon in Denmark, found only in Jutland (EJ, NWJ, WJ). - Not uncommon in southern and central Sweden, found up to Ång. - Scarce in Norway, found in AK, Bø, AAy. - Fairly scarce in southern and central East Fennoscandia, found in Al, Ab, N, St, Ta, Sa, Kb; Vib, Kr. - Widespread in Europe, also recorded from Tunisia, Anatolia, Israel, Moldavia, Georgia, Kazakhstan, Azerbaijan, Altai Mts., Tuva, Mongolia, m. and w. Siberia, Maritime Territory, Korean Peninsula.

Biology. "Auf Waldwiesen" (Kuntze, 1937). "Auf niedrigem Bodenbewuchs, besonders auf *Helianthemum*" (Wagner, 1939). "Bewohnt Trockenrasen in warmen Lagen" (Wagner & Franz, 1961). "Reported from *Pteris* (bracken), from sallow and from oak" (Le Quesne, 1969). Hibernation in the egg-stage, one generation (Schiemenz, 1969b). On deciduous trees and shrubs (Vilbaste, 1974). I found *P. undatus* on *Rhamnus, Crataegus, Prunus padus, Rubus idaeus;* adults in June-September.

312. *Platymetopius major* (Kirschbaum, 1868)
Text-figs. 2145-2150.

Jassus (Platymetopius) major Kirschbaum, 1868b: 147.

Resembling *P. undatus,* on an average slightly larger. Appendages of males pygofer as in Text-fig. 2145, style as in Text-fig. 2146, aedeagus as in Text-figs. 2147-2149, 7th abdominal sternum in female as in text-fig. 2150. Overall length (according to Ribaut, 1952) of male 5.2-5.9 mm, of female 5.9-6.5 mm. - Walter (1978) described the nymphs of this species.

Distribution. Rare in Denmark, found in a few places in EJ: Ørnsø skov and Frijsenborg many years ago. - Very rare in Sweden, one female only found in Öl.: Högsrum, Rönnerum 21.VII.1940 (E. Wieslander). - Not recorded from Norway and East Fennoscandia. - France, Netherlands, German D.R. and F.R., Austria, Italy, Greece, Albania, Bulgaria, Bohemia, Slovakia, Yugoslavia, Poland, Ukraine, Anatolia, Jordan, Iran, Moldavia, w. Siberia.

Biology. "An Eiche. (Ich klopfte sie von *Crataegus*.)" (Kuntze, 1937). "Xerotherme Zikade, die innerhalb der Pflanzenassoziation des Querceto-Carpinetum nur in der trockenen Subassoziation des Querceto-Carpinetum pubescentetosum vorkommt . . ." (Schwoerbel, 1957). Hibernation takes place in the egg-stage (Müller, 1957; Schiemenz, 1969b). One generation (Schiemenz, 1969b).

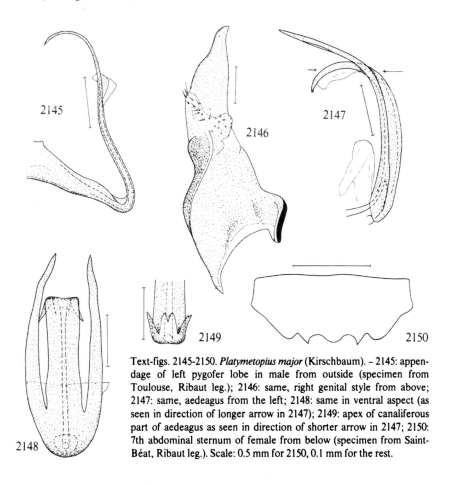

Text-figs. 2145-2150. *Platymetopius major* (Kirschbaum). – 2145: appendage of left pygofer lobe in male from outside (specimen from Toulouse, Ribaut leg.); 2146: same, right genital style from above; 2147: same, aedeagus from the left; 2148: same in ventral aspect (as seen in direction of longer arrow in 2147); 2149: apex of canaliferous part of aedeagus as seen in direction of shorter arrow in 2147; 2150: 7th abdominal sternum of female from below (specimen from Saint-Béat, Ribaut leg.). Scale: 0.5 mm for 2150, 0.1 mm for the rest.

313. ***Platymetopius guttatus*** Fieber, 1869
Plate-fig. 168, text-figs. 2151-2157.

Platymetopius guttatus Fieber, 1869: 202.

Elongate, upper side shining brownish yellow, venter largely whitish yellow. Head

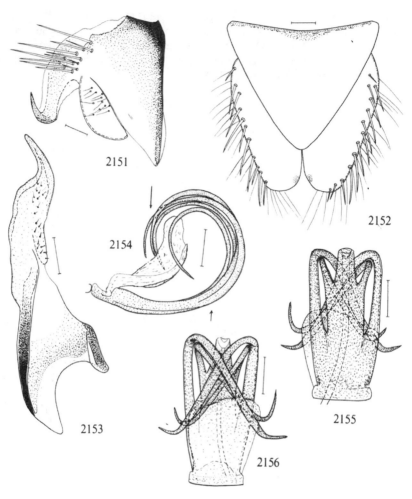

Text-figs. 2151-2157. *Platymetopius guttatus*
Fieber. – 2151: right pygofer lobe in male from
outside; 2152: male genital plates and valve
from below; 2153: right genital style from
above; 2154: aedeagus from the left; 2155:
aedeagus in ventral aspect (in direction of
shorter arrow of 2154); 2156: aedeagus as seen
in direction of longer arrow in 2154; 2157: 7th
abdominal sternum in female from below.
Scale: 0.5 mm for 2157, 0.1 mm for the rest.

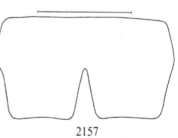

672

frontally obtuse angular. Face sordid yellow, with indistinct darker transverse streaks. Vertex and pronotum whitish mottled. Fore wings considerably longer than abdomen, with many secondary transverse veins, in cells with numerous hyaline milky spots varying in size. Setae of hind tibiae often arising from black dots. Male pygofer lobes as in Text-fig. 2151, genital plates and valve as in Text-fig. 2152, style as in Text-fig. 2153, aedeagus as in Text-figs. 2154-2156, 7th abdominal sternum in female as in Text-fig. 2157. Overall length (♂♀) 5.0-6.5 mm.

Distribution. So far not found in Denmark, nor in East Fennoscandia. – Rare in Sweden, found in Bl.: Karlskrona (Gyllensvärd); Ög.: vicinity of Norrköping (Haglund); Sdm.: Nacka (Tullgren), Botkyrka, Ahlby (Ossiannilsson); Upl.: Hilleshög, Ricksätra (Ossiannilsson), Ekerö (Ossiannilsson). – Rare in Norway, found in AK: Oslo, Kongshavn, and Ormøya (Siebke); Bø: Røyken (Siebke); AAy: Risør (Warloe. – France, German D.R. and F.R., Portugal, Switzerland, Austria, Italy, Bohemia, Hungary, Bulgaria, Yugoslavia, Albania, Greece, Crete, Poland, Cyprus, Anatolia, Ukraine, Iran, Moldavia, Tadzhikistan.

Biology. "Auf *Quercus* in sonniger Lage" (Wagner & Franz, 1961). I collected in all 29 specimens on birch *(Betula),* one specimen on *Rhamnus cathartica.* Adults in July-September.

Genus *Idiodonus* Ball, 1936

Idiodonus Ball, 1936: 57.
    Type-species: *Jassus kennicotti* Uhler, 1864, by original designation.
Orolix Ribaut, 1942: 267.
    Type-species: *Cicada cruentata* Panzer, 1799, by original designation.

Body elongate, structure robust. Head considerably shorter and not wider than pronotum. Transition between vertex and face rounded. Anteclypeus somewhat expanding towards apex. Lateral borders of genital plates in male concave. Genital style short. Connective short and broad. Phallotreme apical. In the Palaearctic region one species.

314. *Idiodonus cruentatus* (Panzer, 1799)
    Plate-fig. 220, text-figs. 2158-2165.

Cicada cruentata Panzer, 1799: 15.

Brownish yellow, shining, more or less densely sprinkled with red dots, also with some black markings. Head on apex with or without a pair of roundish or transversely oval black spots. A black spot varying in size may be present below each antenna. Sutures on face often black. Frontoclypeus usually with brownish transverse streaks. Vertex sometimes with an indistinct dark transverse band. Apical part of fore wing fumose. The red markings may be entirely absent and when the head is strongly black-marked confusion with *Colladonus torneellus* is possible. Thoracic venter largely black, legs

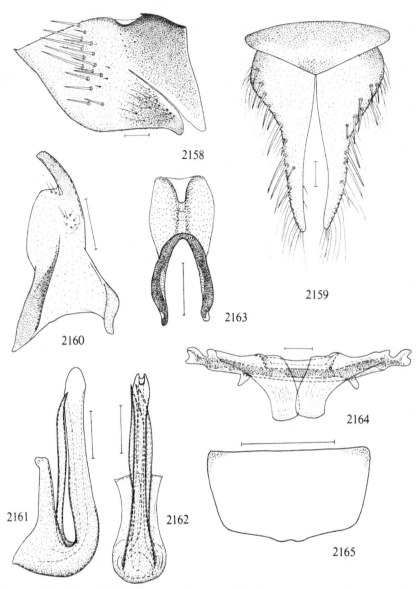

Text-figs. 2158-2165. *Idiodonus cruentatus* (Panzer). – 2158: right pygofer lobe of male from the right; 2159: genital plates and valve from below; 2160: right genital style from above; 2161: aedeagus from the left; 2162: aedeagus in ventral aspect; 2163: connective; 2164: 1st abdominal sternum in male from above; 2165: 7th abdominal sternum in female from below. Scale: 0.5 mm for 2165, 0.1 mm for the rest.

light. Abdomen largely black in males, largely light in females. Male pygofer lobes as in Text-fig. 2158, genital plates and valve as in Text-fig. 2159, style as in Text-fig. 2160, aedeagus as in Text-figs. 2161, 2162, connective as in Text-fig. 2163, 1st abdominal sternum in male as in Text-fig. 2164, 7th abdominal sternum in female longer than the preceding sternum, shaped as in Text-fig. 2165. Overall length of males 4.3-5.3 mm, of females 4.5-5.5 mm. – In nymphs, there are 4 longitudinal rows of hairs on abdominal tergum; dorsal body surface speckled brownish and red; 2 paler semicircular patches often present on sides of abdomen; sides of anteclypeus concave (Vilbaste, 1982). See also Walter (1978).

Distribution. Quite rare in Denmark, found mainly in Jutland and once in LFM: Tykskov. – Common and widespread in Sweden, Sk.-T.Lpm. – Common and widespread in Norway, found in Ø, AK-SFi, NTi, TRy, TRi, Fi. – Common and widespread in southern and central East Fennoscandia, Al, Ab-ObN, Ks; Vib, Kr. – Widespread in Europe; also Morocco, Kazakhstan, Altai Mts., Tuva, Mongolia, m. and w. Siberia, Sakhalin, Kamchatka, Korean Peninsula, Maritime Territory.

Biology. On *Betula* and other deciduous trees, especially in bogs, in July and August (Sahlberg, 1871). In pine bogs with undershrubs, cloudberry-*Sphagnum fuscum* bogs, rich swampy woods and dry *Calluna* heaths (Linnavuori, 1952a). Adults in June-September.

Genus *Colladonus* Ball, 1936

Colladonus Ball, 1936: 57.
Type-species: *Thamnotettix collaris* Ball, 1902, by original designation.
Hypospadianus Ribaut, 1942: 264.
Type-species: *Cicada torneella* Zetterstedt, 1828, by original designation.

Structure of body, head, anteclypeus and wings more or less as in *Idiodonus*. Sides of genital plates in male not distinctly concave. Genital style long, connective narrowly Y-shaped. Phallotreme situated near middle of dorsal side of shaft. In the Palaearctic region one species.

315. *Colladonus torneellus* (Zetterstedt, 1828)
Plate-fig. 221, text-figs. 2166-2173.

Cicada torneella Zetterstedt, 1828: 528.

Brownish yellow with black markings, shining. Head at apex with two transverse black spots, behind these a black or brownish transverse band which may be interrupted medially or disintegrated. Dark transverse streaks on frontoclypeus, if present, not reaching the median line. A black spot below each antenna, frontal suture usually black. In dark specimens also the genae may be dark or dark-spotted. Pronotum often blackish along fore border, caudal part usually greyish white. Scutellum often with a black transverse band enclosing the scuto-scutellar suture. Veins of fore wings light, sometimes indistinctly dark-bordered, claval suture blackish brown. Apical part of fore wing fumose. Thoracic venter largely black. Legs light, femora often with dark longitudinal streaks. Abdomen in male black with light lateral and segment borders, genital plates partly light, female abdomen with light colour more extended. Male

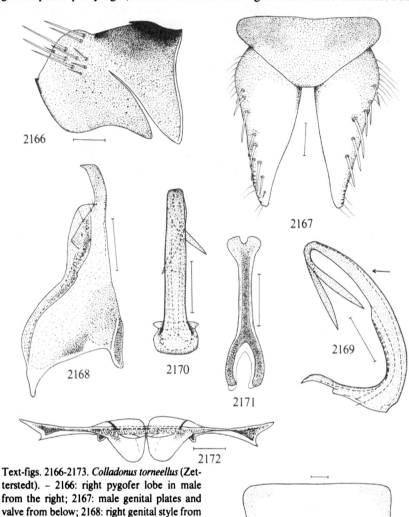

Text-figs. 2166-2173. *Colladonus torneellus* (Zetterstedt). – 2166: right pygofer lobe in male from the right; 2167: male genital plates and valve from below; 2168: right genital style from above; 2169: aedeagus from the left; 2170: aedeagus in ventral aspect (in direction of arrow in 2169); 2171: connective; 2172: 1st abdominal sternum in male from above; 2173: 7th abdominal sternum in female from below. Scale: 0.1 mm.

676

pygofer lobes as in Text-fig. 2166, genital plates and valve as in Text-fig. 2167, genital style as in Text-fig. 2168, aedeagus as in Text-figs. 2169, 2170, connective as in Text-fig. 2171, 1st abdominal sternum in male as in Text-fig. 2172, 7th abdominal sternum in female as in Text-fig. 2173. Overall length of male 4.4-4.9 mm, of female 4.9-5.2 mm.

Distribution. Very rare in Denmark, found in NEJ: Allerup Bakker 27.IX.1880 and 8.IX.1912 (A. Chr. Thomsen). – Sweden: rare in the south, more common in the north, found in Sm., Ög., Vg., Upl., Dlr., Hls., Hrj., Vb., Ly.Lpm., P.Lpm., Lu.Lpm., T.Lpm. – Scarce but widespread in Norway, found in HEn, Bø, TRi, Fi, Fø. – Widespread in East Fennoscandia, fairly scarce and sporadic in the southern part. – England, Scotland, France, Netherlands, Belgium, Switzerland, Austria, Italy, German D.R. and F.R., Hungary, Moravia, Slovakia, Romania, Poland, Estonia, Latvia, Lithuania, n., m. and s. Russia, Ukraine, Moldavia, Altai Mts., Tuva, Mongolia, Manchuria, n., m. and w. Siberia, Kamchatka, Maritime Territory.

Biology. On *Salix* species (Sahlberg, 1871). "Auf Gebüsch und Bäumen, meist in der subarktischen Region" (Lindberg, 1932a). In bogs with *Myrica, Betula nana* etc. (Ossiannilsson, 1947b). Among lush vegetation in pine forest, mainly *Vaccinium (myrtillus, uliginosum, vitis idaea),* some *Betula nana* (Holgersen, 1945). "Hibernation takes place in the larval stage. The species lives on grasses in mesophilous undergrowth of dense deciduous woods" (Lauterer, 1980). Adults in May-September.

Genus *Lamprotettix* Ribaut, 1942

Lamprotettix Ribaut, 1942: 267.
    Type-species: *Cicada octopunctata* Schrank, 1796, by original designation.

Body structure as in *Idiodonus*. Anteclypeus long, distinctly expanding towards apex. Lateral border of male genital plates straight or convex. Styles well developed. Connective short and broad. Phallotreme subapical, situated on the ventral side of aedeagus. One species.

316. *Lamprotettix nitidulus* (Fabricius, 1787)
    Plate-fig. 169, text-figs. 2174-2181.

Cicada nitidula Fabricius, 1787: 273.
Cicada octopunctata Schrank, 1796: 211.
Cicada splendidula Fabricius, 1803: 79.

Dirty yellow with black, fuscous and milky white markings, shining. Head at apex with a pair of black spots, caudally of these on vertex another pair of semicircular or trapezoidal spots, each of which are often connected with the frontal suture by a narrow black streak. A black spot present on each gena above and below antenna. Frontoclypeus with black transverse streaks, these more or less fusing in lower part but

677

usually interrupted by a light median streak. Anteclypeus entirely or largely black. Pronotum with some black dots near fore border, behind these six more or less distinct brownish longitudinal bands. Scutellum usually cream-coloured with a large black spot

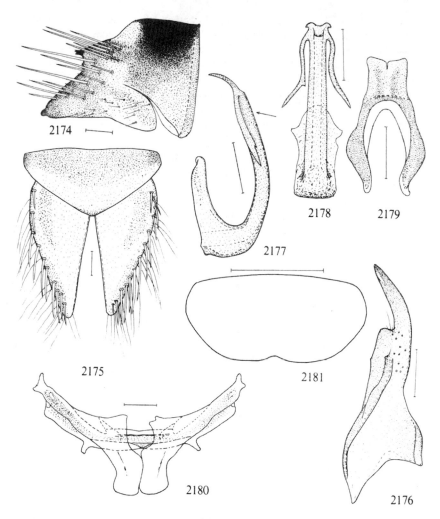

Text-figs. 2174-2181. *Lamprotettix nitidulus* (Fabricius). – 2174: right pygofer lobe in male from the right; 2175: genital plates and valve from below; 2176: right genital style from above; 2177: aedeagus from the left; 2178: aedeagus in ventral aspect (in direction of arrow in 2177); 2179: connective; 2180: 1st abdominal sternum in male from above; 2181: 7th abdominal sternum in female from below: Scale: 0.5 mm for 2181, 0.1 mm for the rest.

on lateral corners, between these two small roundish dots. Fore wings brownish yellow, veins partly light, partly dark-bordered; in proximal and distal ends of the cells the dark colour expands, filling up the space between veins. Transverse veins and distal ends of claval veins usually conspicuously pale and enclosed in milky or at least light spots. Thoracic venter largely black, legs yellowish white, posterior tibiae with black points. Abdominal tergum black with light lateral borders, venter strongly varying in colour: largely black in some males to largely light in some females. Male pygofer lobes as in Text-fig. 2174, genital plates and valve as in Text-fig. 2175, style as in Text-fig. 2176, aedeagus as in Text-figs. 2177, 2178, connective as in Text-fig. 2179, 1st abdominal sternum in male as in Text-fig. 2180, 7th abdominal sternum in female as in Text-fig. 2181. Overall length of males 4.9-5.4 mm, of females 5.2-6.1 mm.

Distribution. Scarce in Denmark, found in SJ, EJ, LFM, SZ, NEZ. – Scarce in Sweden, Sk.-Upl. – Not found in Norway and East Fennoscandia. – England, Wales, Ireland, France, Belgium, Netherlands, German D.R. and F.R., Austria, Switzerland, Italy, Greece, Bohemia, Moravia, Hungary, Romania, Poland, Latvia, Estonia, Lithuania, Ukraine.

Biology. " An Ulmen, in Waldlichtungen" (Kuntze, 1937). "Sur les arbres (Chêne, Hêtre, Orme, Charme, Frêne, Aulne), sur les plantes basses au pied des Sapins" (Ribaut, 1952). Adults in July-October.

Genus *Allygus* Fieber, 1872

Allygus Fieber, 1872: 13.
Type-species: *Cicada atomaria* Fabricius, 1794, by subsequent designation by Van Duzee, 1916.
or: *Cicada mixta* Fabricious, 1794, by subsequent designation by Van Duzee, 1917 (see Note).
Syringius Emeljanov, 1966: 101 (subgenus).
Type-species: *Allygus syrinx* Dlabola, 1961, by original designation.

Note. In an application recently sent to the International Commission on Zoological Nomenclature I propose that the Commission suppress all designations of type-species made for the genus *Allygus* prior to that made by Van Duzee, 1917. This is necessary for keeping current usage of the names *Allygus* and *Allygidius*. If the Commission rejects my proposal, the name *Allygidius* will be a junior objective synonym of *Allygus*, and the present genus will be called *Syringius* Emeljanov. The subgenus *Allygus* sensu Ribaut will be without a name.

Body elongate, structure robust. Head as wide as pronotum, about half as long as the latter, frontally rounded obtusely angular, medially little longer than near eyes. Fore wings always well developed, longer than abdomen, with partly anastomosing whitish secondary veins. Pygofer lobes in male and aedeagus without appendages (in *Allygus* s.str.). Genital styles small. 7th abdominal sternum in female normal, concealing base of ovipositor. In Denmark and Fennoscandia four species, all belonging to subgenus *Allygus* s.str.

Key to species of *Allygus*

1   Vertex with two narrow black streaks extending obliquely caudad from each ocellus towards coronal suture (Text-fig. 2206). Aedeagus in lateral aspect slender (Text-fig. 2210). 7th abdominal sternum in female with a broad excision (Text-fig. 2212) .... 320. *modestus* Scott

–   Pigmentation pattern of vertex different. Aedeagus stouter. 7th abdominal sternum in female nearly straight or slightly concave (Text-figs. 2189, 2197) 2

2 (1) Frontoclypeus above (below ocelli) with a pair of large clavate black spots medially not fused with remaining black markings of frontoclypeus (Text-fig. 2199). Aedeagus short and stout, in lateral aspect abruptly narrowing towards apex (Text-figs. 2203-2205) .... 319. *maculatus* Ribaut

–   Black markings of frontoclypeus different, usually more or less coalescing (as in Text-fig. 2190). If the uppermost spots below ocelli are not fused with the rest, then small 3

3 (2) Male pygofer long (Text-fig. 2182), extending caudally of apices of genital plates. Genital plates long, median margin as long as basal line along genital valve (Text-fig. 2183), macrosetae situated on or quite near lateral margins. Aedeagus comparatively slender (Text-figs. 2186-2188). 7th abdominal sternum in female considerably longer than 6th sternum, caudal margin almost straight (Text-fig. 2189) .... 317. *mixtus* (Fabricius)

–   Male pygofer short, not extending caudally of genital plates (text-fig.. 2191). Genital plates short, median margin shorter than basal line; macrosetae situated inside lateral margins of plates (Text-fig. 2192). Aedeagus stouter (Text-figs. 2195, 2196). 7th abdominal sternum in female only slightly longer than 6th sternum, distinctly and evenly concave (Text-fig. 2197) .... 318. *communis* (Ferrari).

317. *Allygus mixtus* (Fabricius, 1794)
Plate-fig. 170, text-figs. 2182-2190.

Cicada mixta Fabricius, 1794: 39.
Jassus mixtus var. γ *corisipennis* Ferrari, 1882: 139.
Allygus alticola Horváth, 1903: 474 (sec. spec. typ.) **n. syn.**

Brownish yellow to dirty yellow, immature specimens often greenish, with black and milky markings, shining. Vertex along fore border usually with 6 small black dots,

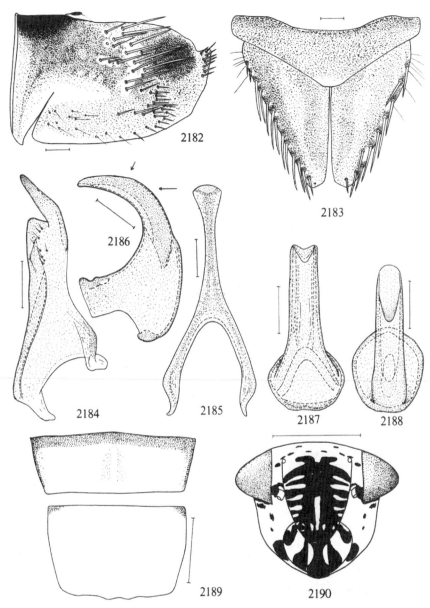

Text-figs. 2182-2190. *Allygus mixtus* (Fabricius). – 2182: left male pygofer lobe from the left; 2183: genital valve and plates from below; 2184: right genital style from above; 2185: connective in dorsal aspect; 2186: aedeagus from the left; 2187: aedeagus in ventral aspect (seen in direction of longer arrow in 2186); 2188: same in direction of shorter arrow in 2186; 2189: 6th and 7th abdominal sterna in female from below; 2190: face of male. Scale: 0.5 mm for 2189, 1 mm for 2190, 0.1 mm for the rest.

behind these a somewhat larger black spot and a small dot near caudal lateral corners. Often with a short streak on each side between lower margin of ocellus and eye. Transverse streaks and sutures, a spot below each antenna, and a longitudinal band on anteclypeus, black (Text-fig. 2190). Pronotum with an irregular transverse row of black dots behind fore border, caudally of these some dark spots and streaks arranged in 5 or 7 irregular longitudinal bands. The surface between these bands is partly whitish or greyish. Scutellum dark dotted. Fore wings dark mottled, with many milky secondary transverse veins and some milky spots. Venter brownish yellow, dark spotted, fore and middle femora each with one proximal and one distal dark transverse band, hind tibiae with dark spots. Pygofer lobes in male as in Text-fig. 2182, genital plates and valve as in Text-fig. 2183, normally brownish, genital styles as in Text-fig. 2184, connective as in Text-fig. 2185, aedeagus as in Text-figs. 2186-2188, 6th and 7th abdominal sterna in female as in Text-fig. 2189. Overall length of males 5.4-6.1 mm, of females 5.6-6.8 mm. – Abdomen of nymphs with hairs present only on hind angles of tergite VIII; pattern of body rather variable (Vilbaste, 1982).

Distribution. Common and widespread in Denmark; I have seen specimens from SJ, EJ, NWJ, NEJ, F, LFM, SZ, NEZ. – Common in southern and central Sweden, found Sk.-Hls. – Records from Norway not revised (Ø, AK, Bø, VE, AAy, AAi, Ry, Ri, HOi). – Moderately common in southern and central East Fennoscandia, found Al, Ab-Kb; Vib, Kr (revised by Huldén, in litt.). – Since *mixtus* and *communis* have been confused until recently, earlier records cannot be accepted as reliable. – England (Le Quesne, in litt.); France (Ribaut, 1952); German F.R. (Remane, in litt.); Italy! (Ferrari), Ukraine! (Horváth).

Biology. On *Quercus, Alnus glutinosa*. "Larvae green, feeding on grasses" (Le Quesne, 1969). Adults in July-October.

318. *Allygus communis* (Ferrari, 1882)
Text-figs. 2191-2197.

Jassus mixtus var. α *communis* Ferrari, 1882: 139.
Jassus mixtus var. β *margaritinus* Ferrari, 1882: 139.

Resembling *mixtus*. Dark markings on vertex often obsolete or absent. Male pygofer lobes as in Text-fig. 2191, male genital plates and valve (Text-fig. 2192) usually pale, genital styles as in Text-fig. 2193, connective as in Text-fig. 2194, aedeagus as in Text-figs. 2195, 2196, 6th and 7th abdominal sterna in female as in Text-fig. 2197. Overall length of male 5.7-6.3 mm, of female 5.9-6.5 mm.

Distribution. Widespread in Denmark, found in EJ: Vejle (O. Jacobsen); SZ: Risenfeld Skove, Vordingborg, and Marienlyst Skov, Vordingborg (Jacobsen); NEZ: København (Schiødte), Charlottenlund (Schlick), Tisvilde, Hornbæk, and Hillerød (Jacobsen). – Widespread and fairly common in southern Sweden, found in Sk., Bl., Öl., Gtl., Ög., Vg., Boh., Sdm., Upl. – So far not recorded from Norway and East

Fennoscandia. – Widespread in Europe (Remane, in litt.), N. Germany; Poland (Nast, in litt.); Italy.

Biology. I found adults on *Quercus* and *Betula,* July-September.

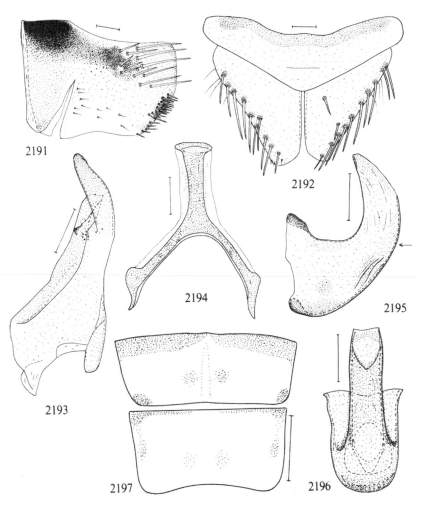

Text-figs. 2191-2197. *Allygus communis* (Ferrari). – 2191: left pygofer lobe in male from the left; 2192: male genital plates and valve from below; 2193: right genital style from above; 2194: connective in ventral aspect; 2195: aedeagus from the left; 2196: aedeagus in ventral aspect (in direction of arrow in 2195); 2197: 6th and 7th abdominal sterna in female from below. Scale: 0.5 mm for 2197, 0.1 mm for the rest.

683

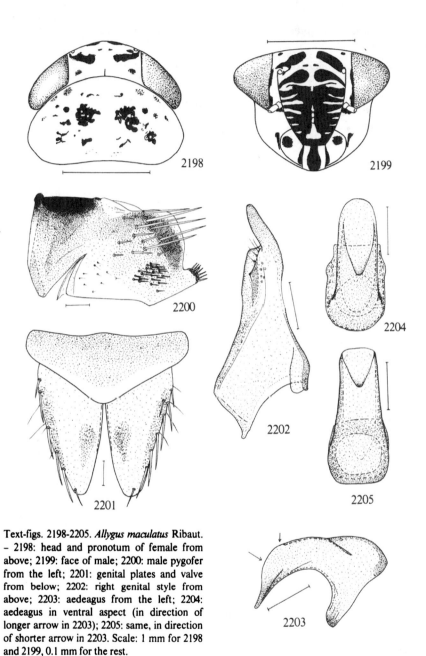

Text-figs. 2198-2205. *Allygus maculatus* Ribaut. - 2198: head and pronotum of female from above; 2199: face of male; 2200: male pygofer from the left; 2201: genital plates and valve from below; 2202: right genital style from above; 2203: aedeagus from the left; 2204: aedeagus in ventral aspect (in direction of longer arrow in 2203); 2205: same, in direction of shorter arrow in 2203. Scale: 1 mm for 2198 and 2199, 0.1 mm for the rest.

684

319. *Allygus maculatus* Ribaut, 1952
Text-figs. 2198-2205.

Allygus maculatus Ribaut, 1952: 204.

Size and general aspect as in *mixtus*, differing by pigmentation of face and structure of male genitalia. Head and pronotum from above as in Text-fig. 2198, face as in Text-fig. 2199, male pygofer lobes as in Text-fig. 2200, genital plates and valve as in Text-fig. 2201, genital styles as in Text-fig. 2202, aedeagus as in Text-figs. 2203-2205. 7th abdominal sternum in female as in *mixtus* (according to Ribaut, l.c.).

Distribution. Very rare in Denmark, only one female taken in SJ: Sønderborg by Wüstnei. – Very rare in Sweden, only one male found in Bl.: Ysane 7.VIII.1939 (Ossiannilsson). – Not found in Norway and East Fennoscandia. – France, German D.R. and F.R., Italy, Poland, Romania.

Biology. In the "Querceto-Carpinetum alnetosum" (Schwoerbel, 1957).

320. *Allygus modestus* Scott, 1876
Text-figs. 2206-2212.

Allygus modestus Scott, 1876: 172.
Jassus mixtus var. ♂ *juvenis* Ferrari, 1882: 139.

Resembling *mixtus*, differing by characters given in the key. Head from above as in Text-fig. 2206, male pygofer lobe as in Text-fig. 2207, genital plates and valve as in Text-fig. 2208, genital style as in Text-fig. 2209, aedeagus as in Text-figs. 2210, 2211, 7th abdominal sternum in female as in Text-fig. 2212. Overall length of males 6.2-6.9 mm, of females 6.8-7.5 mm (according to Le Quesne, 1969).

Distribution. Not found in Sweden, Norway and East Fennoscandia. – Very rare in Denmark, one male only found in SJ: Sønderborg, by Wüstnei. – England, France, Spain, German D.R. and F.R., Netherlands, Austria, Italy, Greece, Moravia, Slovakia, Romania, Yugoslavia, Poland, Ukraine, Morocco, Tunisia.

Biology. "An Ulmen in Waldlichtungen" (Kuntze, 1937). On oak, elm, alder, lime, hawthorn (Ribaut, 1952). "Adult generally on trees, but larva on grasses" (Le Quesne, 1969).

Genus *Allygidius* Ribaut, 1948

Allygidius Ribaut, 1948: 58.
Type-species: *Cicada atomaria* Fabricius, 1794, by original designation.

Body robust. Head narrower than pronotum. Anteclypeus parallel-sided or a little wider towards apex. Transition between vertex and face rounded. Vertex little longer medially than near eyes. Fore wings with numerous secondary transverse veins. Male

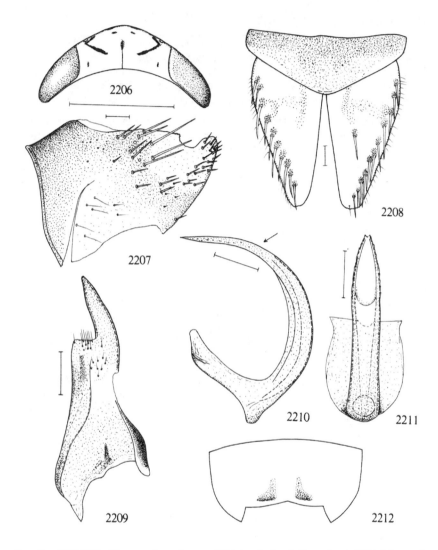

Text-figs. 2206-2212. *Allygus modestus* Scott. – 2206: head of male from above; 2207: left pygofer lobe in male from the left; 2208: male genital plates and valve from below; 2209: right genital style from above; 2210: aedeagus from the left; 2211: aedeagus in ventral aspect (as seen in direction of arrow in 2210); 2212: 7th abdominal sternum in female (after Le Quesne, 1969). Scale: 1 mm for 2206, 0.1 mm for the rest.

pygofer with a long appendage. Genital plates not covering genital chamber, their macrosetae in disorder. Shape of genital styles aberrant. Connective short and broad. Aedeagus without appendages, phallotreme large, terminal. 7th abdominal sternum in female medially short, not covering base of ovipositor. In Denmark and Fennoscandia one species.

321. *Allygidius commutatus* (Fieber, 1872)
   Text-figs. 2213-2220.

Cicada reticulata Thunberg, 1784: 21 (nec Linné, 1758).
Allygus commutatus Fieber, 1872: 13.

Strongly built. Brownish yellow to whitish yellow, brownish mottled and spotted. A few spots along fore border of head, and behind these a transverse band on vertex which is often medially interrupted, more or less distinct. Transverse dark streaks on

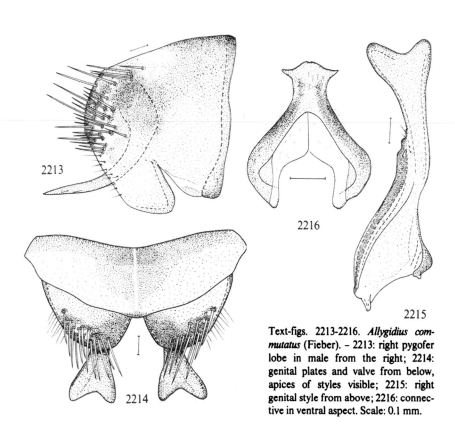

Text-figs. 2213-2216. *Allygidius commutatus* (Fieber). – 2213: right pygofer lobe in male from the right; 2214: genital plates and valve from below, apices of styles visible; 2215: right genital style from above; 2216: connective in ventral aspect. Scale: 0.1 mm.

687

frontoclypeus partly broken up into small dots. Anteclypeus with a T-shaped marking. Black spots below eyes often continuing on propleurum forming a dark longitudinal band. Scutellum dark spotted or mottled. Fore wings dark mottled, veins entirely or partly whitish. Thoracic venter light, posterior tibiae with dark dots. Male pygofer lobes as in Text-fig. 2213, genital plates, valve and apices of styles as in Text-fig. 2214, genital styles as in Text-fig. 2215, connective as in Text-fig. 2216, aedeagus as in Text-figs. 2217, 2218, 2nd and 3rd abdominal sterna in male as in Text-fig. 2219, apical part of abdominal venter in female as in Text-fig. 2220. Overall length of males 6.5-7 mm, of females 6.9-7.5 mm. – Abdomen of nymph with 8 (often incomplete) rows of hairs (Vilbaste, 1982).

Distribution. Fairly common in Denmark, found in SJ, EJ, F, LFM, SZ, NEZ, B. – Fairly common in Sweden, Sk.-Dlr. – Norway: found in AK, Bø, VE, AAy, AAi, VAy, HOi, SFi. – Fairly common in East Fennoscandia, found in Al, Ab-Sa, Sb, Kb; Vib, Kr. – Widespread in Europe, also in Algeria, Tunisia, Moldavia, Georgia, Kazakhstan, Altai Mts., w. Siberia.

Biology. "An Eichen und anderen Laubbäumen" (Kuntze, 1937). Adults in June-September.

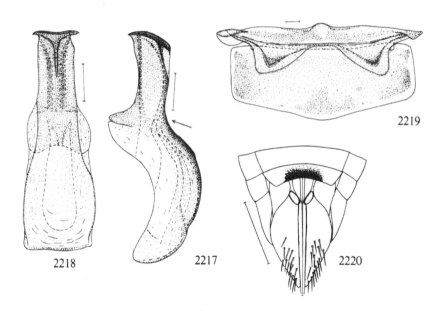

Text-figs. 2217-2220. *Allygidius commutatus* (Fieber). – 2217: aedeagus from the left; 2218: aedeagus in ventral aspect (in direction of arrow in 2217); 2219: 2nd and 3rd abdominal sterna in male from above; 2220: apical part of female abdomen from below. Scale: 1 mm for 2220, 0.1 mm for the rest.

Genus *Graphocraerus* Thomson, 1869

Graphocraerus Thomson, 1869: 57.
  Type-species: *Cicada ventralis* Fallén, 1806, by subsequent designation.

Body broad, robust. Head as long as, or a little shorter than, pronotum, just wider than the latter, behind fore border with an arcuate impression. Frontoclypeus above medially with an indistinct impression. Pronotum finely transversely striate. Fore wings approximately as long as abdomen. Pygofer lobes in male with a pointed appendage. Connective V-shaped. Macrosetae of genital plates in disorder. Aedeagus without appendages, phallotreme ventral, situated proximally of apex. Base of ovipositor not covered by 7th abdominal sternum. A monotypic genus.

322. *Graphocraerus ventralis* (Fallén, 1906)
  Plate-fig. 171, text-figs. 24, 2221-2230.

Cicada ventralis Fallén, 1806: 18.

Light green to yellowish green, shining. (The green colour of the live insects is delicate, dry specimens often being yellow, not green). Head with the following rounded black spots: on vertex two pairs in a transverse row, the lateral pair immediately above ocelli; one pair on frontoclypeus near junction with vertex, and one above each antenna. Pronotum with four black spots arranged in a curve parallel with fore border. Pronotum and scutellum often with a lighter median line. Fore wings often with brownish longitudinal streaks in the cells. Setae on tibiae, femora and hind tarsi arising from well-marked black dots, claws blackish. Male pygofer and anal apparatus as in Text-figs. 24, 2221, genital plates and valve as in text-figs. 24, 2222 (plates in rest coarranged giving a boat-like appearance), genital styles as in Text-figs. 2223, 2224, connective as in Text-fig. 2225, aedeagus as in Text-figs. 2226, 2227, 1st abdominal sternum in male as in Text-fig. 2228, 2nd and 3rd abdominal sterna in male as in Text-fig. 2229, 7th abdominal sternum in female as in Text-fig. 2230. Overall length of males 4.5-5.2 mm, of females 5.6-6.5 mm.

Distribution. Fairly common in Denmark, especially in Jutland. – Common in southern and central Sweden, Sk.-Vstm. – Rare in Norway, only found in Bø: Hofstangen at Hønefoss 1923 (Warloe), Bingen, Modum 1944 (Holgersen). – Common in southern and central East Fennoscandia, Ab-Kb; Vib, Kr. – Widespread in Europe, also found in Tunisia, Anatolia, Moldavia, Kazakhstan, m. and w. Siberia, Mongolia, Tuva, Maritime Territory.

Biology. In "Binnendünen, Sandfeldern, Wiesen" (Kuntze, 1937). In dryish fields, moist sloping meadows, cultivated fields (Linnavuori, 1952a). In "Glatthaferwiesen, Bentgraswiesen" (Marchand, 1953). In the "Arrhenatheretum" (Schwoerbel, 1957). Hibernation takes place in the egg stage, 1 generation (Remane, 1958; Schiemenz, 1969b). On *Poa pratensis* and *Anthoxanthum odoratum* (Schaefer, 1973).

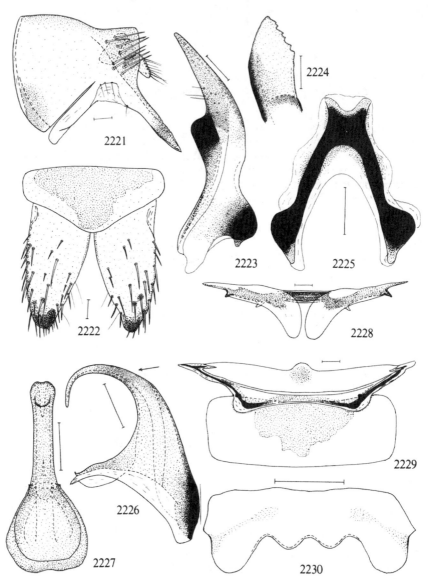

Text-figs. 2221-2230. *Graphocraerus ventralis* (Fallén). – 2221: male pygofer from the left; 2222: male genital plates and valve from below; 2223: right genital style from above; 2224: apical part of right style from inside; 2225: connective in ventral aspect; 2226: aedeagus from the left; 2227: aedeagus in ventral aspect (in direction of arrow of 2226); 2228: 1st abdominal sternum in male from above; 2229: 2nd and 3rd abdominal sterna in male from above; 2230: 7th abdominal sternum in female from below (depressed under coverglass). Scale: 0.5 mm for 2230, 0.1 mm for the rest.

Genus *Rhytistylus* Fieber, 1875

Rhytistylus Fieber, 1875: 404.

Type-species: *Jassus (Athysanus) proceps* Kirschbaum, 1868, by subsequent designation.

Body fairly broad. Head frontally obtusely angular or almost rectangularly produced, above near fore border with a shallow impression. Fore border of head finely, pronotum caudally more coarsely, transversely striate. Anteclypeus rectangular. Transition of frontoclypeus and vertex fairly abrupt. Lateral borders of pronotum very short, carinate. Fore wings a little shorter than abdomen. Male pygofer lobes spinose, without appendages. Genital plates elongate, triangular, with macrosetae in a row along lateral border and in a small group near median border. Connective short, broadly Y-shaped. Aedeagus without a distinct socle, with 2 horn-shaped appendages, phallotreme large, subapical. A monotypic genus.

323. **Rhytistylus proceps** (Kirschbaum, 1868)
Plate-fig. 172, text-figs. 2231-2237.

Jassus (Athysanus) proceps Kirschbaum, 1868b: 105.
Athysanus canescens Douglas & Scott, 1873: 210.

Pale straw-coloured. Frontoclypeus above with an inverted broadly V-shaped black transverse band consisting of more or less confluent streaks (Plate-fig. 172), continuing below eyes and on pro- and mesopleura, often also on lateral margins of abdominal venter. Pronotum often with indistinct lateral longitudinal bands. Mesonotum usually with a dark median longitudinal band visible through the semi-transparent prothoracic tergum. Fore wings entirely light or with dark streaks in the cells. Femora with black longitudinal streaks, posterior femora with a transverse band and a spot near apex, hind tibiae with a black longitudinal streak and black points round bases of setae. Abdomen above with three black longitudinal bands. Male genital plates and valve as in Text-fig. 2231, genital style as in Text-fig. 2232, connective and aedeagus as in Text-figs. 2233, 2234, 1st abdominal sternum in male as in Text-fig. 2235, 2nd and 3rd abdominal sterna in male as in Text-fig. 2236, 7th abdominal sternum in female as in Text-fig. 2237. Overall length of males 3.7-4-4 mm, of females 4.8-5.4 mm. – Abdomen of nymphs with 4 usually incomplete rows of short setae; face with a broad dark transverse band (broader than in adults) continuing on propleurum; thoracal and abdominal tergum with 4 dark longitudinal bands, abdominal venter often largely dark.

Distribution. Denmark: known only from the west-coast of Jutland: NWJ: Hansted (N. P. Kristensen); WJ: Skallingen (Kemner, Trolle). – Scarce in southern Sweden, found in Sk.: Brunnby, Kullen (Ossiannilsson), Dalhem near Helsingborg (Bo Henriksson); Hall.: Slättåkra, Måsarp (Ossiannilsson). – Not found in Norway, nor in East Fennoscandia. – England, France, Belgium, Netherlands, German D.R. and F.R., Portugal, Italy, Bohemia, Poland.

Biology. On *Weingaertneria canescens* and *Nardus stricta* (Kuntze, 1937). In the Corynephoretum (Marchand, 1953). Hibernates in the egg-stage, one generation (Remane, 1958; Schiemenz, 1959b). "Near base of taller grasses, particularly on calcareous soils" (Le Quesne, 1969). "Larvae and imagines abundant on *Carex arenaria*" (Trolle, in litt.).

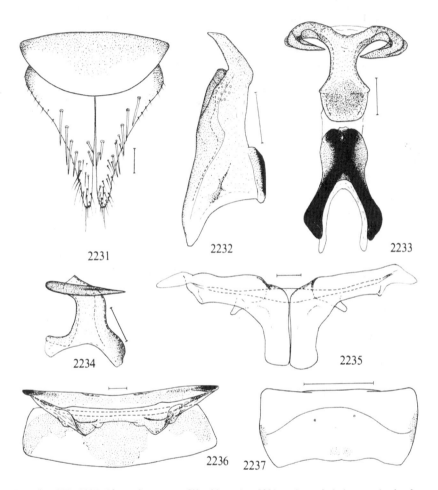

Text-figs. 2231-2237. *Rhytistylus proceps* (Kirschbaum). – 2231: male genital plates and valve from below; 2232: right genital style from above; 2233: aedeagus and connective in ventral aspect; 2234: aedeagus from the left; 2235: 1st abdominal sternum in male from above; 2236: 2nd and 3rd abdominal sterna in male from above; 2237: 7th abdominal sternum in female from below (depressed under coverglass). Scale: 0.5 mm for 2237, 0.1 mm for the rest.

692

Genus *Hardya* Edwards, 1922

Hardya Edwards, 1922: 206.
  Type-species: *Aphrodes melanopsis* Hardy, 1850, by monotypy.

Body moderately elongate, strongly built, Head fairly angularly produced, shorter but wider than pronotum. Ratio length:width of frontoclypeus = 1.6. Anteclypeus somewhat wider towards apex. Transition between vertex and frontoclypeus fairly rounded. Wings somewhat longer than abdomen. Male pygofer lobes apically with a comb-like appendage (Text-fig. 2238). Genital plates with median margins widely divergent, their macrosetae in disorder. Shape of genital style aberrant (Text-fig. 2241). Aedeagus without appendages, phallotreme dorsal. In Denmark and Fennoscandia one species.

324. *Hardya tenuis* (Germar, 1821)
  Plate-fig. 207, text-figs. 2238-2245.

Jassus tenuis Germar, 1821: 92.
Thamnotettix fulvopicta J. Sahlberg, 1871: 212.

Brownish yellow or greyish yellow with black markings, shining. Head on each side of fore border with a transverse curved line, concavity downwards; vertex with markings as in Plate-fig. 207. Face with black transverse streaks and sutures and a black spot laterally of each ocellus and another below each antenna. Anteclypeus with a black median streak. Pronotum along fore border with some dark spots, behind these with six dark longitudinal bands, interspaces usually greyish to whitish. Scutellum yellow with brick-red basal triangles and a brick-red median longitudinal band, the latter often with a pair of small black dots. Veins of fore wings partly black-bordered, the black colour tending to expand into spots in distal ends of the cells. Thoracic venter black with brownish yellow streaks; femora with transverse spots and longitudinal streaks; tibiae black dotted. Abdomen black with light lateral borders and segmental margins, ventrally often with a row of light patches on each side, especially in females. Male pygofer and anal apparatus as in Text-fig. 2238, genital plates and valve as in Text-fig. 2239, genital plate from above as in Text-fig. 2240, genital style as in Text-fig. 2241, aedeagus as in Text-figs. 2242, 2243, 1st abdominal sternum in male as in Text-fig. 2244, 2nd and 3rd abdominal sterna in male as in Text-fig. 2245. Caudal margin of 7th abdominal sternum in female faintly undulated. Overall length of male 2.7-3.8 mm, of female 2.8-4.0 mm.

Distribution. Not yet found in Denmark and Norway. – Scarce, locally common in Sweden, found Sk.-Upl. and Nrk. – Scarce and sporadic in southern and central East Fennoscandia, more common in the eastern part, found in Al, Ab, N, St, Ta, Sa, Kb; Vib, Kr. – Portugal, Spain, France, Belgium, German D.R. and F.R., Netherlands, Austria, Italy, Bohemia, Moravia, Slovakia, Hungary, Romania, Yugoslavia, Poland, Lithuania, s. Russia, Ukraine, Canary Isl., Algeria, Morocco, Tunisia, Georgia, Kazakhstan, Azerbaijan, Altai Mts., Mongolia.

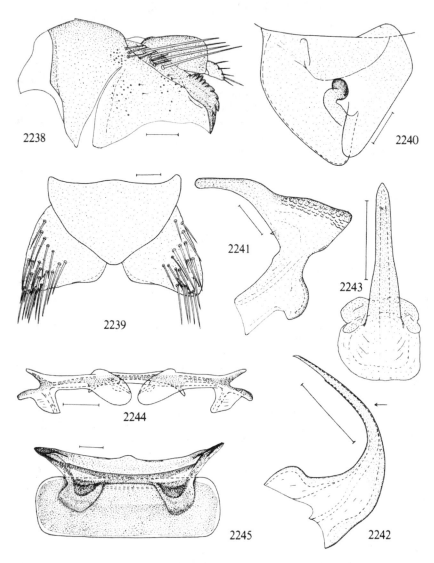

Text-figs. 2238-2245. *Hardya tenuis* (Germar). – 2238: male pygofer and anal apparatus from the left; 2239: male genital plates and valve from below; 2240: left genital plate from above, setae not considered; 2241: right genital style from above; 2242: aedeagus from the left; 2243: aedeagus in ventral aspect (in direction of arrow in 2242); 2244: 1st abdominal sternum in male from above; 2245: 2nd and 3rd abdominal sterna in male from above. Scale: 0.1 mm.

694

Biology. In "Binnendünen, Sandfeldern, Heiden, Hochmooren, Flachmooren, Wäldern; an Gräsern zwischen *Calluna*" (Kuntze, 1937). Xerophilous (Schwoerbel, 1957). "An Gräsern, vorwiegend unter Kiefern" (Wagner & Franz, 1961). Hibernation by adult females (Schiemenz, 1964); one generation (Schiemenz, 1969b). Fairly abundant in a wheat seedling crop (Tullgren, 1925).

Genus *Rhopalopyx* Ribaut, 1939

Rhopalopyx Ribaut, 1939: 267.
   Type-species: *Jassus preyssleri* Herrich-Schäffer, 1838, by original designation.

Elongate to moderately elongate species. Head wider than pronotum, obtusely angular or almost rectangularly produced. Anteclypeus more or less distinctly dilated towards apex. Transition between frontoclypeus and vertex more or less rounded. Anterior tibiae normally with three setae in the inner and four setae in the outer dorsal row. Fore wings apically rounded, as long as abdomen or longer. Male pygofer dorsally deeply excised, each lobe with a process directed caudad-ventrad. Genital plates long, each dorsally near apex of style with a conical process; macrosetae in disorder. Shaft of aedeagus simple, without processes, phallotreme dorsal-subapical. 7th abdominal sternum in female deeply concave, not covering base of ovipositor. In Denmark and Fennoscandia three species.

Key to species of *Rhopalopyx*

1   Vertex and pronotum with black markings 2
–   Vertex and pronotum without black markings .... 327. *vitripennis* (Flor)
2 (1) Male pygofer process long, gradually tapering towards apex,
      apical half without macrosetae (Text-fig. 2246) ....
      .... 325. *preyssleri* (Herrich-Schäffer)
–   Male pygofer process short, distally rounded with a short
      abruptly marked spur, on inside near apex with a bundle of
      macrosetae (Text-figs. 2255, 2256) .... 326. *adumbrata* (C. Sahlberg)

325. ***Rhopalopyx preyssleri*** (Herrich-Schäffer, 1838)
Plate-fig. 173, text-figs. 2246-2254.

Jassus preyssleri Herrich-Schäffer, 1838: 7.

Comparatively broad and robust, light sordid yellow with black and fuscous markings, shining. Head as seen from above obtusely angular, with the following black markings: one spot on apex, another laterally of each ocellus, and one on middle of caudal border. A large black spot present below each antenna. Frontoclypeus often with a diffusely delimited brownish median patch. Anteclypeus medially often with a pair of short parallel longitudinal streaks. Pronotum with a broad fuscous longitudinal band not reaching fore border but continuing on frontal part of scutellum. Fore wings as

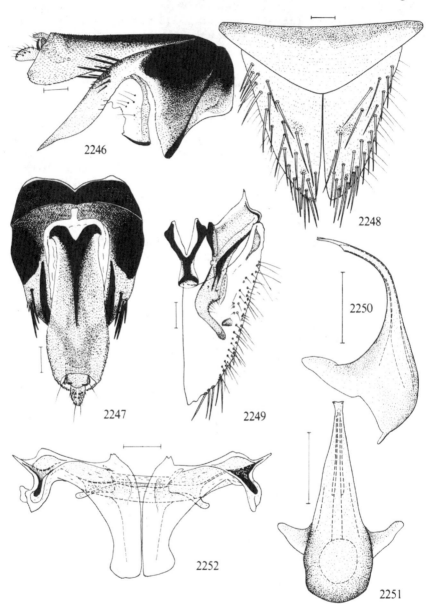

2246

2248

2247

2249

2250

2252

2251

long as or a little longer than abdomen, each with two fuscous longitudinal streaks: one along commissural border, another starting from wing base bordering vein M. Also the claval suture may be dark; apical part of fore wing broadly dark-bordered. Thoracic venter largely light with spots varying in size, legs longitudinally striped. Abdominal tergum with three broad black longitudinal bands, venter with one median band or three longitudinal bands. Male pygofer and anal apparatus as in Text-figs. 2246, 2247, genital plates, valve, connective and styles as in Text-figs. 2248, 2249, aedeagus as in Text-figs. 2250, 2251, 1st abdominal sternum in male as in Text-fig. 2252, 2nd and 3rd abdominal sterna as in Text-fig. 2253, 7th abdominal sternum in female as in Text-fig. 2254. Overall length ($\male\female$) 3.6-3.9 mm. – Nymphs with 4 rows of hairs on dorsal surface of abdomen; vertex without sharply delimited black spots, black spots beneath transition from vertex to frons absent; pattern of dorsal pigmentation distinct, consisting of 3 greyish brown longitudinal bands, lateral bands extending to lateral margins of abdomen; face with grey or grey-brown bands dilating downwards; anteclypeus wholly dark, usually also lora (Vilbaste, 1982). (See also Walter, 1978).

Distribution. Widespread and common in Denmark. – Widespread and common in Sweden, Sk.-Vb. – I have seen only one male from Norway, Ø: Mysen 18.VIII.1960 (Holgersen); earlier records not revised. – Common in southern and central East Fennoscandia, Al, Ab-Om;Vib, Kr. – France, Belgium, Netherlands, German D.R. and F.R., Switzerland, Austria, Bulgaria, Bohemia, Moravia, Slovakia, Hungary, Romania, Poland, Estonia, Latvia, Lithuania, m. and s. Russia, Ukraine, Moldavia, Altai Mts., Kazakhstan, m. Siberia, Tuva, Mongolia.

Biology. Xerophilous (Vilbaste, 1962). Hibernation in the egg-stage, one generation (Schiemenz, 1969a, b). In dry meadows, dry birch forests, in sandy felled areas (Vilbaste, 1974). Abundance maximum (in middle Finland) on 2.VIII (Törmälä & Raatikainen, 1976). Adults in June-September.

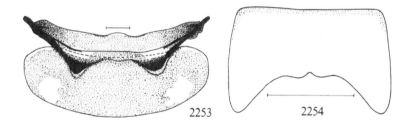

Text-figs. 2246-2254. *Rhopalopyx preyssleri* (Herrich-Schäffer). – 2246: male pygofer and anal apparatus from the right; 2247: same from above; 2248: male genital plates and valve from below; 2249: connective, right genital plate and style from above; 2250: aedeagus from the left; 2251: aedeagus in ventral aspect; 2252: 1st abdominal sternum in male from above; 2253: 2nd and 3rd abdominal sterna in male from above; 2254: 7th abdominal sternum in female from below (depressed under coverglass). Scale: 0.5 mm for 2254, 0.1 mm for the rest.

697

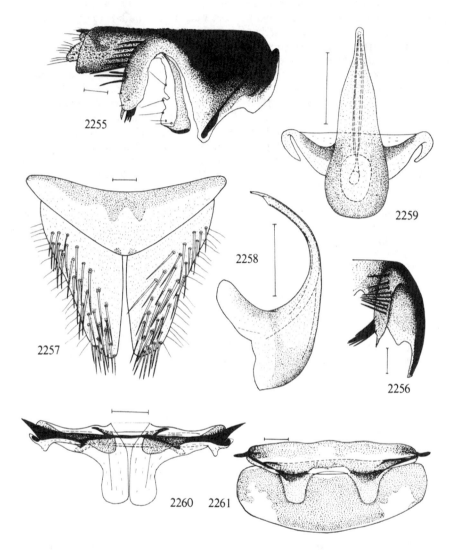

Text-figs. 2255-2261. *Rhopalopyx adumbrata* (C. Sahlb.). – 2255: male pygofer and anal apparatus from the right; 2256: right pygofer lobe in male from behind; 2257: male genital plates and valve from below; 2258: aedeagus from the left; 2259: aedeagus in ventral aspect; 2260: 1st abdominal sternum in male from above; 2261: 2nd and 3rd abdominal sterna in male from above. Scale: 0.1 mm.

698

326. **Rhopalopyx adumbrata** (C. Sahlberg, 1842)
   Text-figs. 2255-2261.

Cicada adumbrata C. Sahlberg, 1842: 91.

Body structure and pigmentation as in *preyssleri*. Male pygofer and anal apparatus as in Text-figs. 2255, 2256, genital plates and valve as in Text-fig. 2257, aedeagus as in Text-figs. 2258, 2259, 1st abdominal sternum in male as in Text-fig. 2260, 2nd and 3rd abdominal sterna in male as in Text-fig. 2261. 7th abdominal sternum in female as in *preyssleri*. Overall length of males 3.6-4.0 mm, of females 3.7-4.1 mm. – Nymphs resembling those of *preyssleri* (Vilbaste, 1982).

   Distribution. Denmark: known only from the west-coast of Jutland, NEJ: Tornby; NWJ: Hansted, and WJ: Skallingen. – Scarce in Sweden, found in Sk., Hall., Ög., Vg., Hls., Med. – Norway: only found in HEs: Magnor 30.VIII.1961 (Holgersen). – East Fennoscandia: distribution incompletely investigated, so far found in Ab, N, St. – England, German D.R. and F.R., Bohemia, Moravia, Slovakia, Romania, Estonia, Lithuania, Latvia, Kazakhstan, Altai Mts.

   Biology. "Eine hygrophile Art, gehört zu den Bewohnern von nassen Wiesen und Mooren" (Vilbaste, 1962). "In damp meadows, in fens" (Vilbaste, 1974). "On grasses, usually on calcareous or other dry hillsides" (Le Quesne, 1969). On roadside verges, only "where *Festuca rubra* L. was present" (Port, 1981). – Adults in July-September.

327. **Rhopalopyx vitripennis** (Flor, 1861)
   Text-figs. 2262-2270.

Jassus (Deltocephalus) vitripennis Flor, 1861a: 255.
Thamnotettix andropogonis Haupt, 1924: 291.
Thamnotettix graminis Haupt, 1935: 204.
Rhopalopyx monticola Ribaut, 1939: 272.
Rhopalopyx parvispinus W. Wagner, 1947: 88.

Elongate, less robust than *preyssleri* and *adumbrata*. Head more angularly produced. Upper side pale green-yellowish, head with a brownish yellow tinge, above along fore border with a brownish yellow inverted V-shaped marking between ocelli, caudally of that often with obsolete brownish yellow spots or a pair of parallel longitudinal brownish yellow streaks running close together. Face with brownish yellow or greyish yellow transverse streaks, median line often light. Below each antenna a small black spot. Pronotum, scutellum, and fore wings usually without dark markings, apical part of fore wing usually partly fumose and with fuscous veins. Venter of thorax largely light; femora with fuscous longitudinal streaks or rows of dots, tibiae striped and spotted with black. Male pygofer (in Swedish specimens) as in Text-fig. 2262, but there is a geographical variation in this species, process of pygofer lobes of varieties *monticola* (Pyrenees) and *parvispinus* (Czechoslovakia, Hungary) shaped more or less as in *adum-*

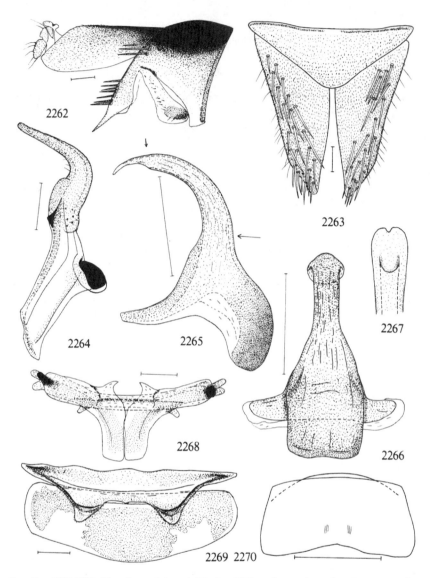

Text-figs. 2262-2270. *Rhopalopyx vitripennis* (Flor). – 2262: male pygofer and anal apparatus from the right; 2263: genital plates and valve from below; 2264: right genital style from above; 2265: aedeagus from the left; 2266: aedeagus in ventral aspect (in direction of longer arrow of 2265); 2267: apex of aedeagus in ventro-terminal aspect (in direction of shorter arrow of 2265); 2268: 1st abdominal sternum in male from above; 2269: 2nd and 3rd abdominal sterna in male from above; 2270: 7th abdominal sternum in female from below (depressed under coverglass). Scale: 0.5 mm for 2270, 0.1 mm for the rest.

brata (but without subapical bundle of setae on inside). (See Wagner, 1968: 47-48). Male genital plates and valve as in Text-fig. 2263, style as in Text-fig. 2264, aedeagus as in Text-figs. 2265-2267, 1st abdominal sternum in male as in Text-fig. 2268, 2nd and 3rd abdominal sterna in male as in Text-fig. 2269, 7th abdominal sternum in female as in Text-fig. 2270. Overall length of male 3.0-4.0 mm, of female 3.3-4.6 mm.

Distribution. Scarce in Denmark, known from F: Svendborg, NEZ: Hillerød, and B: Bølshavn. – Widespread but fairly scarce in Sweden, found in Sk., Sm., Gtl., Ög., Vg., Dlsl., Ång., Vb. – Recorded from Norway by Siebke (1874) but these records have not been checked (Holgersen, 1945). – Fairly scarce and sporadic in southern and central East Fennoscandia, found in Al, Ab, N, Ta, Sb, Kb; Kr. – Widespread in Europe, also in Algeria, Tunisia, Anatolia, Moldavia, Georgia, Kazakhstan, Azerbaijan, Altai Mts., m. Siberia, Mongolia, Maritime Territory.

Biology. In "Binnendünen, Sandfeldern, Heiden" (Kuntze, 1937). Xerophilous, in dryish fields (Linnavuori, 1952a). But Marchand (1953) placed *vitripennis* in the "Klein-seggenwiese" (Cariceto canescentis – Agrostidetum caninae-Ass., Subass. v. *Carex inflata*). Hibernation takes place in the egg-stage (Müller, 1957; Remane, 1958); two generations in central Germany (Remane, 1958; Schiemenz, 1969b). "Stenotope Art der Trockenrasen" (Schiemenz, 1969b). Adults in July and August.

Genus *Paluda* De Long, 1937

Paluda De Long, 1937: 233.
Type-species: *Thamnotettix placidus* Osborn, 19055, by original designation.

As *Rhopalopyx*, but processes of pygofer lobes curved dorsad-frontad. Aedeagus laterally carinate. Phallotreme apical. In Denmark and Fennoscandia one species.

328. *Paluda flaveola* (Boheman, 1845)
Text-figs. 2271-2279.

Thamnotettix flaveola Boheman, 1845b: 157.
Thamnotettix sulphurellus Haupt, 1935: 207, nec Zetterstedt, 1828.

Shape and structure of body as in *Rhopalopyx preyssleri*, light yellow or light orange, shining. Head frontally broadly rounded. Vertex, notum, and fore wings entirely light, veins of fore wings concolorous, apical part somewhat fumose. Sub-antennal black dot small or absent. Femora spotless or with longitudinal streaks consisting of dots arranged in rows, posterior tibiae dark dotted. Abdominal tergum in male black with broadly yellow lateral margins, or yellow with three broad black longitudinal bands, venter yellow with a basal broad black longitudinal band and lateral margins of sternites narrowly black; in the female, the black colour is reduced or absent. Male pygofer and anal apparatus as in Text-figs. 2271, 2272, genital valve, plates, connective and style as in Text-figs. 2273, 2274, aedeagus as in Text-figs. 2275, 2276, 1st abdominal sternum in male as in Text-fig. 2277, 2nd and 3rd abdominal sterna in male as in Text-

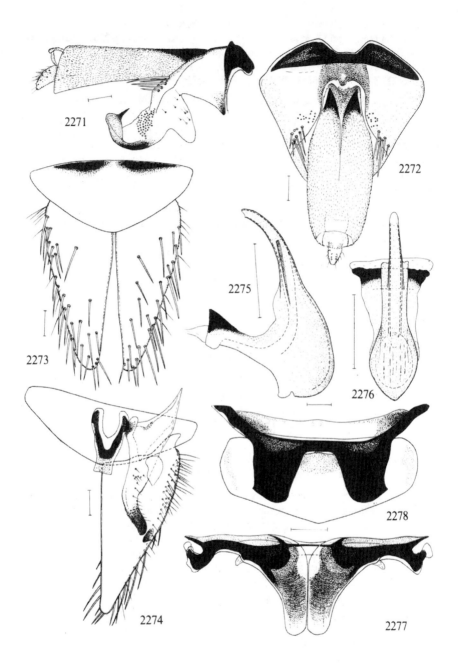

2271

2272

2273

2275

2276

2274

2278

2277

702

fig. 2278, 7th abdominal sternum in female as in Text-fig. 2279. Overall length of male 4.0-4.6 mm, of female 4.4-5.0 mm. – Nymph resembling *Rhopalopyx preyssleri* but pattern indistinct (bands darker on anterior margin of head); lateral bands of abdomen do not extend to lateral margin; face pale (Vilbaste, 1982).

Distribution. Uncommon in Denmark, found in several places in NEJ, NEZ, and B. – Common and widespread in Sweden, Sk.-Nb. – Norway: found in Ø: Kjølberg; AK: L. Langerud, Oslo, and Son; HEs: Magnor; HEn: Snippen, Råna, and Nybergsund (Holgersen leg.). – Common in southern and central East Fennoscandia, found in Al, Ab-Om, ObN; Vib, Kr. – England, France, Netherlands, German D.R. and F.R., Bohemia, Moravia, Slovakia, Poland, Estonia, Lithuania, n. and m. Russia, Ukraine, Tunisia, Tuva, Mongolia.

Biology. In "besonnten Hängen, Wäldern" (Kuntze, 1937). "An *Calamagrostis epigeios* in sandigen Kiefernwäldern" (W. Wagner, in Strübing, 1955). "Its original biotopes are probably fens, peaty meadows and moist meadows in forest clearings. Nowadays the species occurs chiefly in leys, pastures and wastelands. Its wild hostplants are grasses" (Raatikainen & Vasarainen, 1976). Hibernates as eggs (Törmälä & Raatikainen, 1976). Adults in June-September.

Genus *Elymana* De Long, 1936

Elymana De Long, 1936: 218.
  Type-species: *Thamnotettix inornata* Van Duzee, 1892, by original designation.
Solenopyx Ribaut, 1939: 273.
  Type-species: *Cicada sulphurella* Zetterstedt, 1828, by monotypy.

Body elongate, fairly slender. Head as seen from above shorter but wider than pronotum. Anteclypeus somewhat dilated towards apex. Antennae long, fore wings long, narrow (Text-fig. 2281). Anterior tibiae with one seta in the inner, four in the outer dorsal row. Male pygofer dorsally deeply excised. Genital plates in male without a process on dorsal side, macrosetae arranged in a row parallel with lateral margin (Text-fig. 2284). Aedeagus without appendages, apically forked into two short prongs.

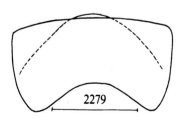

Text-figs. 2271-2279. *Paluda flaveola* (Boheman). – 2271: male pygofer and anal apparatus from the right; 2272: same from above; 2273: genital plates and valve from below; 2274: valve, connective, right genital plate and style from above; 2275: aedeagus from the left; 2276: aedeagus in ventral aspect; 2277: 1st abdominal sternum in male from above; 2278: 2nd and 3rd abdominal sterna in male from above; 2279: 7th abdominal sternum in female from below (depressed under coverglass). Scale: 0.5 mm for 2279, 0.1 mm for the rest.

703

7th abdominal sternum in female with caudal border only faintly concave, covering base of ovipositor. In Denmark and Fennoscandia two species.

Key to species of *Elymana*

1   Male pygofer lobes each with a row of small denticles on dorsal side (Text-fig. 2283). Apical prongs of aedeagus not curved dorsad (Text-fig. 2287) .... 329. *sulphurella* (Zetterstedt)
-   Male pygofer lobes each with a comb of stout denticles on dorsal side (Text-figs. 2289, 2290). Apical prongs of aedeagus curved dorsad (Text-fig. 2292) .... 330. *kozhevnikovi* (Zachvatkin)

329. *Elymana sulphurella* (Zetterstedt, 1828)
Text-figs. 2280-2288.

Cicada virescens Fabricius, 1794: 46 (nec Gmelin, 1790).
Cicada sulphurella Zetterstedt, 1828: 534.

Greenish yellow, shining. Head as seen from above rounded produced, above without markings or with indistinct markings consisting of two medially interrupted, faintly darker, transverse streaks, the anterior one angular, parallel with fore border, being the uppermost pair of frontoclypeal transverse streaks. Frontoclypeus with very light greyish yellow or brownish transverse streaks (not considered in Text-fig. 2280). A black spot present below each antenna. Pronotum and scutellum unicolorous. Fore wings (Text-fig. 2281) considerably longer than abdomen, almost transparent with greenish yellow veins, apical part fumose. Thoracic venter with black spots caudally of fore and middle coxae. Abdominal tergum black with yellow lateral margins and narrow yellow segmental borders, venter largely light with a basal black median band. Tibiae black spotted. Male anal apparatus as in Text-fig. 2282, pygofer as in Text-figs. 2282, 2283, genital plates and valve as in Text-fig. 2284, connective and style as in Text-fig. 2285, aedeagus as in Text-figs. 2286, 2287, 2nd and 3rd abdominal sterna in male as in Text-fig. 2288. Overall length of males 4.0-4,8 mm, of females 4.5-5.4 mm. – Nymphs on abdominal tergum with 4 longitudinal rows of hairs; hind tibiae dark spotted, vertex as wide as long; anteclypeus slightly dilated towards tip; greyish or pale brownish lateral bands usually present on dorsum (Vilbaste, 1982). See also Walter (1978).

Distribution. Common and widespread in Denmark and Sweden (Sk.-Ång.). – I have seen male specimens from Ø, AK, Os, Bø, Bv, VE, AAy, VAy, VAi, Ry, HOy, SFy and MRy in Norway. – Very common in southern and central East Fennoscandia, found in Al, Ab-Om; Vib, Kr. – Widespread in Europe, also in Algeria, Morocco, Cyprus, Anatolia, Kazakhstan, Kirghizia, Altai Mts., m. and w. Siberia, Manchuria, Korean Peninsula, Japan.

Biology. On seashores, and in dryish fields, moist sloping meadows, peaty meadows, cultivated fields (Linnavuori, 1952a). Feeds on *Holcus mollis* (Morcos, 1953). Hibernation takes place in the egg stage (Morcos, 1953; Müller, 1957; Remane, 1958;

Schiemenz, 1969b; Raatikainen, 1971). One generation (Remane, 1958; Schiemenz, 1969b). "Was very frequent and abundant in grass leys and frequently migrated to cereal crops" (Raatikainen & Vasarainen, 1973). "In cages it reproduced on *Phleum pratense* and fed on grasses, e.g. oats" (Raatikainen & Vasarainen, 1976). Adults in June-September.

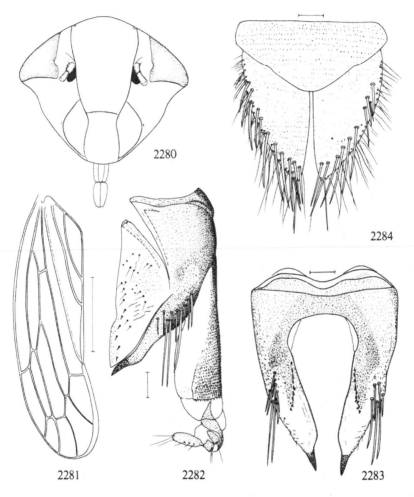

Text-figs. 2280-2284. *Elymana sulphurella* (Zetterstedt). – 2280: face of female; 2281: left fore wing of male; 2282:male pygofer and anal apparatus from the left; 2283: male pygofer from above; 2284: male genital plates and valve from below. Scale: 1 mm for 2281, 0.1 mm for the rest.

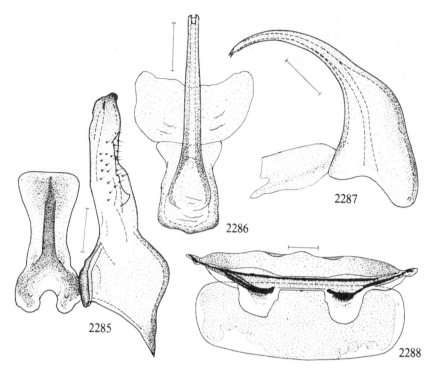

Text-figs. 2285-2288. *Elymana sulphurella* (Zetterstedt). – 2285: connective and left genital style from above; 2286: aedeagus in ventral aspect; 2287: aedeagus from the left; 2288: 2nd and 3rd abdominal sterna in male from above. Scale: 0.1 mm.

330. *Elymana kozhevnikovi* (Zachvatkin, 1938)
Text-figs. 2289-2294.

Limotettix kozhevnikovi Zachvatkin, 1938: 285-287.
Elymana ikumae auct., nec Matsumura, 1911.

Resembling *E. sulphurella*. Male pygofer as in Text-figs. 2289, 2290, anal apparatus as in Text-fig. 2289, genital style as in Text-fig. 2291, aedeagus as in Text-figs. 2292, 2293, 2nd and 3rd abdominal sterna in male as in Text-fig. 2294. Overall length (♂♀) 5.0-5.4 mm. Nymphs as in *sulphurella* (Vilbaste, 1982).

Distribution. Not yet recorded from Denmark and Norway. – Scarce in Sweden, found in Öl., Gtl., Upl., Vstm., Dlr., Gstr., Hls. – East Fennoscandia: several times found in Ab, N, Sa, Kb, but probably with a wider distribution. – German D.R. and F.R., Austria, Poland, Estonia, Latvia, Lithuania, n. and m. Russia, Altai Mts., Tuva, Mongolia, Kazakhstan, m. Siberia, Sakhalin, Maritime Territory, Korean Peninsula.

Biology. "Bewohner der Feldschicht der Laubwälder und bevorzugt deutlich feuchtere, doch nicht von Seggenvegetation eingenommene Stellen" (Kontkanen, 1947). In damp meadows; in shore meadows (Linnavuori, 1952a). "In den Nordostalpen

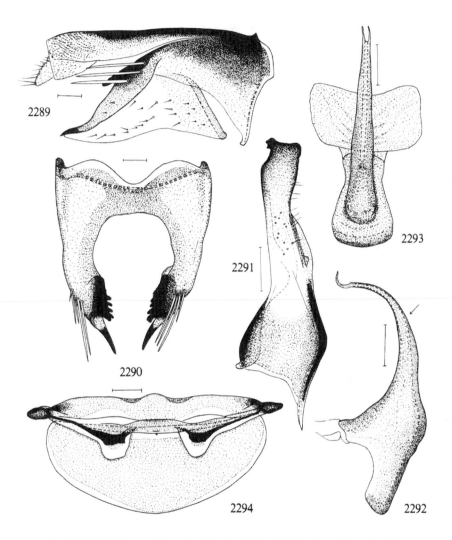

Text-figs. 2289-2294. *Elymana kozhevnikovi* (Zachvatkin). – 2289: male pygofer and anal apparatus from the left; 2290: male pygofer from above; 2291: left genital style from above; 2292: aedeagus from the left; 2293: aedeagus in ventral aspect (in direction of arrow in 2292); 2294: 2nd and 3rd abdominal sterna in male from above. Scale: 0.1 mm.

ist die Art ein typischer Bewohner der Föhrenheidewälder und Felsenheiden vorwiegend auf Dolomit" (Wagner & Franz, 1961). In woods and shrubs, but also in sand areas (Vilbaste, 1974). Adults in July-September.

Genus *Cicadula* Zetterstedt, 1838

Cicadula Zetterstedt, 1838: 296.
    Type-species: *Cicada quadrinotata* Fabricius, 1794, by subsequent designation.
Cyperana De Long, 1936: 218.
    Type-species: *Jassus melanogaster* Provancher, 1872, by original designation.
Henriana Emeljanov, 1964: 412 (as subgenus).
    Type-species: *Jassus frontalis* Herrich-Schäffer, 1835, by original designation.

Body elongate, more or less as in *Elymana*. Head wider than pronotum. Anteclypeus dilated towards apex. Antennae long, fore wings long and narrow. Chaetation of anterior tibiae as in *Elymana*. Male genital plates short, apically rounded, macrosetae in one row, more or less abruptly turning off mediad (Text-figs. 2296, 2333, 2345). Pygofer in male dorsally deeply excised, anal tube dorsally well sclerotized. In Denmark and Fennoscandia eleven species.

Key to species of *Cicadula*

1   Head at fore border with a transverse row of semilunar or elongate black spots, two on frontoclypeus, one on each gena (Plate-fig. 176). Pygofer lobes and aedeagus without processes (Subgenus *Cyperana* De Long) 10
–   Head on fore border without a transverse row of elongate black spots, with different black markings or without markings 2
2 (1) Pygofer lobes in male without processes. Aedeagus with a pair of appendages parallel with shaft, arising considerably proximally of apex (Text-figs. 2347, 2348). Phallotreme apical. Ground-colour of female brownish yellow (Subgenus *Henriana* Emeljanov) .... 341. *frontalis* (Herrich-Schäffer)
–   Pygofer lobes in male each with two spine-shaped processes on dorsal side (Text-fig. 2295). Appendages of aedeagus, if present, arising from apex of shaft. Phallotreme ventral, subapical. Ground-colour of females greenish yellow, light yellow, or orange yellow (Subgenus *Cicadula* s.str.) 3
3 (2) Head as seen from above rounded rectangular apically, always with four large black spots (Plate-fig. 174) 4
–   Head apically rounded obtusely angular, above without spots or with four smaller spots, of which the caudal is usually smaller; sometimes also a pair of short longitudinal streaks at anterior end of coronal suture (Plate-fig. 175) 8

4 (3) Aedeagus without appendages (Text-figs. 2313-2315) .... 335. *saturata* (Edwards)
–   Aedeagus with slender apical appendages 5
5 (4) Socle of aedeagus with a marked gibbosity on ventral side
      (Text-fig. 2299) .... 331. *quadrinotata* (Fabricius)
–   Socle of aedeagus without a marked ventral gibbosity 6
6 (5) Base of aedeagus on dorsal side carinate 7
–   Base of aedeagus on dorsal side not carinate. Appendages
      of aedeagus long (Text-figs. 2303-2305) .... 332. *persimilis* (Edwards)
7 (6) Basal carina of aedeagus usually evenly curved, not denti-
      culate. Apical appendages of aedeagus short (Text-figs. 2310,
      2311) .... 334. *longiventris* (J. Sahlberg)
–   Basal carina of aedeagus usually irregularly dilated, denti-
      culate. Appendages longer (Text-figs. 2307-2309) .... 333. *albingensis* W. Wagner
8 (3) Aedeagus apically with four appendages (Text-fig. 2329).
      Head above often spotless or spots small .... 338. *flori* (J. Sahlberg)
–   Aedeagus apically with three processes 9
9 (8) Base of aedeagus on dorsal side with a more or less distinct
      gibbosity, shaft near base carinate or somewhat compres-
      sed. Appendages of aedeagus longer (Text-fig. 2321). 1st ab-
      dominal sternum in male as in Text-fig. 2324 .... 337. *nigricornis* (J. Sahlberg)
–   Base of aedeagus on dorsal side not gibbose, nor carinate.
      Apical appendages of aedeagus shorter (Text-figs. 2317-
      2319). 1st abdominal sternum in male as in Text-fig. 2320 ....
      ....336. *quinquenotata* (Boheman)
10 (1) Male pygofer lobes with dorsal margin straight or slightly
      convex (Text-fig. 2332). Apex of aedeagus (distally of phal-
      lotreme) in lateral aspect narrow (Text-fig. 2334). Smaller,
      overall length not above 5.2 mm .... 339. *intermedia* (Boheman)
–   Dorsal margin of male pygofer lobes slightly concave. Apex
      of aedeagus in lateral aspect wider (Text-fig. 2339). Larger,
      overall length exceeding 5.6 mm .... 340. *ornata* (Melichar)

331. ***Cicadula (Cicadula) quadrinotata*** (Fabricius, 1794)
       Text-figs. 2295-2302.

Cicada quadrinotata Fabricius, 1794: 43.
Cicada strigipes Zetterstedt, 1828: 532.

Orange yellow or greenish yellow, shining. Head as seen from above rounded obtusely angular, ratio median length:width between eyes 0.56 (♀) – 0.62 (♂). On transition of frontoclypeus and vertex a pair of large rounded or polygonal black spots, vertex caudally of each ocellus with a black spot usually not or little smaller; below each eye a black patch varying in size; frontal suture black; black transverse streaks on frontoclypeus present or absent. Pronotum, scutellum and fore wings usually entirely

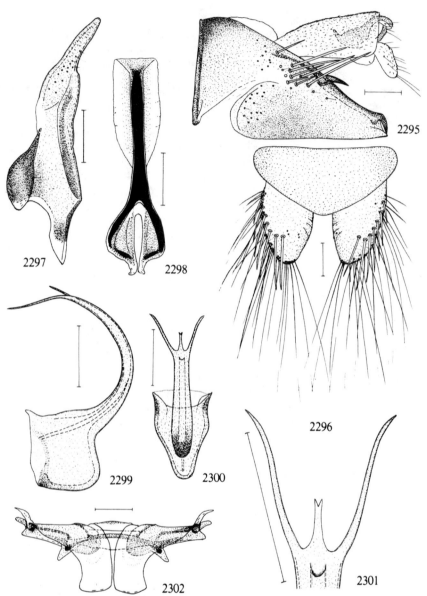

Text-figs. 2295-2302. *Cicadula quadrinotata* (Fabricius). – 2295: male pygofer and anal apparatus from the left; 2296: genital plates and valve in ventral aspect; 2297: left genital style from above; 2298: connective from above; 2299: aedeagus from the left; 2300: aedeagus in ventro-terminal aspect; 2301: apex of aedeagus in ventral aspect; 2302: 1st abdominal sternum in male from above. Scale: 0.1 mm.

710

light, veins yellow, apiçal part fumose. Sometimes there are dark patches and longitudinal streaks on pronotum and fore wings. Thoracic venter and abdomen in male black, abdomen with narrow light lateral borders and segmental margins; in females the light colour may be more extended than in males. Legs unicolorous light or with dark longitudinal streaks, hind tibiae dark spotted. Male pygofer and anal apparatus as in Text-fig. 2295, genital plates and valve as in Text-fig. 2296, style as in Text-fig. 2297, connective as in Text-fig. 2298, aedeagus as in Text-figs. 2299-2301, 1st abdominal sternum in male as in Text-fig. 2302. 7th abdominal sternum in female broadly and shallowly concave, medially with a small obtusely angular projection. Overall length of males 3.9-5.1 mm, of females 4.2-5.2 mm. – Nymphs with 4 rows of hairs on abdominal dorsum; vertex without sharply delimited black spots; a row of black spots present beneath transition from vertex to frons; body with 3 brown longitudinal bands, of which the median one is bisected by a narrow pale line; dorsal pattern of pigmentation consisting of bands, these contrasting, lateral band of vertex (along eye) distinct along its entire length; black spots usually present along sides of anclypeus (Vilbaste, 1982). See also Walter (1978).

Distribution. Common in Denmark. – Common in Sweden, Sk.-T.Lpm. – Common in Norway, Ø, AK-TRi. – Common in East Fennoscandia, Al, Ab-LkW; Vib, Kr. – Widespread in the Palaearctic region.

Biology. In "Salzstellen, Stranddünen, Hochmooren, Flachmooren, Waldlichtungen, Wiesen" (Kuntze, 1937). In "Bentgraswiesen, Calthawiesen, extremnassen Carexwiesen" (Marchand, 1953). Hibernation in egg-stage (Müller, 1957, Remane, 1958), 2 generations (Remane, 1958). Adults in May-September.

332. *Cicadula (Cicadula) persimilis* (Edwards, 1920)
Plate-fig. 174, text-figs. 2303-2306.

Limotettix persimilis Edwards, 1920: 57.

Resembling *C. quadrinotata*. Aedeagus as in Text-figs. 2303-2305, 1st abdominal sternum in male as in Text-fig. 2306. Overall length of males 4.2-4.7 mm, of females 4.4-5.0 mm. – Nymphs as in *C. quadrinotata* but very pale; bands on dorsum, if present, pale brown, band along eyes always absent, face entirely pale (Vilbaste, 1982). See also Walter (1978).

Distribution. Widespread and common in Denmark. – Widespread, fairly common in southern and central Sweden, found Sk.-Jmt. – Norway: so far found in AK, Os, Bø, VE, TEy, TEi, VAy, STi. – Fairly scarce in East Fennoscandia, found in Al, Ab, N, Ta, Kb; Kr. – England, Scotland, Wales, France, Netherlands, German D.R. and F.R., Austria, Switzerland, Italy, Bohemia, Moravia, Slovakia, Bulgaria, Poland, Estonia, Latvia, Lithuania, m. Russia, Kazakhstan; Nearctic region.

Biology. In "Waldlichtungen, Wiesen" (Kuntze, 1937). "Auf trocknen Hangwiesen" (Kontkanen, 1938). "Auf den Glatthaferwiesen (Arrhenatheretum elatioris)"

(Marchand, 1953). Hibernates in the egg-stage, 2 generations in Central Europe (Schiemenz, 1969b). In meadows (dry and moist) (Vilbaste, 1974). Adults in July-September.

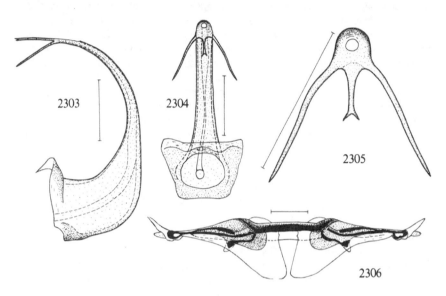

Text-figs. 2303-2306. *Cicadula persimilis* (Edwards). – 2303: aedeagus from the left; 2304: aedeagus in dorsal aspect; 2305: apex of aedeagus in terminal aspect; 2306: 1st abdominal sternum in male from above. Scale: 0.1 mm.

333. *Cicadula (Cicadula) albingensis* W. Wagner, 1940
Text-figs. 2307-2309.

Cicadula albingensis W. Wagner, 1940: 110.

Resembling *C. quadrinotata*. Caudal black spots on vertex larger than those on junction of frontoclypeus and vertex. Aedeagus as in Text-figs. 2307-2309. Overall length of males 4.5-5.0 mm, of females 4.8-5.2 mm. – Nymphs, see Walter (1978).

Distribution. So far not found in Denmark and Sweden. – Norway: recorded from a few places in AK, Bø, and VE. – East Fennoscandia: found in N: Strömfors, Vahterpää 30.VII.1975 by Albrecht. – France, German D.R. and F.R., Austria, Bohemia, Poland, Lithuania, Estonia, Altai Mts.

Biology. "Auf einer feuchten Wiese mit viel *Carex*" (Wagner, 1940). Makes great demands as regards humidity (Schwoerbel, 1957). Adults in July-September.

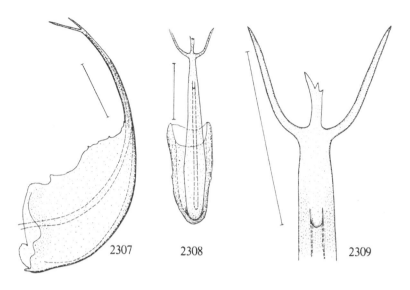

Text-figs. 2307-2309. *Cicadula albingensis* Wagner. – 2307: aedeagus from the left; 2308: aedeagus in ventral aspect; 2309: apex of aedeagus in ventral aspect. Scale: 0.1 mm.

334. *Cicadula (Cicadula) longiventris* (J. Sahlberg, 1871)
Text-figs. 2310-2312.

Limotettix longiventris J. Sahlberg, 1871: 231.
Cicadula rubroflava Linnavuori, 1952b: 74.
(Synonymy according to Vilbaste, in litt.).

Resembling *C. quadrinotata*. Caudal pair of black spots on vertex large, sometimes wider than interspace. Aedeagus as in Text-figs. 2310, 2311, 1st abdominal sternum in male as in Text-fig. 2312. Length (according to Linnavuori, 1952b): ♂ 3.67 mm, ♀ 3.78 mm. – Nymphs resembling those of *persimilis* but bands darker; bands along eyes interrupted or absent (Vilbaste, 1982).

Distribution. So far not found in Denmark, Sweden and Norway. – Rare in East Fennoscandia, found in N: Strömfors and Sibbo (Albrecht); Sa: Joutseno (Linnavuori, Thuneberg). – The male type of *longiventris* was found at Mjatusow at Swir in Russian Karelia. – Estonia, Altai Mts., Tuva, Mongolia.

Biology. "On a small swampy patch in a mixed wood. Here grew *Sphagnum* and sparse *Carex globularis* and *C. goodenowii*" (Linnavuori, 1952b). In fens (Vilbaste, 1974). Adults in July-September.

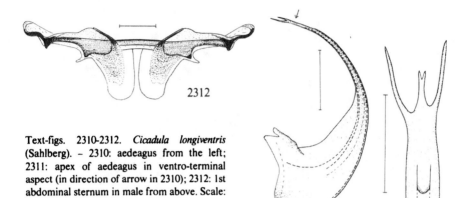

Text-figs. 2310-2312. *Cicadula longiventris* (Sahlberg). – 2310: aedeagus from the left; 2311: apex of aedeagus in ventro-terminal aspect (in direction of arrow in 2310); 2312: 1st abdominal sternum in male from above. Scale: 0.05 mm for 2311, 0,1 mm for the rest.

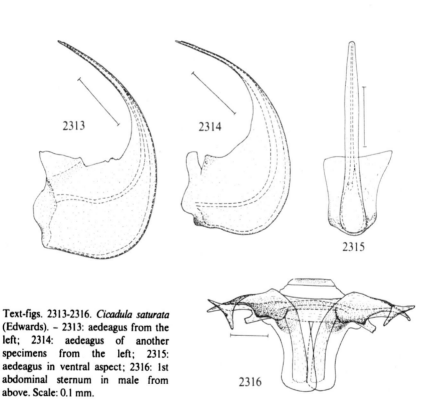

Text-figs. 2313-2316. *Cicadula saturata* (Edwards). – 2313: aedeagus from the left; 2314: aedeagus of another specimens from the left; 2315: aedeagus in ventral aspect; 2316: 1st abdominal sternum in male from above. Scale: 0.1 mm.

714

335. *Cicadula (Cicadula) saturata* (Edwards, 1915)
Text-figs. 2313-2316.

Limotettix saturata Edwards, 1915: 208.

Resembling *quadrinotata*, ground-colour in male usually orange. Caudal black spots on vertex varying in size. Aedeagus as in Text-figs. 2313-2315, 1st abdominal sternum in male as in Text-fig. 2316, 7th abdominal sternum in female as in *quadrinotata*. Overall length of male 4.2-4.7 mm, of female 4.7-5.3 mm.

Distribution. Denmark: so far known only from EJ: Moesgård and Kasted Mose (Trolle). – Widespread and fairly common in Sweden, Sk.-Vb. – Norway: found in AK, HEs, HEn, Bø, TEy, TEi. – Fairly scarce and sporadic in southern and central East Fennoscandia, found in Ab, N, Ta, Sa, Kb, Om; Kr. – England, Scotland, Netherlands, France, German D.R. and F.R., Bohemia, Poland, Estonia, Latvia, Lithuania, n. and m. Russia, Altai Mts.

Biology. In tall-sedge bogs (Linnavuori, 1952). In moist meadows, in woodland clearings (Vilbaste, 1974). "Auf Sumpfwiesen und Mooren, im *Eriophorum*-Bestand; an *Carex*" (Schiemenz, 1976). Adults in July-October.

336. *Cicadula (Cicadula) quinquenotata* (Boheman, 1845)
Plate-fig. 175, text-figs. 2317-2320.

Thamnotettix quinquenotata Boheman, 1945b: 159.
Thamnotettix ribauti Kontkanen, 1937: 148.

Upper side orange-yellow to greenish yellow, shining. Head considerably shorter than pronotum, on junction of frontoclypeus with vertex almost always with a pair of reniform or oval transverse black spots, the convex side upwards. In addition there is often a smaller black dot caudally of each ocellus, sometimes also a small spot consisting of two short longitudinal streaks at anterior end of coronal suture (Plate-fig. 175). Face more or less as in *quadrinotata*. Pronotum, scutellum and fore wings usually unicolorous, the latter with anal part fumose. Sometimes there are indistinctly delimited dark spots on pronotum and dark longitudinal bands in fore wing cells. Thoracic venter and abdomen in male black, the latter with lateral and segmental margins narrowly yellow; in females the light colour may be more extending. Legs yellow with black dots on tibiae, with or without black longitudinal streaks on outside of fore and middle tibiae. Aedeagus as in Text-figs. 2317-2319, 1st abdominal sternum in male as in Text-fig. 2320. Overall length of males 4.1-4.9 mm, of females 4.4-4.9 mm. – Nymphs resembling those of *quadrinotata* (Vilbaste, 1982).

Distribution. Rare in Denmark, found only in NWJ: Hansted 26.VIII.1962 and 3.VIII.1963 (N. P. Kristensen). – Fairly common and widespread in Sweden, Sk.-Lu.Lpm. – Not recorded from Norway. – East Fennoscandia: found in Al, Ab, N, Ta, Sa, Oa, Kb; Vib, Kr. – General distribution imperfectly known owing to possible confusion with the following species. – England. France, German F.R. and D.R. Austria, Poland, Estonia, Latvia, Bohemia, n. Russia.

Biology. "Auf *Carex*-Weissmooren recht häufige und in grosser Menge auftretende Art; erscheint in der zweiten Julihälfte" (*"ribauti"*, Kontkanen, 1938). Tyrphophilous; in tall-sedge bogs, short-sedge bogs, wet "rimpi" bogs, quagmire marshes (Linnavuori, 1952a). "Auf Hochmooren an *Carex* und *Eriophorum*" (Wagner & Franz, 1961). Adult in July-September.

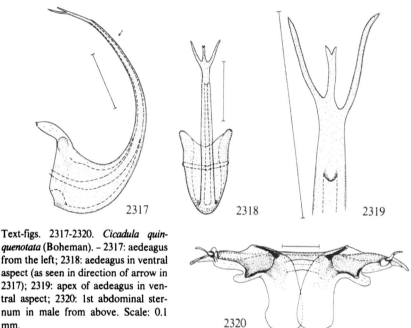

Text-figs. 2317-2320. *Cicadula quinquenotata* (Boheman). – 2317: aedeagus from the left; 2318: aedeagus in ventral aspect (as seen in direction of arrow in 2317); 2319: apex of aedeagus in ventral aspect; 2320: 1st abdominal sternum in male from above. Scale: 0.1 mm.

337. *Cicadula (Cicadula) nigricornis* (J. Sahlberg, 1871)
Text-figs. 2321-2324.

Limotettix nigricornis J. Sahlberg, 1871: 232.

Resembling *C. quinquenotata*, on an average slightly larger. Orange-coloured specimens dominating; vertex often without black markings. Aedeagus as in Text-figs. 2321-2323, 1st abdominal sternum in male as in Text-fig. 2324. Overall length of male 4.4-4.9 mm, of females 4.8-5.4 mm. – Nymphs often with thorax and tergites VII-VIII brown, but rarely entirely so (Vilbaste, 1982).

Distribution. Not recorded from Denmark and Norway. – Scarce in Sweden, found in Sm. (Boheman); Ög.: Rystad, Bjursholmen (Ossiannilsson); several localities in Upl.

(Ossiannilsson); Hls.: Edsbyn (Henriksson). – East Fennoscandia: Vib: Kivinebb (J. Sahlberg); Swir: Mjatusow (J. Sahlberg), Syväri, Uslanka (Platonoff). – Latvia, Lithuania, Estonia.

Biology. In fens and moist meadows (Vilbaste, 1974). In a moist meadow with *Calamagrostis canescens* (Ossiannilsson). Adults in July and August.

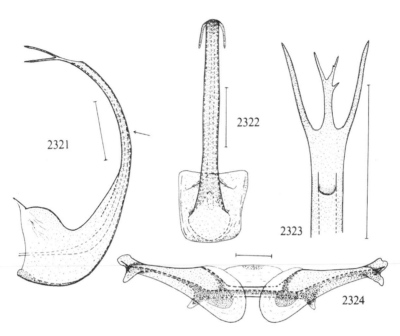

Text-figs. 2321-2324. *Cicadula nigricornis* (Sahlberg). – 2321: aedeagus from the left; 2322: aedeagus in ventral aspect (in direction of arrow in 2321), 2323: apex of aedeagus in ventral aspect; 2324: 1st abdominal sternum in male from above: Scale: 0.1 mm.

338. *Cicadula (Cicadula) flori* (J. Sahlberg, 1871)
   Text-figs. 2325-2331.

Limotettix flori J. Sahlberg, 1871: 239.

Upper side orange yellow to greenish yellow, shining. Head as seen from above shorter than pronotum, with or without black spots: one pair on junction of vertex and frontoclypeus (Text-figs. 2325, 2326), and another behind each ocellus; if present, these spots are usually quite small. A spot of varying size may be present below each antenna. Sutures on face partly black. Antennae long. Fore wings long and narrow (Text-fig. 2327), apical part fumose. Thoracic venter and abdomen coloured as in *quinquenotata*.

717

Legs yellow, posterior tibiae black spotted. Aedeagus as in Text-figs. 2328, 2329, 1st abdominal sternum in male as in Text-fig. 2330, 7th abdominal sternum in female as in Text-fig. 2331. Overall length of males 4.3-5.0 mm, of females 4.7-5.6 mm.

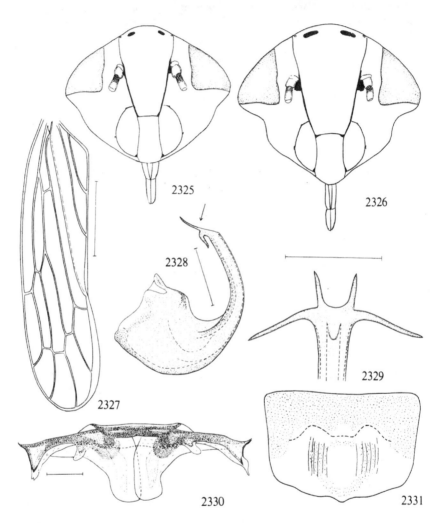

Text-figs. 2325-2331. *Cicadula flori* (Sahlberg). – 2325: face of male; 2326: face of female; 2327; left fore wing of male; 2328: aedeagus from the left; 2329: apex of aedeagus in terminal aspect (in direction of arrow in 2328); 2330: 1st abdominal sternum in male from above; 2331: 7th abdominal sternum in female from below (depressed under coverglass). Scale: 1 mm for 2327, 0.1 mm for 2328-2330.

Distribution. Denmark: known from a few localities in EJ: Moesgård and Ny Nørup, NEZ: Tisvilde, and B: Stakkelemose. – Scarce in Sweden, found in Bl., Sm., Öl., Gtl., Vg., Sdm., Upl., Vstm., Vrm., Hls. – Not recorded from Norway. – Fairly scarce and sporadic in southern and central East Fennoscandia, Al, Ab, N, Ka, St, Sa, Oa; Vib, Kr. – France, Switzerland, German D.R. and F.R., Bohemia, Slovakia, Hungary, Poland, Estonia, Latvia, Lithuania, m. and s. Russia, Ukraine, Kazakhstan, Israel, Altai Mts., m. Siberia, Tuva, Mongolia, Maritime Territory.

Biology. On seashores, in the *Phragmites* zone, the *Scirpus Tabernaemontani-S. Maritimus* zone, and the *Juncus Gerardi-Festuca* zone (Linnavuori, 1952a). Hibernates in the egg-stage, in Central Europe 2 generations (Remane, 1958). "Auf Sumpfwiesen an *Carex*" (Schiemenz, 1976). I found *C. flori* on moist meadows with *Calamagrostis canescens*. Adults in July-September.

339. *Cicadula (Cyperana) intermedia* (Boheman, 1845)
   Plate-fig. 176, text-figs. 2332-2337, 2343.

Thamnotettix intermedia Boheman, 1845b: 159.
Limotettix lunulifrons J. Sahlberg, 1871: 236.

Upper side orange yellow or greenish yellow, shining. Head as seen from above shorter than pronotum, on junction of vertex with face with four black spots in a transverse row (Plate-fig. 176), the median pair on frontoclypeus and semilunar with the convex side above, the lateral pair on upper part of genae and shaped as the median spots, only smaller. A small black spot is often present caudally of each ocellus; at frontal end of the coronal suture there are often two small black streaks, as in strongly pigmented specimens of *quinquenotata*. Markings of face for the rest as in *flori*. Fore wings apically more or less distinctly fumose. Colour of thoracic venter and abdomen as in *quinquenotata*. Fore and middle legs with black longitudinal streaks on backside of femora and outside of tibiae. Posterior tibiae black dotted. Male pygofer and anal apparatus as in Text-fig. 2332, genital plates and valve as in Text-fig. 2333, aedeagus as in Text-figs. 2334, 2335, style as in Text-fig. 2336, 1st abdominal sternum in male as in Text-fig. 2337. 7th abdominal sternum in female largely black, with characteristic surface sculpture (Text-fig. 2343). Overall length of male 3.9-4.5 mm, of female 4.3-4.8 mm.

Distribution. So far not found in Denmark. – Sweden: rare in the south (Öl. only), common in the northern part, Vstm.-T.Lpm. – Common and widespread in Norway, AK-Fi. – East Fennoscandia: widespread, comparatively scarce, commoner in the north, N-Li; Lr. – England, Scotland, Ireland, Estonia, n. and m. Russia, Tuva, Altai Mts., Kamchatka; Nearctic region.

Biology. "Auf seggenreichen Weissmooren" (Kontkanen, 1949b). "On boggy, often water-soaked ground with *Juncus, Carex* etc., on the beach of lakes and ponds, from July 28 to September 20" (Holgersen, 1949). I have seen an adult collected as late as 17.X.

719

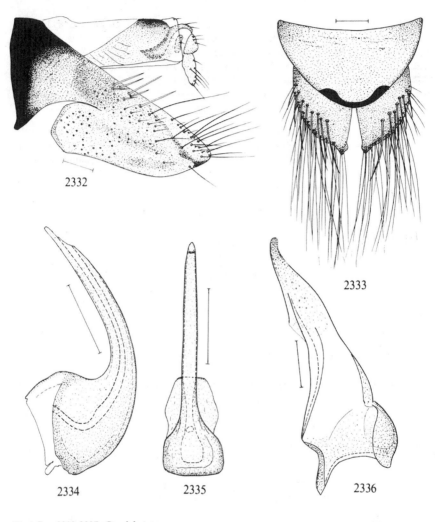

Text-figs. 2332-2337. *Cicadula intermedia* (Boheman). – 2332: male pygofer and anal apparatus from the left; 2333: genital plates and valve from below; 2334: aedeagus from the left; 2335: aedeagus in ventral aspect; 2336: right style from above; 2337: 1st abdominal sternum in male from above. Scale: 0.1 mm.

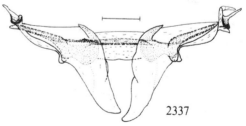

720

340. *Cicadula (Cyperana) ornata* (Melichar, 1900)
     Text-figs. 2338-2342.

Thamnotettix ornatus Melichar, 1900: 36.
Thamnotettix stramineus Sanders & De Long, 1917: 90.
Cicadula intermedia W. Wagner, 1939: 187, nec Boheman, 1845.
Cicadula ossiannilssoni Kontkanen, 1947b: 171 (n.n.).

Resembling *intermedia,* distinctly larger. Colour of upper side normally orange yellow in both sexes. Transverse row of black spots on junction of vertex with face often coalescent. Dorsal margin of male pygofer lobe faintly concave. Genital style as in Text-fig. 2338, aedeagus as in Text-figs. 2339, 2340, 1st abdominal sternum in male as in Text-fig. 2341. 7th abdominal sternum in female largely black, with characteristic sur-

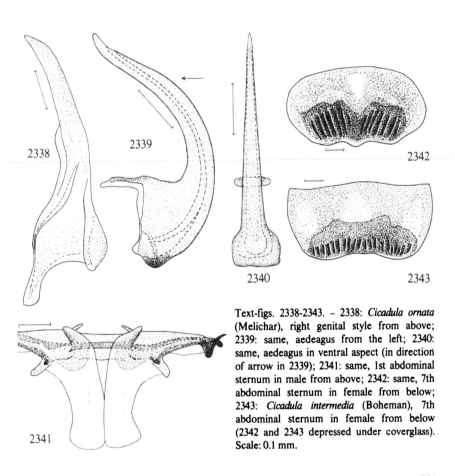

Text-figs. 2338-2343. – 2338: *Cicadula ornata* (Melichar), right genital style from above; 2339: same, aedeagus from the left; 2340: same, aedeagus in ventral aspect (in direction of arrow in 2339); 2341: same, 1st abdominal sternum in male from above; 2342: same, 7th abdominal sternum in female from below; 2343: *Cicadula intermedia* (Boheman), 7th abdominal sternum in female from below (2342 and 2343 depressed under coverglass). Scale: 0.1 mm.

face sculpture (Text-fig. 2342). Overall length of males 5.0-5.4 mm, of females 5.4-5.8 mm. – Nymphs with chaetotaxy as in *quadrinotata,* longitudinal bands on dorsum faint or absent; width of head in N$_5$ greater than 1.1 mm (Vilbaste, 1982).

Distribution. Not found in Denmark and Norway. – Replacing *intermedia* in the south of Sweden, Sk.-Hls., fairly common. – Fairly scarce and sporadic in East Fennoscandia, found in Ab, Ta, Kb. – Estonia, Latvia, Tuva, Mongolia, Altai Mts., m. Siberia, Kamchatka; Nearctic region.

Biology. With *C. intermedia* (Kontkanen, 1949b). In wet peaty meadows (Linnavuori, 1952a). On *Carices* (Vilbaste, 1974). Adults in July-September.

341. *Cicadula (Henriana) frontalis* (Herrich-Schäffer, 1835)
Plate-fig. 217, text-figs. 2344-2350.

Jassus frontalis Herrich-Schäffer, 1835: 70.
Thamnotettix antennata Boheman, 1845b: 35.

Male orange yellow or brownish yellow, female brownish yellow, shining. Head as seen from above rounded angular, little shorter than abdomen, often spotless, especially in females, often with two usually small spots on junction of vertex and frontoclypeus. In strongly pigmented specimens these spots may be comparatively large, accompanied by a small spot on each gena at the same level, and by a small spot caudally of each ocellus. Coronal suture in male sometimes strongly red; a black spot below each antenna present or absent, sutures of face partly black. Antennae very long (Plate-fig. 217). Lateral triangular spots on scutellum often orange-coloured. Fore wings long, usually unicolorous brownish yellow or (in males) with fuscous streaks in the cells. Venter of thorax black spotted, legs unicolorous with black longitudinal streaks, at least posterior tibiae black spotted. Abdomen in male usually largely black with light lateral and segmental borders, in female often largely light. Male pygofer and anal apparatus as in Text-fig. 2344, genital plates and valve as in Text-fig. 2345, style as in Text-fig. 2346, aedeagus as in Text-figs. 2347, 2348, 1st-3rd abdominal sterna in male as in Text-figs. 2349, 2350. Overall length of males 4.5-5.4 mm, of females 5.4-6.2 mm.

Distribution. Widespread but rather uncommon in Denmark, found in SJ, EJ, F, SZ, NEZ. – Scarce in southern Sweden, found in Sk., Sm., Öl., Vg., Sdm., Upl. – Not recorded from Norway. – East Fennoscandia: only found in Al: Finström, Pålsböle 20-26.VIII.1943 (Håkan Lindberg). – England, Ireland, France, Spain, Netherlands, Belgium, German D.R. and F.R., Switzerland, Italy, Bohemia, Hungary, Poland, Estonia, Latvia, s. Russia, Moldavia, Kazakhstan, Uzbekistan, Tuva, Mongolia, Altai Mts., Kirghizia.

Biology. "In Salzstellen, Flachmooren, Wäldern, Waldlichtungen; an Riedgräsern" (Kuntze, 1937). "Stellt hohe Ansprüche an die Feuchtigkeit" (Schwoerbel, 1957). Hibernation takes place in the egg-stage (Müller, 1957, Schiemenz, 1976). Two generations in Central Europe (Schiemenz, 1976). Adults in July-October.

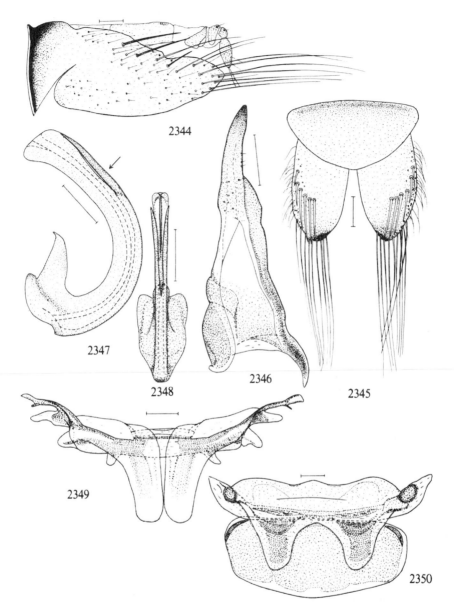

Text-figs. 2344-2350. *Cicadula frontalis* (Herrich-Schäffer). – 2344: male pygofer and anal apparatus from the left; 2345: genital plates and valve from below; 2346: left genital style from above; 2347: aedeagus from the left; 2348: aedeagus in ventral aspect (in direction of arrow in 2347); 2349: 1st abdominal sternum in male from above; 2350: 2nd and 3rd abdominal sterna in male from above. Scale: 0.1 mm.

Genus *Mocydiopsis* Ribaut, 1939

Mocydiopsis Ribaut, 1939: 274.
   Type-species: *Jassus attenuatus* Germar, 1821, by original designation.

Body comparatively robust. Head just wider than pronotum. Transition between vertex and frontoclypeus fairly abrupt. Anteclypeus dilated towards apex. Chaetotaxy of anterior tibiae as in *Elymana*. Fore wings apically narrowly rounded, median apical cell elongate, asymmetrical, apical membrane narrow but distinct. Male pygofer dorsally deeply incised. Anal tube long, well sclerotized, without appendages. Genital plates short, rounded, their macrosetae arranged in a row along lateral margin. Connective Y-shaped. Aedeagus with socle well developed, shaft slender, phallotreme ventral. In Denmark and Fennoscandia two species.

Key to species of *Mocydiopsis*

1   Fore wings with at least one, usually 2 or 3, dark spots in cla-
      vus along commissural border (Plate-fig. 177). Each lateral
      appendage of aedeagus distinctly longer than socle (Text-fig.
      2354). Phallotreme situated considerably proximally of point
      of attachment of appendages .... 342. *attenuata* (Germar)
–   Clavus spotless. Lateral appendages of aedeagus shorter than
      socle (Text-fig. 2362). Phallotreme situated quite near point of
      attachment of appendages .... 343. *parvicauda* Ribaut.

342. *Mocydiopsis attenuata* (Germar, 1821)
   Plate-fig. 177, text-figs. 2351-2356.

Jassus attenuatus Germar, 1821: 91.

Cream-coloured, shining. Head above with two brownish yellow longitudinal bands converging towards apex. Transverse streaks and sutures of face, a spot below each antenna and a longitudinal spot on anteclypeus, brownish yellow to fuscous. Pronotum with 6 brownish yellow longitudinal bands. Lateral spots on scutellum brownish yellow or orange yellow. Veins of fore wings partly dark-bordered, the dark colour here and there extending into spots. 1-3 dark spots present in clavus at commissural border. Median apical cell dark. Venter of thorax and abdomen with large black patches. Legs striped with black. Anal tube of male almost as long as rest of abdominal dorsum. Male genital plates and valve as in Text-fig. 2351, style as in Text-fig. 2352, aedeagus as in Text-figs. 2353-2355, 1st abdominal sternum in male as in Text-fig. 2356. Caudal border of 7th abdominal sternum of female strongly concave, medially almost straight. Overall length of males 3.7-4.4 mm, of females 3.9-4.6 mm. – For description of nymphs see Walter (1978).

Distribution. Widespread but rare in Denmark, found only in Jutland. – Very rare in Sweden, found only in Bl.: Torhamn 15.VI.1962 (1♀, Gyllensvärd leg.) and in Öl.: Stenåsa 21.VI.1935 (1♀, Ossiannilsson leg.). – Not found in Norway and East Fennoscandia. – England, France, Netherlands, German D.R. and F.R., Switzerland,

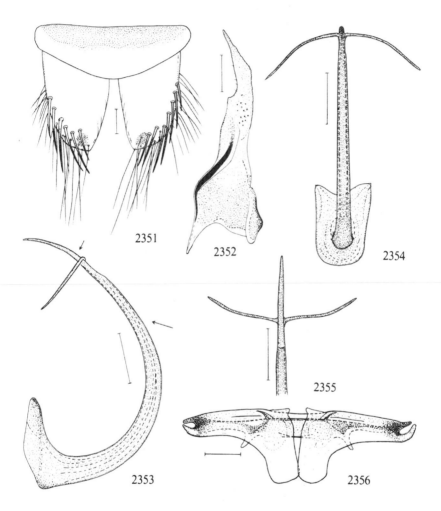

Text-figs. 2351-2356. *Mocydiopsis attenuata* (Germar) (Specimen from Luxemburg, Cobben leg.). – 2351: male genital plates and valve from below; 2352: right genital style from above; 2353: aedeagus from the left; 2354: aedeagus in ventral aspect (in direction of longer arrow in 2353); 2355: apex of aedeagus in ventro-terminal aspect (in direction of shorter arrow in 2353); 2356: 1st abdominal sternum in male from above. Scale: 0.1 mm.

Austria, Italy, Luxemburg, Albania, Bulgaria, Bohemia, Moravia, Slovakia, Hungary, Romania, Yugoslavia, Poland, Algeria, Tunisia, Ukraine, Georgia, Moldavia.

Biology. Probably a litoral species (Gravestein, 1953). "In Küstendünen im Ammophiletum . . ., an relativ trockenwarmen Kalkhängen . . . und auch in grasdurchsetzten Calluneten. Weiter südlich scheint die Art feuchtere Grasbestände (Wiesen) zu bevorzugen. . . . Beide Geschlechter überwintern, und die Paarung erfolgt höchstwahrscheinlich (auf alle Fälle aber Ei-Entwicklung und -Ablage) erst im Frühjahr" (Remane, 1961a). – "Eurytope Art der Trockenrasen, wobei der Schwerpunkt des Auftretens im Bereich X (= xerotherme Biotope) liegt" (Schiemenz, 1969b).

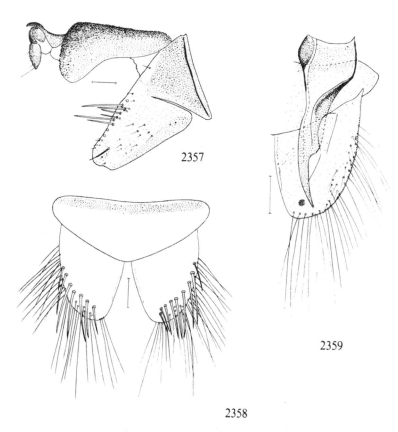

Text-figs. 2357-2359. *Mocydiopsis parvicauda* Ribaut. – 2357: male pygofer and anal apparatus from the right; 2358: genital plates and valve from below; 2359: right genital plate and right style from above (ventral macrosetae not considered). Scale: 0.1 mm.

343. *Mocydiopsis parvicauda* Ribaut, 1939
Text-figs. 2357-2364.

Mocydiopsis parvicauda Ribaut, 1939: 277.

Resembling *attenuata,* dark markings much reduced, clavus always spotless. Anal tube in male considerably shorter than rest of abdominal tergum. Male pygofer with anal apparatus as in Text-fig. 2357, genital valve, plates, and style as in Text-figs. 2358, 2359, connective as in Text-fig. 2360, aedeagus as in Text-figs. 2361, 2362, 1st abdominal sternum in male as in Text-fig. 2363, 2nd abdominal sternum in male as in Text-fig. 2364. 7th abdominal sternum in female as in *attenuata.* Overall length of male 3.7-4.4 mm, of female 4.2-4.5 mm.

Distribution. Not found in Denmark, Norway and East Fennoscandia. – Rare in Sweden, found only in Upl.: Värmdö, Lövbergaviken 15.IX.1958 (Coulianos), and

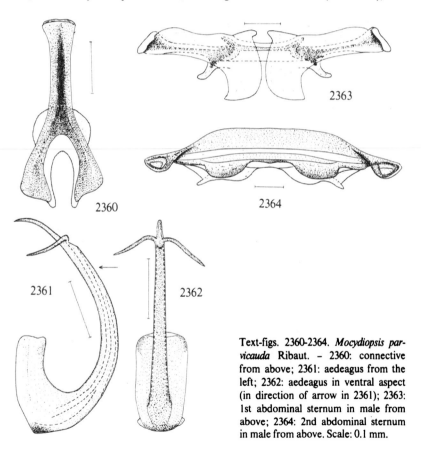

Text-figs. 2360-2364. *Mocydiopsis parvicauda* Ribaut. – 2360: connective from above; 2361: aedeagus from the left; 2362: aedeagus in ventral aspect (in direction of arrow in 2361); 2363: 1st abdominal sternum in male from above; 2364: 2nd abdominal sternum in male from above. Scale: 0.1 mm.

727

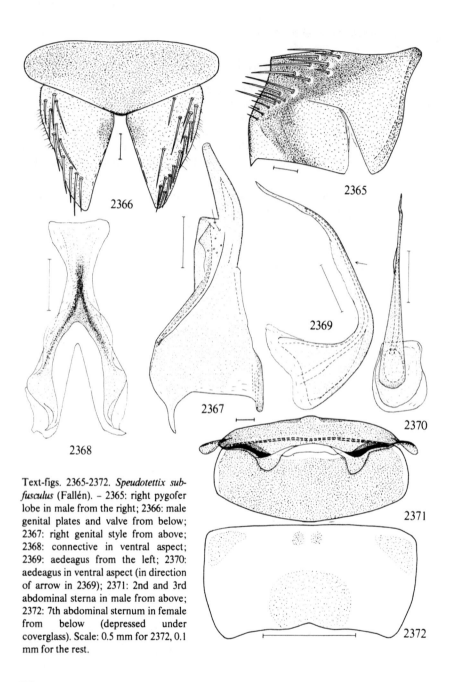

Text-figs. 2365-2372. *Speudotettix subfusculus* (Fallén). – 2365: right pygofer lobe in male from the right; 2366: male genital plates and valve from below; 2367: right genital style from above; 2368: connective in ventral aspect; 2369: aedeagus from the left; 2370: aedeagus in ventral aspect (in direction of arrow in 2369); 2371: 2nd and 3rd abdominal sterna in male from above; 2372: 7th abdominal sternum in female from below (depressed under coverglass). Scale: 0.5 mm for 2372, 0.1 mm for the rest.

Ekerö 1.X.1959 and 11.IX.1960 (Ossiannilsson). – England, France, Belgium, Netherlands, German D.R. and F.R., Austria, Italy, Bohemia, Bulgaria, Iran.

Biology. "In *Calluna*-Beständen, die mit Gräsern durchwachsen sind" (Wagner & Franz, 1961). "Sowohl in sauren, grasdurchsetzten Calluneten, in ± kultivierten Grasflächen verschiedener Art und auch am Rande von Kalkhängen" (Remane, 1961a). "Imaginal-Überwinterer, 1 Generation" (Schiemenz, 1969b). Eurytopic species in both xeric and mesic biotopes (Schiemenz, l.c.).

Genus *Speudotettix* Ribaut, 1942

Speudotettix Ribaut, 1942: 261.
  Type-species: *Cicada subfuscula* Fallén, 1806, by original designation.

Body elongate, robust. Head frontally angular, much shorter than pronotum. Anteclypeus tapering towards apex. Fore wings considerably longer than abdomen. Chaetotaxy of anterior tibiae as in *Elymana*. Anal tube in male not sclerotized dorsally. Connective Y-shaped. Male pygofer lobes without processes. Genital plates triangular, their setae arranged along lateral border (not always in a row). Aedeagus asymmetrical, shaft twisted, phallotreme on left side. In Europe one species.

344. *Speudotettix subfusculus* (Fallén, 1806)
  Text-figs. 2365-2372.

Cicada subfuscula Fallén, 1806: 30.

Greyish brown-yellow, shining. Head on fore border between frontal sutures with a fuscous transverse band consisting of two arcuate lines, concave sides ventrad. Caudally of this band there is a similar band extending between ocelli, often interrupted medially. Behind this a third transverse band is usually present; it is fuscous or lighter, widest on middle and almost straight. Frontally of hind border of vertex a pair of light spots may be present. Frontoclypeus with black transverse streaks, antennal pit and sutures on face black, anteclypeus light or fuscous with or without light spots. Dark markings of face often confluent. Pronotum with or without a row of black spots along fore border. Scutellum with brownish lateral spots and two smaller fuscous spots between these. Fore wings greyish brown with light veins; cells often darkened by a diffuse black pigment usually more dense along veins. In dark specimens, this fuscous pigment may extend over large parts of pronotum and scutellum. Venter of thorax black with sordid yellow sclerite margins. Legs sordid yellow with indistinctly delimited fuscous spots and streaks sometimes extending over the major part of the surface of the legs. Abdomen in male black with sordid yellow or orange yellow lateral and segmental borders, genital plates largely light; abdomen in female strongly varying in colour. Male pygofer lobes as in Text-fig. 2365, genital valve and plates as in Text-fig. 2366 (median margins not diverging in live specimens); style as in Text-fig. 2367, connective as in

729

Text-fig. 2368, aedeagus as in Text-figs. 2369, 2370, 2nd and 3rd abdominal sterna in male as in Text-fig. 2371, 7th abdominal sternum in female as in Text-fig. 2372. Overall length of male 5.0-5.7 mm, of female 5.1-6.0 mm. – Nymph with hairs present only on abdominal tergite VIII – two on hind margin and two on hind angles (Vilbaste, 1982; see also Walter, 1978).

Distribution. Common in Denmark. Common in Sweden (Sk.-T.Lpm.), in Norway (Ø, AK-Fn), and in East Fennoscandia (Ab-Li; Kr, Lr). – Widespread in the Palaearctic region.

Biology. In woods (Kuntze, 1937). Hibernation takes place in the larval stage (Müller, 1957), Schiemenz, 1976). "An Waldgräsern" (Wagner & Franz, 1961). "In deciduous and mixed forest in the early stages of forest succession, and on forest undergrowth. Its host plants include *Alnus incana, Betula* spp., *Picea abies, Sorbus aucuparia, Salix* spp., *Lonicera xylosteum, Tilia cordata, Populus tremula, Juniperus communis . . . Vaccinium uliginosum* and *V. myrtillus . . .*" (Raatikainen & Vasarainen, 1976). Adults in May-September.

Genus *Hesium* Ribaut, 1942

Hesium Ribaut, 1942: 261.
Type-species: *Cicada biguttata* Fallén, 1806, by original designation.

Body elongate, robust. Head little longer than half length of pronotum. Frontoclypeus wide, only slightly longer than wide. Anteclypeus dilated towards apex. Anterior tibiae with one seta in the inner, five in the outer dorsal row. Pygofer lobes in male without macrosetae, each apically with a T-shaped appendage. Apophysis of style long, slender. Connective Y-shaped, short and broad. Phallotreme ventral, placed at a long distance proximally of apex. 7nd abdominal sternum in female with a deep median incision. A monotypic genus.

345. *Hesium domino* (Reuter, 1880)
Plate-fig. 178, text-figs. 2373-2380.

Cicada biguttata Fallén, 1806: 27 (nec Fabricius, 1781).
Athysanus domino Reuter, 1880: 212.
Hesium falleni Metcalf, 1955: 265 (n.n.).

Brownish yellow, shining. Vertex with an indistinct vestige of a darker transverse band or two spots. Face often with fuscous transverse streaks. Lateral spots on scutellum a little darker brownish. Veins of fore wings more or less distinctly light, indistinctly dark-bordered, transverse veins somewhat more broadly light; a small milky spot present on distal end of each claval vein. Hind wings fumose, veins dark. Thoracic venter black spotted, legs light, hind tibiae black spotted. Abdomen in male black, lateral and segmental borders and genital segment light, or abdomen largely light as in female.

Male pygofer as in Text-fig. 2373, genital valve and plates as in Text-fig. 2374 (but median borders of plates not divergent in live specimens at rest), style as in Text-fig. 2375, connective as in Text-fig. 2376, aedeagus as in Text-figs. 2377, 2378, 2nd and 3rd

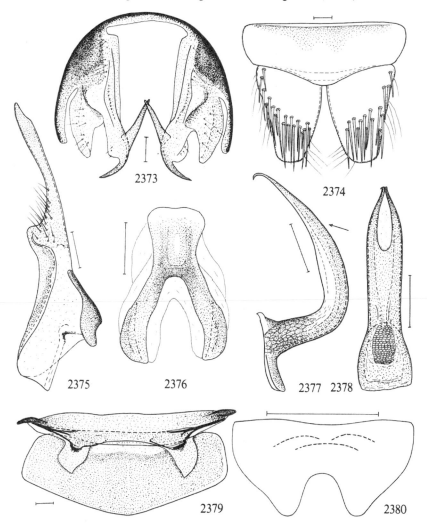

Text-figs. 2373-2380. *Hesium domino* (Reuter). – 2373: male pygofer from behind; 2374: genital plates and valve from below; 2375: right genital style from above; 2376: connective in ventral aspect; 2377: aedeagus from the left; 2378: aedeagus in ventral aspect (in direction of arrow in 2377); 2379: 2nd and 3rd abdominal sterna in male from below; 2380: 7th abdominal sternum in female from below (depressed under coverglass). Scale: 1 mm for 2380, 0.1 mm for the rest.

731

abdominal sterna in male as in Text-fig. 2379. 7nd abdominal sternum in female as in Text-fig. 2380. Overall length (♂♀) 5.6-7.0 mm. – Nymphs: 4 longitudinal rows of dorsal hairs, dorsal body surface uniformly pale, ground colour yellowish green or pale green, chaetation of anterior tibiae as in adults; N5 longer than 5 mm (Vilbaste, 1982).

Distribution. Not found in Denmark. – Fairly common in southern and central Sweden, Sk.-Med. – Norway: found in Ø, AK, Os, On, Bø, VE, TEi, SFi. – Fairly scarce and sporadic in East Fennoscandia, recorded from Ab, N, Ta, Sa, Kb; Vib, Kr; Swir. – France, Belgium, German D.R. and F.R., Austria, Italy, Albania, Bohemia, Slovakia, Bulgaria, Hungary, Romania, Poland, Latvia, Lithuania, Estonia, n. and m. Russia, Ukraine, Moldavia, Armenia, Georgia.

Biology. In "besonnten Hängen" (Kuntze, 1937). "In dry meadows, on wood margins, on sand dunes, sometimes also on trees and shrubs, especially in the autumn" (Vilbaste, 1974). Adults in July-September.

Genus *Thamnotettix* Zetterstedt, 1838

Thamnotettix Zetterstedt, 1838: 292.
    Type-species: *Cicada prasina* Fallén, 1806 (nec Fabricius), by subsequent designation.
Thamnus Fieber, 1866: 505.
    Type-species: *Cicada confinis* Zetterstedt, 1828, by subsequent designation.
Loepotettix Ribaut, 1942: 264 (subgenus).
    Type-species: *Jassus dilutior* Kirschbaum, 1868, by original designation.

Body elongate, robust. Head considerably shorter than pronotum, frontally rounded obtusely angular. Frontoclypeus broad, little longer than wide in upper part, anteclypeus tapering towards apex. Wings long. Anterior tibiae (in our species) with one seta in the inner, four in the outer dorsal row. Male pygofer lobes with macrosetae. Macrosetae of male genital plates arranged in a row along lateral border. Genital styles small. Connective short and broad. Phallotreme ventral. In Denmark and Fennoscandia two species.

Key to species of *Thamnotettix*

1 Ocelli smaller, their distance from eye considerably greater
   than their diameter. Genital plates of male longer (Text-fig.
   2382). Phallotreme situated on apex of a tube arising from
   ventral side of aedeagal shaft (Text-fig. 2384) (*Thamnotettix*
   s.str.) .... 346. *confinis* (Zetterstedt)
– Ocelli larger, their distance from eye approximally equal-
   ling their diameter. Genital plates of male shorter (Text-fig.
   2389) Phallotreme situated on ventral side of shaft (Text-figs.
   2391, 2392) (Subgenus *Loepotettix*) .... 347. *dilutior* (Kirschbaum)

346. *Thamnotettix (Thamnotettix) confinis* (Zetterstedt, 1828)
Text-figs. 2381-2387.

Cicada prasina Fallén, 1806: 27 (nec Fabricius, 1794).
Cicada confinis Zetterstedt, 1828: 527.

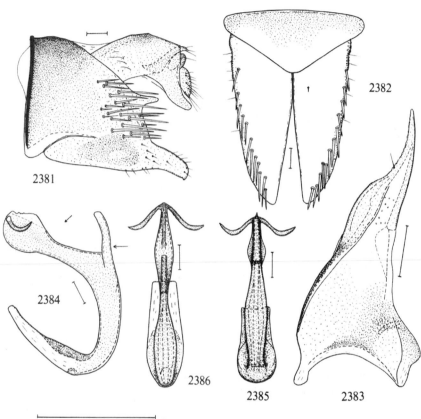

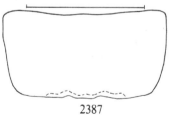

Text-figs. 2381-2387. *Thamnotettix confinis* (Zetterstedt). – 2381: male pygofer with anal apparatus from the left; 2382: genital plates and valve from below; 2383: right style from above; 2384: aedeagus from the left; 2385: aedeagus in ventral aspect (in direction of longer arrow in 2384); 2386: same in direction of shorter arrow in 2384; 2387: 7th abdominal sternum in female from below (depressed under coverglass). – Scale: 1 mm for 2387, 0.1 mm for the rest.

Jassus simplex Herrich-Schäffer, 1834c: 7.
Thamnotettix tincta Zetterstedt, 1838: 293.
Thamnotettix stupidula Zetterstedt, 1838: 294.
Athysanus prominulus Reuter, 1880: 213.
Thamnotettix schlueteri Haupt, 1917: 253.

Upper side greenish yellow with a fumose tinge, shining. Fore wings considerably longer than abdomen. In *f. typica,* upper side and face without distinct dark markings, apical part of fore wing fumose, abdomen and thoracic venter in male largely black with abdominal segmental borders and genital plates yellowish, in female lighter; tibiae dark dotted. Sometimes especially head, pronotum and scutellum are more or less densely sprinkled with sanguine dots (*f. tincta* Zett.). In *f. stupidula* Zett. there is a tendency of brownish pigment developing on upper side and face. In this form, some dark markings are visible on vertex, viz. a pair of small dots on apex, caudally of these a somewhat larger pair, on each side between these and each ocellus a transverse streak, more caudally a transverse band interrupted medially; transverse streaks on frontoclypeus fuscous, often coalescing into a large patch; pronotum with more or less distinct longitudinal bands; fore wings with brownish longitudinal streaks along veins. These two forms are on an average somewhat smaller than the typical form. Male pygofer and anal apparatus as in Text-fig. 2381, genital valve and plates as in Text-fig. 2382 (median margins of plates not diverging in live specimens); genital style as in text-fig. 2383, aedeagus as in Text-figs. 2384-2386. 7th abdominal sternum in female as in Text-fig. 2387. Overall length of male 4.9-6.5 mm, of female 5.5-7.7 mm. – Nymphs with 4 complete rows of hairs on abdomen, dorsal body surface uniformly pale brownish, ventral surface clearly darker; chaetotaxy of anterior tibiae as in adults, hind tibiae usually with dark spots around bases of spines; vertex clearly wider than long; anteclypeus narrow slightly towards tip (Vilbaste, 1982). Abdomen sometimes sanguine. With an ivory-coloured median longitudinal line. See also Walter (1978).

Distribution. Common in Denmark, Sweden, Norway, and East Fennoscandia; *f. stupidula* Zett. in the northern part of Fennoscandia. – Widespread in the Palaearctic, also present in the Nearctic region.

Biology. In moist sloping meadows, peaty meadows, rich swampy woods, rich moist grass-herb woods, moist *Oxalis-Myrtillus* spruce woods (Linnavuori, 1952a). "Vorwiegend in der Feldschicht, jedoch ziemlich oft auch polyphag auf kleineren Bäumen und auf Sträuchern, auf letzteren Imagines 1.VI.-28.VIII" (Nuorteva, 1952). "In Laubwäldern . . . mehr im Schatten und auf feuchteren Stellen" (Schwoerbel, 1957). "Particularly in grass-herb forests and moist herb forests, and in peaty or marshy meadows; also occurring in forest clearings and even in fields" (Raatikainen & Vasarainen, 1976). Adults in May-September. – F. *stupidula:* on *Betula nana,* birches, willows and herbs (Lindberg, 1932).

347. *Thamnotettix (Loepotettix) dilutior* (Kirschbaum, 1868)
Text-figs. 2388-2392.

Jassus (Thamnotettix) dilutior Kirschbaum, 1868b: 92.

Brownish yellow, sometimes sprinkled with sanguine dots. 1st apical cell in fore wing often divided by one or two secondary transverse veinlets. Frontoclypeus with indistinct brownish transverse streaks. Fore wings often with a few fuscous spots, veins whitish, apical part fumose. Posterior tibiae with setae arising from black dots. Male pygofer and anal apparatus as in Text-figs. 2388, genital plates and valve as in Text-fig. 2389, styles as in Text-fig. 2390, aedeagus as in Text-figs. 2391, 2392. Caudal margin of 7th abdominal sternum in female straight. Overall length (♂♀) 5.9-6.7 mm. – Nymphs resembling those of *confinis* (Walter, 1978).

Distribution. Widespread but rare in Denmark, only found in the southern part, known from SJ, F, LFM, and SJ. – Very rare in southern Sweden, two specimens only found in Sk.: Brunnby, Kullen 31.VII.1961 and 30.VII.1962 (Ossiannilsson). – Not found in Norway and East Fennoscandia. – England, Ireland, France, Portugal, Belgium,

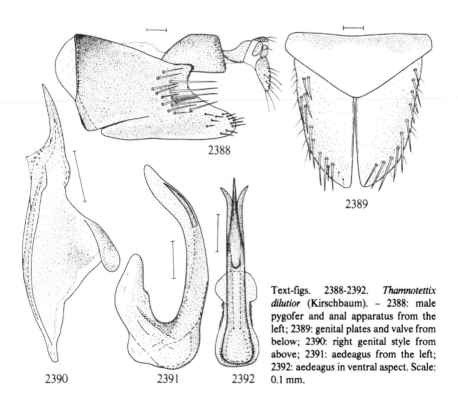

2388

2389

2390   2391   2392

Text-figs. 2388-2392. *Thamnotettix dilutior* (Kirschbaum). – 2388: male pygofer and anal apparatus from the left; 2389: genital plates and valve from below; 2390: right genital style from above; 2391: aedeagus from the left; 2392: aedeagus in ventral aspect. Scale: 0.1 mm.

735

Netherlands, German D.R. and F.R., Austria, Italy, Albania, Bohemia, Hungary, Romania, Yugoslavia, Poland, Ukraine, Algeria, Morocco, Tunisia, Georgia, w. Siberia.

Biology. In "besonnten Hängen" (Kuntze, 1937). "Besonders auf Eichengebüsch (Wagner & Franz, 1961). "Surtout sur les Chênes" (Ribaut, 1952). "Usually on trees, especially oaks, but sometimes on lower vegetation" (Le Quesne, 1969). The two Swedish specimens were also found on *Quercus*. Adults in June-October (Le Quesne, l.c.).

Genus *Pithyotettix* Ribaut, 1942

Pithyotettix Ribaut, 1942: 261.
   Type-species: *Cicada abietina* Fallén, 1806, by original designation.

Body elongate, robust. Head as seen from above shorter than pronotum, obtusely angular. Anteclypeus parallel-sided, a little narrower at lower end. Wings longer than body. Anterior tibiae with 3-4 setae in the inner, 4 in the outer dorsal row. Anal tube in male dorsally sclerotized, pygofer lobes with macrosetae. Macrosetae of genital plates in disorder. Styles well developed. Connective short, Y-shaped or X-shaped. Aedeagus asymmetrical, phallotreme terminal. In Europe one species.

348. *Pithyotettix abietinus* (Fallén, 1806)
   Plate-fig. 179, text-figs. 2393-2398.

Cicada abietina Fallén, 1806: 28.

Sordid whitish yellow or greenish white with brown and fuscous markings, often sprinkled with sanguine dots, shining. Head in strongly marked specimens on fore border between frontal sutures and apex of head on each side with an arched line, concavity downwards, caudally of these lines on vertex one or three pairs of indistinctly limited brownish spots. These markings are absent in most specimens. Frontoclypeus usually with brownish transverse streaks interrupted by a light longitudinal band widening towards anteclypeus, sometimes largely dark. Antennal pits black. Pronotum in very dark specimens with brownish shades frontally. Scutellum sometimes with a brownish median longitudinal band. Head, pronotum and scutellum sometimes densely sprinkled with sanguine dots, sometimes without such dots. Fore wings with characteristic markings (Plate-fig. 179): clavus usually light brown, distal ends of claval veins and a spot around each of these ends whitish, claval suture fuscous bordered on inside. Subcostal cell largely light brownish, proximal third of the rest of corium white; distally of this area there follows a broad light brownish field distally delimited by a broad white transverse band; apical part of fore wing mainly light brown with whitish veins. These markings are more or less distinct. In the brownish parts, veins may be partly dark-bordered, the dark streaks often extending into spots. Thoracic venter fuscous spotted, femora without markings or dark striped and spotted, tibial setae arising from fuscous

736

dots. Abdominal tergum black with light lateral and segmental borders, venter light or with a broad black longitudinal band, laterally black spotted. Male pygofer lobes as in Text-fig. 2393, genital valve and plates as in Text-fig. 2394, (in living specimens not diverging), genital styles as in Text-fig. 2395, aedeagus as in Text-figs. 2396, 2397, 7th abdominal sternum in female as in Text-fig. 2398. Overall length ($\male\female$) 5.3-6.0 mm. – Nymphs with hairs present only on tergite VIII, 4 on hind margin, two on hind angles. Whole insects speckled brown or red; if no speckling, a dark double longitudinal band is present along the whole body (Vilbaste, 1982).

Distribution. Fairly common in Denmark and in Sweden up to Lu. Lpm. – Norway: found in AK, Os, Bø, AAy, AAi, STi, NTi. – Fairly common in major part of East

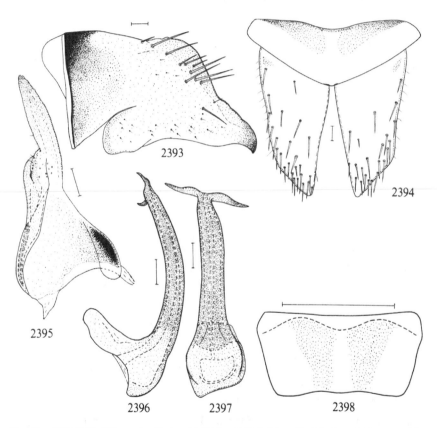

Text-figs. 2393-2398. *Pithyotettix abietinus* (Fallén). – 2393: left pygofer lobe in male from outside; 2394: male genital plates and valve from below; 2395: right genital style from above; 2396: aedeagus from the left; 2397: aedeagus in ventral aspect; 2398: 7th abdominal sternum in female from below (depressed under coverglass). Scale: 1 mm for 2398, 0.1 mm for the rest.

Fennoscandia, Ab-ObN, LkW; Kr, Lr. – France, Spain, Belgium, Netherlands, German D.R. and F.R., Switzerland, Austria, Italy, Bohemia, Moravia, Slovakia, Bulgaria, Hungary, Romania, Yugoslavia, Poland, Estonia, Latvia, Lithuania, n. and m. Russia, Algeria, m. and w. Siberia.

Biology. On *Picea abies*. Hibernates in the egg stage (Müller, 1957). I found young larvae in April, copulating adults 20.VI. Adults 24.V. – 11.IX.

Genus *Perotettix* Ribaut, 1942

Perotettix Ribaut, 1942: 260.

Type-species: *Thamnotettix pictus* Lethierry, 1880, by original designation.

As *Thamnotettix*, chaetotaxy of anterior tibiae as in *Pithyotettix*. Face elongate. Anal tube short, well sclerotized. Male pygofer asymmetrical, with macrosetae. Macrosetae of genital plates in disorder. Connective short and broad, Y-shaped. Genital styles well developed. Aedeagus asymmetrical, phallotreme terminal. In Fennoscandia one species.

349. *Perotettix orientalis* (Anufriev, 1971)
Text-figs. 2399-2406.

Pithyotettix (Perotettix) orientalis Anufriev: 1971: 113.

I have seen only two males of this very rare species. The following is a translation from the Russian text in Anufriev, 1978: 140. "Reddish to dark brown. Vertex and face yellowish, often with small red dots. Pronotum frontally brownish, caudally greyish, scutellum reddish to yellowish brown. Fore wings light brownish, often with faintly marked light spots. 5.9-6 mm." – General aspect of the Swedish specimens suggestive of *Speudotettix subfusculus*. Fore body reddish brown, above with some black markings as in Text-fig. 2399. Anteclypeus parallel-sided, with two approximately parallel black longitudinal streaks. Frontoclypeus below unspotted, upper 3/4 with black transverse stripes, median line pale. Caudal 2/3 of pronotum whitish. Fore wings with black stripes along commissural border and part of longitudinal veins, with whitish spots on distal ends of claval veins, in basal part of cubital cell, near distal end of median cell, and on proximal ends of 2nd subapical cell and 1st apical cell. Thoracic venter and abdomen largely black, segmental borders of abdomen narrowly brick-red. Legs with black longitudinal stripes, femora with a black transverse band near apex. Tibial setae arising from black spots. Male genital plates and valve as in Text-fig. 2400, pygofer lobes as in Text-figs. 2401, 2402, styles as in Text-fig. 2403, connective as in Text-fig. 2404, aedeagus as in Text-figs. 2405, 2406.

Distribution. Sweden: Vrm.: Gräsmark, 31.VIII.1978 one male (Waldén); Hls.: Edsbyn, Ullungen 21.VII.1980, one male (Bo Henriksson). These are the only European records so far. – Maritime Territory.

Biology. In conifer forests (Anufriev, 1978).

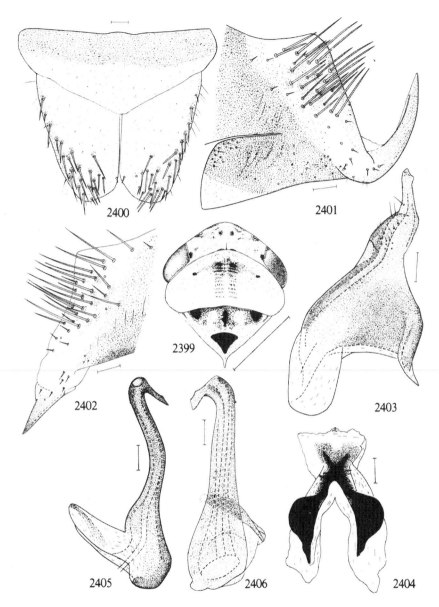

Text-figs. 2399-2406. *Perotettix orientalis* (Anufriev) (Swedish specimen). – 2399: fore body of male from above; 2400: genital plates and valve from below; 2401: left pygofer lobe in male from the left; 2402: right pygofer lobe in male from the right; 2403: right genital style from above; 2404: connective from above; 2405: aedeagus from the left; 2406: aedeagus from behind. Scale: 1 mm for 2399, 0.1 mm for the rest.

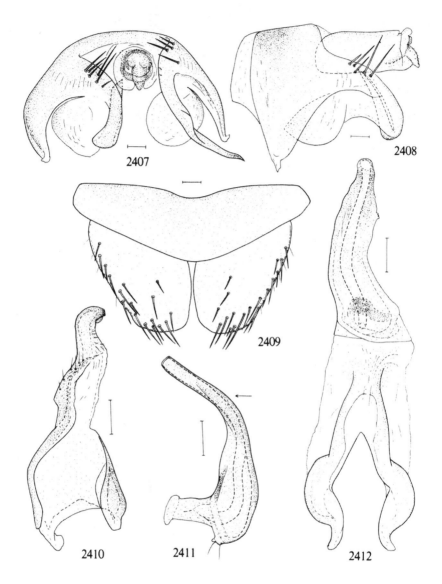

Text-figs. 2407-2412. *Colobotettix morbillosus* (Melichar). – 2407: male pygofer and anal apparatus from behind; 2408: same from the left; 2409: genital plates and valve in ventral aspect; 2410: right genital style from above; 2411: aedeagus from the left; 2412: aedeagus and connective in ventral aspect (aedeagus as seen in direction of arrow in 2411). Scale: 0.1 mm.

Genus *Colobotettix* Ribaut, 1948

Colobotettix Ribaut, 1948: 58.
  Type-species: *Thamnotettix morbillosus* Melichar, 1896.

Allied to *Perotettix* by the asymmetry of male pygofer and aedeagus, by the terminal position of the phallotreme, and by the elongate face. Connective basally asymmetrical, fused with base of aedeagus. A monotypic genus.

350. *Colobotettix morbillosus* (Melichar, 1896)
  Text-figs. 2407-2412.

Thamnotettix morbillosus Melichar, 1896: 293.

The following is essentially a translation of Melichar's original description in German (♀ only). "Much resembling *Idiodonus cruentatus,* but body including fore wings wider, the latter not much longer than abdomen, vertex obtusely angular, yellowish, with light red dots, also face red sprinkled, antennal pits not black. Pronotum red sprinkled on anterior half only, caudal half dull greyish green, transversely striolate. Scutellum yellow, red spotted. Fore wings unicolorous brownish yellow or sordid olive green, with concolorous, little prominent veins, not red sprinkled. Apices of claval veins white. Venter yellow, red sprinkled as well as legs. Posterior tibiae with black dots at bases of macrosetae, their apices as well as those of tarsal segments and claws black, abdominal venter basally blackish. 7th abdominal sternum twice as long as the preceding one, medially shallowly concave, with rounded lateral corners. Length of female 4.5 mm." – Male pygofer as in Text-figs. 2407, 2408, genital plates and valve as in Text-fig. 2409, style as in Text-fig. 2410, aedeagus as in Text-figs. 2411, 2412, connective united with aedeagus (Text-fig. 2412). Apodemes of 1st abdominal sternum in male small, each a little longer than wide and somewhat wider than interspace between them. Apodemes of 2nd abdominal sternum short, inconspicuous. Overall length (♂♀) 4.5-5 mm.

Distribution. Not in Denmark, Sweden, and Norway. – East Fennoscandia: found by Linnavuori in Ab: Raisio 10.VI.1947; Tb: Jyväskylä 25.VII.1949; Sb: Kiuruvesi, Heinäkylä 31.VII.1951. – France, German D.R. and F.R., Austria, Bohemia, Moravia, Poland.

Biology. On *Picea* and *Abies* (Linnavuori, 1950, 1952b). On conifers (Wagner & Franz, 1961).

Genus *Macustus* Ribaut, 1942

Macustus Ribaut, 1942: 261.
  Type-species: *Cicada grisescens* Zetterstedt, 1828, by original designation.

Head a little wider than pronotum. Face broad. Anteclypeus narrower towards apex, in lateral aspect convex. Ocelli small. Junction of vertex and frontoclypeus rounded. Anterior tibiae with 1-4 setae in the inner, 4 in the outer dorsal row. Wing dimorphous. Male pygofer dorsally deeply incised, lobes with an apical claw. Anal tube sclerotized laterally, not dorsally. Styles well developed. Macrosetae of genital plates arranged in an irregular row along lateral margin. Connective Y-shaped. Aedeagus with socle small but well marked, shaft slender, asymmetrical, phallotreme lateral, situated far from apex. A monotypic genus.

351. *Macustus grisescens* (Zetterstedt, 1828)
Plate-fig. 180, text-figs. 2413-2419.

Cicada grisescens Zetterstedt, 1828: 530.

Brownish yellow, shining. Wing dimorphous. Fore wings in macropterous males considerably longer than abdomen, in macropterous females about as long as the latter, in sub-brachypterous males just longer, in sub-brachypterous females considerably shorter than abdomen. Head above with two brownish transverse bands, the anterior one extending between ocelli, obtusely angular, apex directed frontad, sides arcuate; the caudal band wider, almost straight, laterally usually not reaching eyes, lateral end often forming a loop-like figure. Transverse streaks and sutures on face, a spot below each antenna, and one on anteclypeus, black, fuscous or brown. Pronotum with six more or less distinct brownish longitudinal bands. Veins of fore wings light, often dark-bordered, cells sometimes more or less filled up with dark colour; apical cells in macropters usually entirely dark. In sub-brachypterous females, the fore wings are usually unicolorous light. In specimens from northern Fennoscandia, the upper side may be largely dark. Thoracic venter dark spotted, femora dark striped, posterior tibiae with setae arising from black dots. Abdomen in males black with light lateral and segmental borders, in females light, tergum and venter each with three more or less distinctly delimited dark longitudinal streaks. Male pygofer lobes as in Text-fig. 2413, genital plates and valve as in Text-fig. 2414, styles as in Text-fig. 2415, aedeagus as in Text-figs. 2416-2418, 7th abdominal sternum in female as in Text-fig. 2419. Overall length: macropterous males about 4.75 mm, macropterous females 5.0-5.4 mm, sub-brachypterous males 3.9-4.8 mm, sub-brachypterous females 4.25-5.5 mm. British specimens appear to be larger, on an average (Le Quesne, 1969). – Nymph with hairs present only on tergite VIII, viz. 4 on hind margin, 2 on hind angle; abdomen with numerous more or less rounded pale dots; frons with clear arch-lines, separate in upper region; tip of anteclypeus narrows abruptly (Vilbaste, 1982). Good descriptions and excellent reproductions of instars III-V are given in Walter (1975: 258-260).

Distribution. Common and widespread in Denmark, Sweden, Norway, and East Fennoscandia. – Widespread in the Palaearctic region, also present in the Nearctic.

Biology. "Auf Mooren zwischen *Calluna* and *Molinia*" (Wagner, 1935a). "Auf *Carex*-und *Eriophorum*-Wiesen und Mooren, auf niedrigen, grasbewachsenen Wiesen" (Lindberg, 1932). "Auf Hochmooren" (Kuntze, 1937; Wagner & Franz, 1961). "Par-

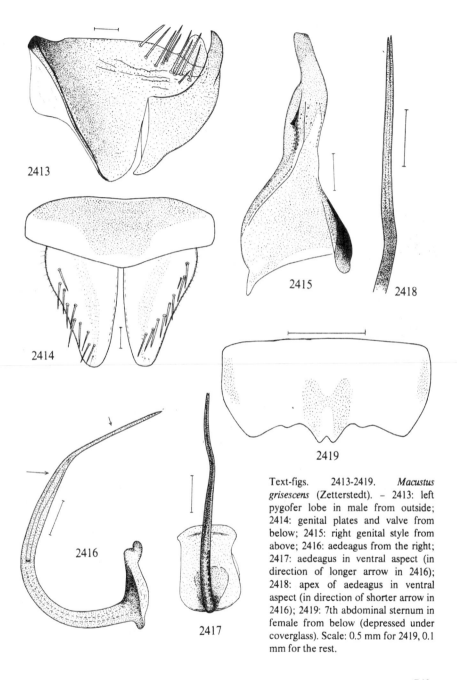

Text-figs. 2413-2419. *Macustus grisescens* (Zetterstedt). – 2413: left pygofer lobe in male from outside; 2414: genital plates and valve from below; 2415: right genital style from above; 2416: aedeagus from the right; 2417: aedeagus in ventral aspect (in direction of longer arrow in 2416); 2418: apex of aedeagus in ventral aspect (in direction of shorter arrow in 2416); 2419: 7th abdominal sternum in female from below (depressed under coverglass). Scale: 0.5 mm for 2419, 0.1 mm for the rest.

743

ticularly in swampy forests, treeless fens, and swampy meadows" (Raatikainen & Vasarainen, 1976). "Larval-Überwinterer mit 1 Generation" (Schiemenz, 1976). Adults in May-August.

Genus *Doliotettix* Ribaut, 1942

Doliotettix Ribaut, 1942: 266.
Type-species: *Cicada pallens* Zetterstedt, 1828, by original designation.

Body fairly broad, robust. Head somewhat wider than pronotum. Frontoclypeus a little longer than maximal width. Anteclypeus narrower towards apex. Anterior tibiae with 1 seta in inner, 4 in outer dorsal row. Tergum of male pygofer deeply incised, lobes with an apical process. Anal tube not sclerotized dorsally. Genital plates triangular, their macrosetae arranged in a row along lateral border. Connective long, Y-shaped. Aedeagus without appendages, phallotreme ventro-terminal. A monotypic genus.

352. *Doliotettix lunulatus* (Zetterstedt, 1838)
Text-figs. 2420-2429.

Cicada pallens Zetterstedt, 1828: 522 (♀) (nec Gmelin, 1790).
Thamnotettix lunulata Zetterstedt, 1838: 295 (♂).
Thamnotettix lapponicus Haupt, 1917: 252 (♂).

Greenish yellow, shining. Owing to the strong sexual dimorphism, the sexes have sometimes been placed in different genera. In the male, the head is shorter than pronotum, above usually with the following fuscous markings: frontally on each side between ocellus and median line two transverse streaks in an arched row, caudally of these a transverse medially interrupted band which laterally reaches outside ocelli. Frontoclypeus at least in upper part with fuscous transverse streaks, the uppermost of which are often visible from above, sutures of face partially black, anteclypeus with two parallel dark longitudinal streaks. Thoracic venter dark spotted. Fore wings considerably longer than abdomen, apically fumose. Femora with longitudinal and transverse bands, tibiae with setae arising from black dots. Abdomen largely black with light lateral and segmental borders. In females, head almost as long as pronotum, the dark transverse bands present in males partly absent or indistinct, especially the caudal band usually missing. Fore wings not longer than abdomen, venter of thorax and abdomen largely light, the latter with a black longitudinal band. Male pygofer lobes as in Text-fig. 2420, genital plates and valve as in Text-fig. 2421 (but plates not diverging at rest), genital styles as in Text-fig. 2422, connective as in Text-fig. 2423, aedeagus as in Text-figs. 2424-2426, 1st abdominal sternum in male as in Text-fig. 2427, 2nd and 3rd abdominal sterna in male as in Text-fig. 2428, 7th abdominal sternum in female as in Text-fig. 2429. Overall length of males 3.9-4.3 mm, of females 4.2-4.8 mm. – In nymphs, abdomen with 4 longitudinal rows of hairs; vertex distinctly wider than long; abdomen with pale midline; dark points present on sides of longitudinal bands of meso- and

744

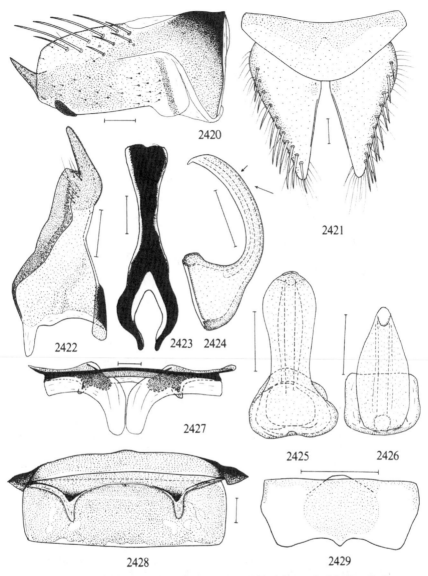

Text-figs. 2420-2429. *Doliotettix lunulatus* (Zetterstedt). – 2420: right pygofer lobe in male from outside; 2421: male genital plates and valve from below; 2422: right genital style from above; 2423: connective in ventral aspect; 2424: aedeagus from the left; 2425: aedeagus in ventral aspect (in direction of longer arrow in 2424); 2426: same in direction of shorter arrow in 2424; 2427: 1st abdominal sternum in male from above; 2428: 2nd and 3rd abdominal sterna in male from above; 2429: 7th abdominal sternum in female from below. Scale: 0.5 mm for 2429, 0.1 mm for the rest.

metathorax; anteclypeus with converging sides; pattern of vertex rather clear and characteristic (Vilbaste, 1982).

Distribution. Not found in Denmark. – Scarce in the south of Sweden, common in the northern part: Sm., Ög., Vg., Sdm.-T.Lpm. – Widespread and common in Norway, AK-Fø. – Widespread in East Fennoscandia, commoner in the north, found in Ab-Li; Kr, Lr. – German D.R. and F.R., Bohemia, Slovakia, Romania, Poland, Estonia, Latvia, Lithuania, n. and m. Russia, Altai Mts., Tuva, e. and w. Siberia, Kamchatka, Maritime Territory.

Biology. "Auf verschiedenen grasreichen Biotopen" (Kontkanen, 1938). "Auf Wiesen und im spärlichen Birkenwalde mit Gräsern, *Trollius, Geranium* usw." (Lindberg, 1932a). "In damp meadows, in clover fields, etc." (Vilbaste, 1974). "In cages it fed and reproduced on oats, wheat, rye, *Phleum pratense, Poa annua, P. nemoralis, P. pratensis, Anthoxanthum odoratum, Agrostis tenuis, Agropyron repens, Deschampsia flexuosa, D. caespitosa, Alopecurus pratensis, Festuca ovina* and *Phalaris arundinacea*, and lived for more than a month on *Luzula pilosa* and *Tripleurospermum inodorum*" (Raatikainen & Vasarainen, 1976). Hibernates at the nymphal stage (Törmälä & Raatikanen, 1976). Adults in late May-September.

Genus *Athysanus* Burmeister, 1838

Athysanus Burmeister, 1838b: [14].
Type-species: *Cicada argentata* Fabricius, 1794, by subsequent designation.

Body broad, strongly built. Head short, wider than pronotum, frontally rounded. Frontoclypeus broad, partly dorsal. Junction of vertex and frontoclypeus rounded. Anterior tibiae with 5 setae in inner, 5 in outer dorsal row. Apical membrane of fore wing strongly reduced or absent. Anal tube in male sclerotized dorsally and laterally. Male pygofer dorsally incised, with macrosetae. Genital plates triangular, their macrosetae in disorder. Connective Y-shaped, short. Styles strong, shape ordinary. Aedeagus symmetrical, phallotreme ventral or dorsal, socle well developed. In Denmark and Fennoscandia two species.

Key to species of *Athysanus*

1 Larger (length about 7 mm), straw-coloured, markings of fore wings consisting of narrow dark longitudinal streaks (Plate-fig. 181) .... 353. *argentarius* Metcalf
– Smaller (length 4.3-5.5 mm), usually largely black or brownish, fore wings spotted (Plate-fig. 182) .... 354. *quadrum* (Boheman)

353. *Athysanus argentarius* Metcalf, 1855
Plate-fig. 181, text-figs. 27, 28, 2430-2434.

Cicada argentata Fabricius, 1794: 38 (nec Olivier, 1790).
Athysanus argentarius Metcalf, 1955: 265 (n.n.).

Straw-coloured, shining. Head above with a black transverse streak almost parallel with caudal margin, extending from eye to eye. Frontoclypeus with brownish yellow transverse streaks. Pronotum on each side with or without two fuscous lateral longitudinal streaks not reaching fore border. Scutellum with a narrow black median streak and a short longitudinal streak on each lateral corner. Fore wings just longer than abdomen; claval suture, commissural border, and narrow longitudinal streaks in cells, brownish. Venter of thorax light. Femora with dark longitudinal streaks and transverse spots. Dorsum and venter of abdomen each with three black longitudinal stripes often more or less reduced in females. Male pygofer as in Text-fig. 2430, genital valve and plates as in Text-fig. 2431 (plates in repose joining medially suggesting the front of a boat), genital style as in Text-fig. 2432, aedeagus as in Text-figs. 2433, 2434, caudal part of abdominal venter in female as in Text-fig. 27. Overall length (♂♀) 6.6-8.0

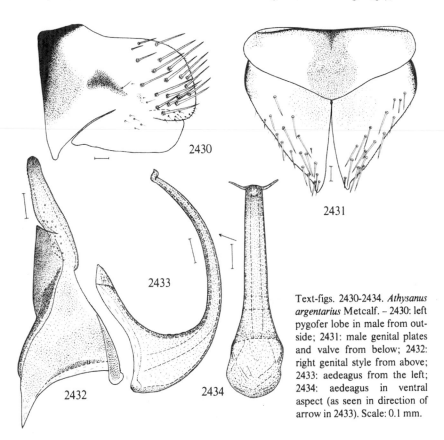

Text-figs. 2430-2434. *Athysanus argentarius* Metcalf. – 2430: left pygofer lobe in male from outside; 2431: male genital plates and valve from below; 2432: right genital style from above; 2433: aedeagus from the left; 2434: aedeagus in ventral aspect (as seen in direction of arrow in 2433). Scale: 0.1 mm.

747

mm. – Abdominal terga of nymphs with 4 complete rows of setae, vertex distinctly wider than long, 3 narrow brownish longitudinal bands present on abdomen, 2 of which continue onto thorax (Vilbaste, 1982). Walter (1975) provided good descriptions and excellent illustrations of instars III-V.

Distribution. "*A. argentarius* has been spreading in eastern Denmark since the turn of the century – now common in B, LFM, SZ, NWZ and NEZ. Also found on minor islands like Hesselø (N. P. Kristensen). Not yet in F, but in 1978 it was found by me on the island of Sprogø in the great Belt (Storebælt)" (Trolle, in litt.). – Fairly common in southern and central Sweden, Sk.-Hls. – Rare in Norway, only found in the Oslo area (Munster, Warloe). – East Fennoscandia: found in the southern and central parts, Al, Ab-Kb, fairly common in coastal areas, scarce and sporadic in the inland. – England, France, Netherlands, Belgium, German D.R. and F.R., Austria, Bohemia, Moravia, Slovakia, Hungary, Romania, Poland, Estonia, Latvia, Lithuania, n. and m. Russia, Ukraine, Moldavia, Altai Mts., Tuva, Mongolia, Kazakhstan, Uzbekistan, m. and w. Siberia, Kirghizia; Nearctic region.

Biology. In "Stranddünen, Binnendünen, Sandfeldern, besonnten Hängen, Flachmooren, Wiesen" (Kuntze, 1937). In the "Molinio-Arrhenatheretea" (Marchand, 1953). "Mehr oder weniger eurytop und besiedelt trockene wie auch feuchte Biotope" (Strübing, 1955). Prefers mesophilous conditions (Schiemenz, 1969b). "Mostly in damp meadows but also in clover fields, etc." (Vilbaste, 1974). Adults in June-September.

354. *Athysanus quadrum* Boheman, 1845
Plate-fig. 182, text-figs. 2435-2442.

Athysanus quadrum Boheman, 1845b: 157.

Brownish yellow, shining, with more or less extended black markings. Vertex with a broad black transverse band which may reach eyes, but often only extends to ocelli. In front of this band the uppermost transverse streaks of frontoclypeus are visible, behind the transverse band there is a dark spot near each eye. Sutures of face dark; on lower part of frontoclypeus the transverse streaks may coalesce into a large patch. Anteclypeus with one or two dark longitudinal bands, sometimes largely dark. Pronotum frontally with some dark spots or more extended mottlings, caudally of these with six more or less distinct longitudinal bands strongly varying in width; the surface between these bands is often greyish white. Scutellum more or less dark spotted. Wing dimorphous, fore wings in macropters (♂♀) apically broadly rounded, usually distinctly longer than abdomen, in sub-brachypters (♀) apically almost truncate, reaching somewhere about caudal margin of 8th tergum; with the following black or brownish markings, much extended in males, often reduced in females: an elongate spot in clavus and a smaller spot in claval apex, a large spot in corium at the level of the interspace between claval spots, and another in apical part often covering its entire surface. Interspaces between these spots as well as veins partly milk-coloured. Thoracic venter and abdomen in macropters almost entirely black, in sub-brachypters brown or black spotted, side margins of abdominal segments – especially those near base – con-

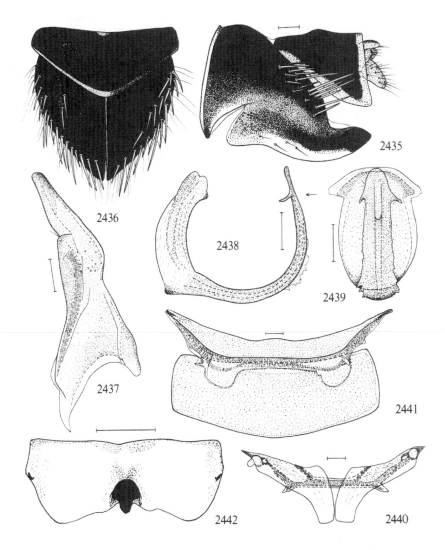

Text-figs. 2435-2442. *Athysanus quadrum* Boheman. – 2435: left pygofer lobe and anal apparatus in male from the left; 2436: genital plates and valve from below; 2437: right genital style from above; 2438: aedeagus from the left; 2439: aedeagus in ventral aspect (as seen in direction of arrow in 2438); 2440: 1st abdominal sternum in male from above; 2441: 2nd and 3rd abdominal sterna in male from above; 2442: 7th abdominal sternum in female from below (depressed under coverglass). Scale: 0.5 mm for 2442, 0.1 mm for the rest.

spicuously ivory white. Fore and middle femora in macropters proximally entirely or largely jet black, in sub-brachypters brown, apices lighter; fore and middle tibiae and tarsi yellow; hind femora striped and/or spotted with black or brown; posterior tibiae with longitudinal stripes and black setigerous spots; 1st and 2nd segments of hind tarsi apically dark. Male pygofer and anal apparatus as in Text-fig. 2435, genital valve and plates as in Text-fig. 2436, styles as in Text-fig. 2437, aedeagus as in Text-figs. 2438, 2439, 1st abdominal sternum in male as in Text-fig. 2440, 2nd and 3rd abdominal sterna in male as in Text-fig. 2441, 7th abdominal sternum in female as in Text-fig. 2442. Overall length of males 4.1-4.9 mm, of sub-brachypterous females 4.9-5.5 mm, of one macropterous female 5.77 mm. – Nymphs with 4 complete rows of hairs on abdominal tergum; vertex distinctly wider than long; abdomen with pale midline; in dark specimens, pronotum wholly dark brown, first abdominal tergites pale, tergites IV-VIII dark, in pale specimens, darkening present only on tergites VI-VII (Vilbaste, 1982).

Distribution. Very rare in Denmark: one male found in NWZ: Grønnehave 3.VIII.-1869 (Budde-Lund). – Fairly scarce, locally common in southern and central Sweden, Sk.-Vstm. – Rare in Norway, found in AK: Oslo (Siebke); On: Dovre (Boheman); Bø: Drammen and Ringerike (Warloe). – Fairly scarce and sporadic in southern East Fennoscandia, found in Ab, N, St, Ta, Sa; Vib. – France, Netherlands, German D.R. and F.R., Austria, Moravia, Slovakia, Hungary, Romania, Poland, Estonia, Latvia, Lithuania, m. Russia, Ukraine, Kazakhstan, Tuva, Mongolia, Altai Mts., m. Siberia, Sakhalin, Maritime Territory.

Biology. In "Flachmooren, Uferzone eines Sees oder Flusslaufes, Wiesen" (Kuntze, 1937). In "extremnassen Carexwiesen" (Marchand, 1953). "Auf feuchten Wiesen mit *Sphagnum* and *Carex*" (Wagner & Franz, 1961). "In damp meadows, in fens, on shores of water bodies" (Vilbaste, 1974). Adults in July-September.

Genus *Stictocoris* Thomson, 1869

Jassus (Stictocoris) Thomson, 1869: 51.
    Type-species: *Cicada lineata* Fabricius, 1789, by subsequent designation.

Body broad, robust. Head frontally rounded, narrower than pronotum. Frontoclypeus broad. Anteclypeus about parallel-sided or a little wider in lower part. Ocelli small. Fore wings approximately as long as abdomen. Fore tibia with one seta in the inner, four in the outer dorsal row. Male anal tube short, not sclerotized dorsally and laterally. Male pygofer lobes with many macrosetae, without appendages. Genital plates long, triangular, their macrosetae arranged in a lateral row. Style ordinary. Connective long, clip-shaped. Aedeagus without a socle, slightly asymmetrical, without appendages, phallotreme subterminal, lateral. 7th abdominal sternum in female caudally with a semicircular scale consisting of the out-turned dorsal lamina of that sternum. A monotypic genus.

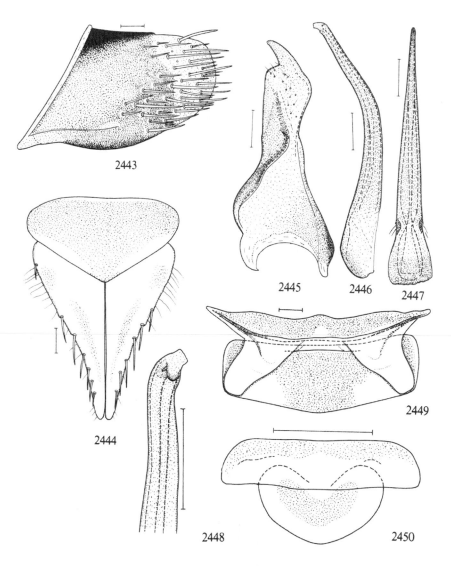

Text-figs. 2443-2450. *Stictocoris picturatus* (C. Sahlberg). – 2443: left pygofer lobe in male from the left; 2444: male genital plates and valve from below; 2445: right genital style from above; 2446: aedeagus from the right; 2447: aedeagus in dorsal aspect; 2448: apex of aedeagus from the left; 2449: 2nd and 3rd abdominal sterna in male from above; 2450: 7th abdominal sternum in female from below (depressed under coverglass). Scale: 0.5 mm for 2450, 0.1 mm for the rest.

355. **Stictocoris picturatus** (C. Sahlberg, 1842)
Plate-fig. 183, text-figs. 2443-2450.

Cicada lineata Fabricius, 1787: 270 (nec Linné, 1758).
Cicada hybneri Gmelin, 1789: 2107 (n.n.) (Nomen oblitum).
Cicada picturata C. Sahlberg, 1842: 89.

Sordid whitish yellow with black or fuscous markings, shining. Head with the following dark markings: on each side a large black spot behind and another below ocellus; between these spots usually a pair of small dots at anterior end of coronal suture; a short transverse band near median part of caudal border of vertex; a pair of large, sometimes confluent spots on centre of frontoclypeus; a large spot below each antenna; lora black-bordered, anteclypeus with a longitudinal band or a T-shaped spot. Pronotum along anterior border with an arched transverse fuscous band or a pair of patches, more caudally 3 black spots, the median one largest. Scutellum with a median black spot. Longitudinal veins of fore wings (not Sc) blackish, fuscous bordered. Thoracic venter largely light, dark spotted, abdominal dorsum black with light lateral and segmental borders, venter light with a dark longitudinal band. Femora with dark transverse bands near apices, tibiae black striped, tarsi partly black. Male pygofer lobes as in Text-fig. 2443, genital valve and plates as in Text-fig. 2444 (apices of plates bent upwards, as in *Macrosteles*), genital style as in Text-fig. 2445, aedeagus as in Text-figs. 2446-2448, 2nd and 3rd abdominal sterna in male as in Text-fig. 2449, 7th abdominal sternum in female as in Text-fig. 2450. Overall length (♂♀) 3.9-4.8 mm. – Nymph with 4 complete rows of dorsal hairs on abdomen; of these, 2 are situated in pale, 2 in dark bands; vertex with two large black spots and an additional pair of spots on transition to frons (Vilbaste, 1982).

Distribution. Not found in Denmark and Norway. – Very rare in Sweden, found in Vg.: Kinnekulle, Österplana hed 2.VIII.1967 (Ossiannilsson); Sdm. (Stål), Huddinge 12.VII.1922 (Håkan Lindberg); "Hlm" (= Stockholm) (in coll. Thomson); Upl.: Skuggan and Bellevue near Stockholm (Boheman). – Moderately common in southern and central East Fennoscandia, found in Al, Ab, N, St, Ta, Kb; Vib, Kr. – France, German D.R. and F.R., Austria, Bohemia, Slovakia, Moravia, Hungary, Romania, Yugoslavia, Italy, Albania, Poland, Estonia, Latvia, Lithuania, n. and m. Russia, Ukraine, Anatolia, Moldavia, Kazakhstan, Uzbekistan, Altai Mts., w. and m. Siberia, Kirghizia, Tuva, Mongolia, China.

Biology. In "besonnten Hängen" (Kuntze, 1937). "Auf trocknen Hangwiesen im Juli und August" (Kontkanen, 1938). "Au pied des plantes basses" (Ribaut, 1952). "Bewohner von Trockenrasen" (Wagner & Franz, 1961). Adults in July and August.

Genus *Ophiola* Edwards, 1922

Ophiola Edwards, 1922: 206.
Type-species: *Cicada striatula* Fallén, 1806, by subsequent designation.

Ophiolix Ribaut, 1942: 264 (subgenus of *Limotettix*).

Type-species: *Thamnotettix paludosus* Boheman, 1845, by original designation.
Scleroracus Oman, 1949: 152.

Type-species: *Athysanus anthracinus* Van Duzee, 1894, by original designation.
Ophiola Metcalf, 1952: 231 (nomenclature).

Body more or less elongate, robust. Transition of vertex and frontoclypeus more or less rounded. Anteclypeus tapering towards apex. Anterior tibiae with one seta in the inner, 3 in the outer dorsal row. Ocelli dorsal. Male pygofer moderately incised. Connective Y-shaped, stem long. Macrosetae of genital plates in disorder or in an irregular row along lateral border, median half of plates without setae. Socle of aedeagus dorsally fused with a laminate appendage. Styles short. 7th abdominal sternum in female medially not concave. In Denmark and Fennoscandia 5 species.

Key to species of *Ophiola*

1   Head much wider than pronotum. Anal tube in male not sclerotized dorsally. Genital plates on ventral side with many
long, fine hairs (Text-fig. 2476) (Subgenus *Ophiolix* Ribaut) ....
.... 360. *paludosa* (Boheman)
–   Head not or just wider than pronotum. Anal tube in
male largely sclerotized dorsally. Genital plates ventrally
not finely pilose (Subgenus *Ophiola* s.str.) 2
2 (1) Vertex with a broad black transverse band (Plate-fig. 185).
Fore wings with black or fuscous longitudinal streaks or
bands in the cells .... 359. *transversa* (Fallén)
–   Markings on vertex different. Veins of fore wings more or
less distinctly dark-bordered, inner surface of cells lighter 3
3 (2) Head as seen from above frontally angularly produced. Shaft
of aedeagus on each side with a pointed tooth directed distad
(Text-figs. 2461, 2462) .... 357. *cornicula* (Marshall)
–   Head frontally more rounded angular. Aedeagus not as above 4
4 (3) Larger, males 3.4-3.9 mm, females 4.2-4.8 mm. Male pygofer
lobes each with an acute, caudad directed tooth (text-fig.
2451) .... 356. *decumana* (Kontkanen)
–   Smaller, males 2.8-3.6 mm, females 3.2-3.7 mm. Male pygofer
lobes each with a rectangular or stump caudal projection
(Text-fig. 2463) .... 358. *russeola* (Fallén).

356. *Ophiola (Ophiola) decumana* (Kontkanen, 1949)
Plate-fig. 184, text-figs. 2451-2457.

Cicada striatula Fallén, 1806: 31 (nec Fabricius, 1794).
Limotettix (Ophiola) russeolus ssp. *decumanus* Kontkanen, 1949a: 89.
Limotettix (Scleroracus) corniculus Ribaut, 1952: 157 (nec Marshall, 1866).

Greyish yellow to brownish yellow to reddish brown, shining, black markings much varying in extension. Head above with four black transverse streaks, the foremost and next of which are interrupted medially. The foremost streak is continued laterally and ventrad in a black stripe along frontal suture; the second streak laterally reaching ocelli. Third transverse band almost straight, reaching eyes; the fourth band irregularly

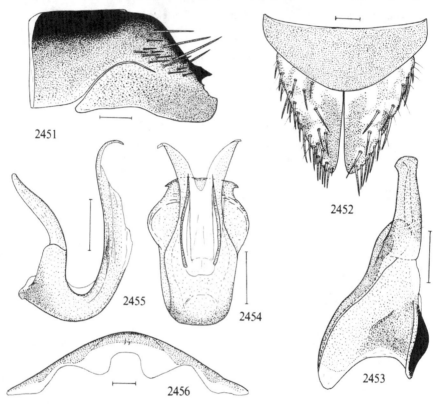

Text-figs. 2451-2457. *Ophiola decumana* (Kontkanen). – 2451: left pygofer lobe in male from outside; 2452: male genital plates and valve from below; 2453: right style from above; 2454: aedeagus in ventral aspect; 2455: aedeagus from the left; 2456: 2nd abdominal tergum in male from behind; 2457: 7th abdominal sternum in female from below (depressed under coverglass). Scale: 0.5 mm for 2457, 0.1 mm for the rest.

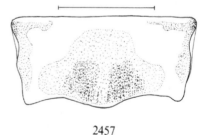

undulated. These bands are often more or less confluent. Face with black transverse streaks on frontoclypeus, black spotted for the rest. Scutellum with black lateral corners and more or less confluent black spots. Fore wings with veins black bordered, interspaces light. Thoracic venter and abdomen largely light, black spotted, or largely black. Femora transversely black banded and spotted, posterior tibiae with black spots and longitudinal black streaks. Male pygofer lobes as in Text-fig. 2451, genital valve and plates as in Text-fig. 2452, styles as in Text-fig. 2453, aedeagus as in Text-figs. 2454, 2455, 2nd abdominal tergum in male as in Text-fig. 2456, 7th abdominal sternum in female as in Text-fig. 2457. Overall length of males 3.4-3.9 mm, of females 4.2-4.8 mm. – Nymphs with hairs present only on tergite VIII; anteclypeus parallel-sided; N5 shorter than 3.5 mm, head not wider than pronotum; frons dark with pale patches or pale with dark patches; abdomen variegated dark/pale; front and middle femora usually ringed (Vilbaste, 1982).

Distribution. Uncommon in Denmark, found mainly in Jutland. – Fairly common in Sweden, Sk.-P.Lpm. – Norway: records of "*striatula*" not revised; found in Fi: Duolbba-javrre 17.VIII.1977 by Huldén. – East Fennoscandia: found in Ab, N, Kb, Ka, St, Ta, Sb, ObN, Ks; Vib. – England, Wales, France, Netherlands, German D.R. and F.R., Bohemia, Moravia, Slovakia, Bulgaria, Romania, Italy, Poland, Estonia, Latvia, Lithuania, Moldavia, Kazakhstan, Altai Mts., Tuva, Mongolia, Kamchatka.

Biology. "An *Vaccinium myrtillus* gefunden . . . aber auch an Standorten, wo diese Pflanze fehlt" (Wagner & Franz, 1961). "In dry meadows, in light pine forests" (Vilbaste, 1974). I found *decumana* among *Vaccinium myrtillus* as well as among *Calluna*. Adults in July-September.

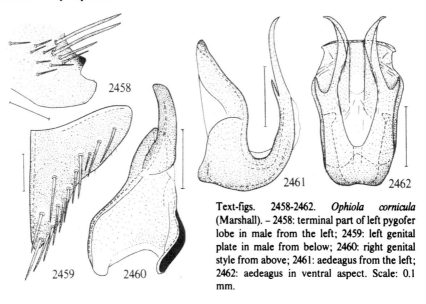

Text-figs. 2458-2462. *Ophiola cornicula* (Marshall). – 2458: terminal part of left pygofer lobe in male from the left; 2459: left genital plate in male from below; 2460: right genital style from above; 2461: aedeagus from the left; 2462: aedeagus in ventral aspect. Scale: 0.1 mm.

357. **Ophiola (Ophiola) cornicula** (Marshall, 1866)
   Text-figs. 2458-2462.

Iassus (?) corniculus Marshall, 1866: 119.
Jassus orichalceus Thomson, 1869: 64.
Limotettix (Ophiola) intractabilis Kontkanen, 1949a: 87.

Resembling *decumana,* head more pointed apically. Apical part of male pygofer lobe as in Text-fig. 2458, genital plate as in Text-fig. 2459, style as in Text-fig. 2460, aedeagus as in Text-figs. 2461, 2462. Overall length of males 3.2-3.5 mm, of females 3.5-4.0 mm.

Distribution. So far not found in Denmark. – Scarce but widespread in Sweden, found in Sk., Hall., Dlr., Ly. Lpm. – Not recorded from Norway. – East Fennoscandia: found in Al, Ab, N, St, Ta, Oa, Tb, Sb, Kb, ObN; Vib, Kr, Lr. – England, Scotland, France, Netherlands, German D.R. and F.R., Austria, Bohemia, Slovakia, Bulgaria, Romania, Estonia, Latvia, Lithuania, n. Russia, Ukraine, m. Siberia, Manchuria, Maritime Territory; Nearctic region.

Biology. In "Binnendünen, Sandfeldern" (Kuntze, 1937). Hibernates in the egg stage (Remane, 1958). "Moorbewohner; kommt oft gemeinsam mit *O. russeola* vor" (Wagner & Franz, 1961). "In fens, pine bogs and moist meadows" (Raatikainen & Vasarainen, 1976). "Ei-Überwinterer mit 1 Generation, Imaginalzeit Ende VI – Anfang X" (Schiemenz, 1976).

358. **Ophiola (Ophiola) russeola** (Fallén, 1826)
   Text-figs. 16, 17, 2463-2468.

Cicada russeola Fallén, 1826: 34.
Jassus plutonius Uhler, 1877: 470.
Thamnotettix striatulellus Edwards, 1894: 102.

Resembling *O. decumana,* differing by smaller body and by the structure of male genitalia. Variation in extension of pigmentation is considerable, almost unpigmented specimens – as the ♀-type of *russeola* – are not uncommon. Some, but not all, of these light specimens are parasitized. – Male pygofer lobes as in Text-fig. 2463, genital valve and plates as in Text-fig. 2464, style as in Text-fig. 2465, aedeagus as in Text-figs. 2466, 2467, 7th abdominal sternum in female as in Text-fig. 2468. Overall length of males 2.8-3.6 mm, of females 3.2-3.7 mm. – Nymphs as in *decumana,* anterior body orange-yellow; abdomen uniformly brownish yellow or with cross-rows of brownish dots (Vilbaste, 1982).

Distribution. Rare in Denmark, found only in EJ: Svejbæk and NEZ: Ryget skov many years ago. – Common in Sweden, Sk.-T.Lpm. – I have examined specimens from HEs, HEn, On and HOi in Norway. – East Fennoscandia: found in Ab, N, Ta, Sb, Kb. – England, Scotland, France, Netherlands, German D.R. and F.R., Austria, Bohemia, Slovakia, Poland, Latvia, Lithuania, Estonia, Tunisia, Mongolia, w. Siberia, Maritime Territory; Nearctic region.

Biology. Dominant in *Calluna* pine heaths (Linnavuori, 1952). Hibernation takes place in the egg stage (Müller, 1957; Schiemenz, 1976). Strübing found and reared the species on *Vaccinium oxycoccus*. "Im Juli eingesammelten Tiere hielten sich bis weit in den Oktober hinein, sie überwinterten als Ei im Freien und im Mai nächsten Jahres schlüpften die Junglarven" (Strübing, 1955). "Eine offenbar heliophile Art, die einerseits in Hochmooren und an heidigen Standorten auf *Calluna vulgaris* und *Oxycoccus spec.*, andererseits in Felsenheiden auf Kalk, vielleicht auf *Erica carnea,* vorkommt" (Wagner & Franz, 1961). One generation (Schiemenz, 1976).

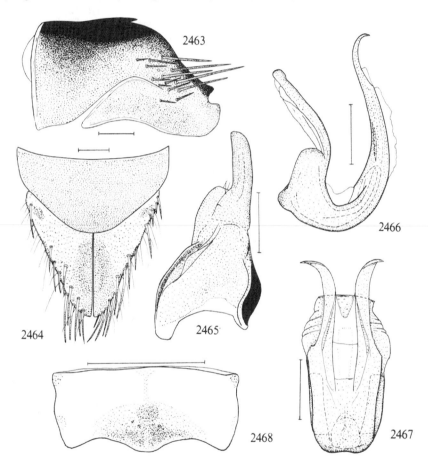

Text-figs. 2463-2468. *Ophiola russeola* (Fallén). – 2463: left pygofer lobe in male from the left; 2464: genital plates and valve from below; 2465: right genital style from above; 2466: aedeagus from the left; 2467: aedeagus in ventral aspect; 2468: 7th abdominal sternum in female from below (depressed under coverglass). Scale: 0.5 mm for 2468, 0.1 mm for the rest.

757

359. *Ophiola (Ophiola) transversa* (Fallén, 1826)
Plate-fig. 185, text-figs. 2469-2474.

Cicada transversa Fallén, 1826: 37.
Cicada lineigera Zetterstedt, 1840: 1076.

Brownish yellow, often with reddish tinge, shining. Transition between vertex and frontoclypeus rounded. Head frontally with two large black spots ventrally consisting of confluent transverse streaks of frontoclypeus. Behind ocelli a broad black transverse band between eyes. Frontoclypeus below the black spots above mentioned usually light with black transverse streaks; black markings sometimes confluent, making face largely black. Pronotum in light specimens with a transverse row of black spots along fore border; in dark specimens, also caudal part of pronotum transversely black mottled. Scutellum largely light with the usual black spots, or black, lateral margins and a narrow median stripe light. Fore wings sometimes almost without black markings,

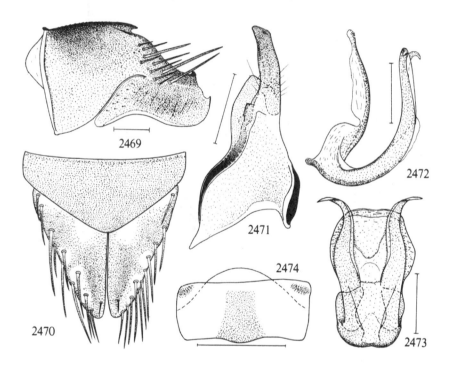

Text-figs. 2469-2474. *Ophiola transversa* (Fallén). – 2469: left pygofer lobe in male from the left; 2470: male genital plates and valve from below; 2471: right genital style from above; 2472: aedeagus from the left; 2473: aedeagus in ventral aspect; 2474: 7th abdominal sternum in female from below (depressed under coverglass). Scale: 0.5 mm for 2474, 0.1 mm for the rest.

usually with longitudinal black streaks more or less completely filling cells. Thoracic venter and abdomen largely black, legs light, spotted and transversely banded with black. Male pygofer lobes as in Text-fig. 2469, genital valve and plates as in Text-fig. 2470, style as in Text-fig. 2471, aedeagus as in Text-figs. 2472, 2473, 7th abdominal sternum in female as in Text-fig. 2474. Overall length of males 2.7-3.2 mm, of females 3.3-3.7 mm. – Nymphs as those of *decumana* but almost entirely dark brown, including front and middle femora (Vilbaste, 1982).

Distribution. Scarce in Denmark, found in a few localities in SJ, EJ and NEZ. – Sweden: common in the southern provinces, scarce in central part, found Sk.-Boh., Nrk.-Vb. – Norway: recorded only from Bø: Drammen and Ringerike (Warloe). – East Fennoscandia: found in the southern and central parts, Al, Ab-Om; Vib, Kr; common in the eastern part, for the rest scarce and sporadic. – France, German D.R. and F.R., Austria, Bulgaria, Poland, Estonia, Latvia, Lithuania, n. and m. Russia, Ukraine, Azerbaijan, Kazakhstan, Altai Mts., Tuva, Mongolia, m. and w. Siberia.

Biology. In "Besonnten Hängen" (Kuntze, 1937). "Auf ± trocknen Wiesen im Juli und August" (Kontkanen, 1938). "Bewohnt Trockenrasen, im Norden Trockenrasen zwischen *Calluna*" (Wagner & Franz, 1961). "Stenotope Art der Trockenrasen" (Schiemenz, 1969b). "In dry meadows, on dry hill-slopes, in juniper stands, etc." (Vilbaste, 1974). Adults in July and August.

360. *Ophiola (Ophiolix) paludosa* (Boheman, 1845)
     Plate-fig. 186, text-figs. 2475-2480.

Thamnotettix paludosa Boheman, 1845a: 34, 1845b: 158.

In general aspect more resembling *Limotettix* spp. Head wider than prothorax, frontally broadly rounded, transition between vertex and face rounded. Uppermost transverse streaks of frontoclypeus visible from above. Vertex caudally of these with a narrow transverse band between eyes, near caudal border with an additional transverse band or a transverse row of black spots. These markings as well as the transverse streaks on frontoclypeus often confluent. Anteclypeus below usually with a triangular black spot. Genae with lower part black. Extension of black markings on pronotum varying; a light median line always remains in dark specimens. Scutellum with black lateral spots, between these additional black markings. Veins of fore wings partly dark, usually dark-bordered; claval suture and commissural border almost always distinctly dark, even in very light specimens. Thoracic venter and abdomen in males largely black, abdomen in females partly yellow. Legs yellow, spotted, striped and banded with black. Pygofer lobes in male as in Text-fig. 2475, genital valve and plates as in Text-fig. 2476, aedeagus as in Text-figs. 2478, 2479, style as in Text-fig. 2477, 7th abdominal sternum in female as in Text-fig. 2480. Overall length (♂♀) 4.0-4.8 mm.

Distribution. Not found in Denmark. – Scarce and sporadic in Sweden, found in Sk., Sm., Upl., Dlr., Hls., Med., Ång., Vb., Nb. – Scarce and sporadic in Norway, found in

AK, Bø, AAy, VAy, Ry. – Common in southern and central East Fennoscandia, found Al, Ab-ObN; Vib, Kr. – Latvia, Estonia, n. Russia, Altai Mts., Tuva, m. Siberia.

Biology. On *Juncus*-spp. (Sahlberg, 1871). "Auf feuchten Wiesen" (Lindberg, 1947). "Bewohnte am Ufer des Lieksajoki die unmittelbar landwärts von der Wasserlinie gelegenen *Carex* – *Equisetum limosum*-Siedlungen" (Kontkanen, 1949b). Adults in July-September.

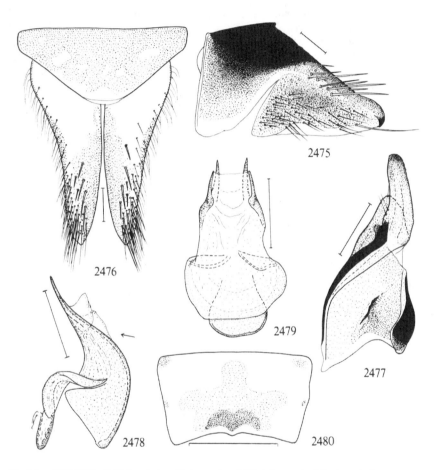

Text-figs. 2475-2480. *Ophiola paludosa* (Boheman). – 2475: left pygofer lobe in male from outside; 2476: male genital plates and valve from below; 2477: right genital style from above; 2478: aedeagus from the left; 2479: aedeagus in ventral aspect (in direction of arrow in 2478); 2480: 7th abdominal sternum in female from below (depressed under coverglass). Scale: 0.5 mm for 2480, 0.1 mm for the rest.

Genus *Limotettix* J. Sahlberg, 1871

Limotettix J. Sahlberg, 1871: 224.
  Type-species: *Cicada striola* Fallén, 1806, by subsequent designation.
Drylix Edwards, 1922: 206.
  Type-species: *Cicada striola* Fallén, 1806, by subsequent designation.

Elongate, fairly robust. Head much wider than pronotum, frontally rounded, transition between vertex and frontoclypeus rounded. Anteclypeus tapering towards apex. Chaetation of anterior tibiae as in *Ophiola*. Male anal tube short, dorsal sclerotization little extended. Macrosetae of genital plates in disorder, also present on median half of plates. Styles long. Connective Y-shaped. Socle of aedeagus dorsally with a laminate appendage not fused with socle. 7th abdominal sternum in female deeply concave, not covering base of ovipositor. In Europe four species.

Key to species of *Limotettix*
(After Vilbaste, 1973; modified)

1 Both sexes (colour characters) 2
– Males only (genital structure characters) 5
2 (1) Light interocellar band reduced to small triangular spots.
    Upper two (rarely three) arch-lines of frontoclypeus always
    fused .... 362. *atricapillus* (Boheman)
– Light interocellar band distinct for its whole extent (usually
    narrowly interrupted beside ocelli only, more rarely also
    medially). Upper arch-lines usually separate (fused only in
    exceptionally dark males) 3
3 (2) Ground-colour ochraceous (in some males the upper side is
    greenish yellow). Lower part of frontoclypeus usually conspi-
    cuously light. (Genal sutures and sides of anteclypeus usually
    broadly darkened) .... 364. *ochrifrons* Vilbaste
– Ground-colour greenish yellow. Arch-lines of frontoclypeus
    usually distinct for their whole extent 4
4 (3) Upper pair of arch-lines not or only somewhat broader than
    the following ones (only in exceptionally dark males some-
    times broader). Fore wings often with brownish pattern or
    entirely brown with light veins. Sutures of anteclypeus and
    lora broadly darkened .... 361. *striola* (Fallén)
– Upper pair of arch-lines distinctly broader than the rest.
    Fore wings usually unicolorous light. Sutures of anteclypeus
    and lora narrowly darkened .... 363. *sphagneticus* Emeljanov
5 (1) Lateral margins of genital plates convex (Text-figs. 2488, 2499) 6
– Lateral margins of genital plates more or less straight (Text-
    figs. 2482, 2494) 7
6 (5) Caudal outline of aedeagus (in lateral aspect) rounded (Text-

fig. 2491). Genital plates along median line approximately as
long as wide at base (Text-fig. 2488) .... 362. *atricapillus* (Boheman)
– Caudal outline of aedeagus in lateral aspect angular (Text-
fig. 2501). Genital plates along median line shorter than wide
at base (Text-fig. 2499) .... 364. *ochrifrons* Vilbaste
7 (5) Apex of style abruptly cut (Text-fig. 2483). Genital plates
(along median margin) as long as wide as base (Text-fig. 2482) 361. *striola* (Fallén)
– Apical part of style as in Text-fig. 2495. Genital plates along
median border longer than wide at base (Text-fig. 2494) ....
.... 363. *sphagneticus* Emeljanov.

361. *Limotettix striola* (Fallén, 1806)
Plate-fig. 187, text-figs. 2481-2487.

Cicada striola Fallén, 1806: 31.
Jassus frenatus Germar, 1821: 86.

Greyish yellow, shining. Head medially slightly longer than near eyes, above between
eyes with a black transverse band rather uniform in width, frontally touching ocelli,
laterally curving ventrad, following inner margin of eye. In front of this band there is
another band consisting of a lying 3-shaped marking laterally continuing into the black
border of frontoclypeus. Frontoclypeus with the usual transverse streaks, here medially
coalescing into a double longitudinal band. These markings may be more or less confluent. Anteclypeus with a T-shaped black marking. Pronotum usually light with a
transverse row of small black dots on anterior half, often dark mottled; pronotum may
be largely black in very dark specimens. Scutellum entirely light or with black markings. Fore wings distinctly longer than abdomen, often entirely light, only claval suture
being narrowly dark bordered. Sometimes also inner margin of commissure narrowly
dark. In many specimens there are dark streaks along the veins; in very dark individuals the fore wings may be entirely black with light veins. Thoracic venter black
spotted, anterior and middle femora with transverse black spots, femora also
longitudinal striped, posterior tibiae black spotted. Male pygofer lobes as in Text-fig.
2481, genital valve and plates as in Text-fig. 2482 (but plates with median margins not
diverging). Styles as in Text-fig. 2483, aedeagus as in Text-figs. 2484, 2485, 1st
abdominal sternum in male as in Text-fig. 2486, 7th abdominal sternum in female as in
Text-fig. 2487. Overall length of males 3.6-4.4 mm, of females 4.0-5.0 mm. – Nymphs
with hairs present only on abdominal tergite VIII, 4 on hind margin, 2 on hind angles;
abdomen uniformly dark or with cross-rows of dark spots; head considerably wider
than pronotum; distinct arch-lines present on frons (Vilbaste, 1982).

Distribution. Common in Denmark, found in most districts. – Common in Sweden,
Sk.-Nb. – Norway: found in Ø, AK, Bø, VAy, Ry, HOy. – Very common and
widespread in East Fennoscandia. – Widespread in the Palaearctic region, also in the
Nearctic.

Biology. In "Salzstellen, Flachmooren" (Kuntze, 1937). "Auf *Carex*-reichen Moorwiesen und auf Weissmooren von Juli bis August" (Kontkanen, 1938). In most seashore biotopes, in wet "rimpi" bogs and quagmire marshes (Linnavuori, 1952a). Hibernation in the egg-stage, in Central Europe 2 generations (Remane, 1958). "An *Scirpus paluster*" (Wagner & Franz, 1961). Adults in July-September.

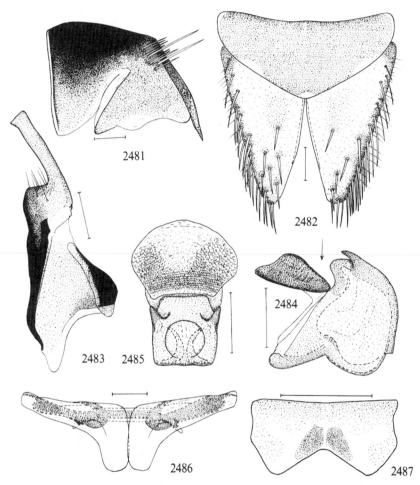

Text-figs. 2481-2487. *Limotettix striola* (Fallén). – 2481: left pygofer lobe in male from the left; 2482: genital plates and valve from below; 2483: right genital style from above; 2484: aedeagus and laminate appendage from the left; 2485: same in morphological ventral aspect (as seen in direction of arrow in 2484); 2486: 1st abdominal sternum in male from above; 2487: 7th abdominal sternum in female from below (depressed under coverglass). Scale: 0.5 mm for 2487, 0.1 mm for the rest.

763

362. *Limotettix atricapillus* (Boheman, 1845)
   Text-figs. 2488-2493.

Thamnotettix atricapilla Boheman, 1845b: 158.
Limotettix nigrifrons Haupt, 1935: 198.

Resembling *striola,* differing by characters given in the key. Head as seen from above

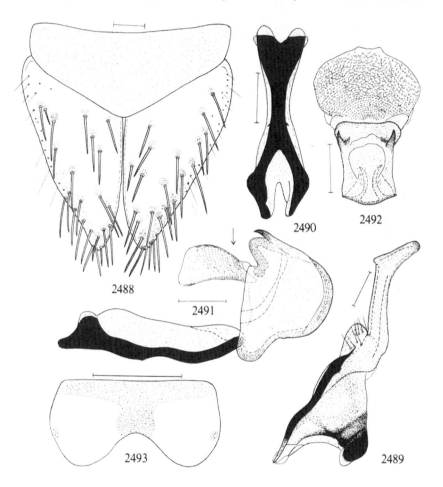

Text-figs. 2488-2493. *Limotettix atricapillus* (Boheman). – 2488: genital plates and valve from above; 2489: right genital style from above; 2490: connective in ventral aspect; 2491: aedeagus with laminate appendage and connective from the left; 2492: same in morph. ventral aspect (as seen in direction of arrow in 2491); 2493: 7th abdominal sternum in female from below (depressed under coverglass). Scale: 0.5 mm for 2493, 0.1 mm for the rest.

medially not longer than near eyes. Male genital valve and plates as in Text-fig. 2488, genital style as in Text-fig. 2489, connective as in Text-figs. 2490, 2491, aedeagus as in Text-figs. 2491, 2492, 7th abdominal sternum in female as in Text-fig. 2493. Overall length of males 4.2-4.6 mm, of females 4.6-5.2 mm.

Distribution. Not found in Denmark. – Rare in Sweden, found in Sk. (according to Thomson, 1869); Sm.: Anneberg (Boheman); Gtl. (specimen in coll. Thomson), Tingstäde 3.VIII.1935 (Ossiannilsson); Ög. (Wahlberg); Hls.: Edsbyn, V. Homnatjärn 22.VIII.1980 (Bo Henriksson). – Norway: only found in VAy: Saevik near Farsund 12.VIII.1944 (Holgersen). – Scarce and sporadic in southern and central East Fennoscandia, recorded from Al, N, Ka, Ta, Sa, Kb; Kr. – England, German D.R., Poland, Estonia, Lithuania.

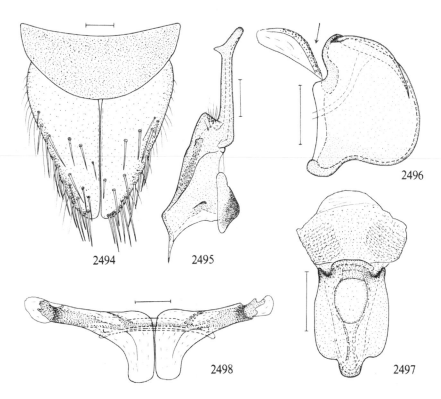

Text-figs. 2494-2498. *Limotettix sphagneticus* Emeljanov. – 2494: male genital plates and valve from below; 2495: right genital style from above; 2496: aedeagus and laminate appendage from the left; 2497: same from above (morph. ventral aspect, as seen in direction of arrow in 2496); 2498: 1st abdominal sternum in male from above. Scale: 0.1 mm.

Biology. "Auf feuchten Mooswiesen mit *Carex lasiocarpa*" (Lindberg, 1947). "On damp ground with *Sphagnum, Juncus, Menyanthes, Carex* etc." (Holgersen, 1949). "Eine Art der offenen seggenreichen Weissmoore. . . . Vorherrschend . . . auf einem relativ nassen, eutrophen *Carex vesicaria*-Weissmoor, rezedent trat sie dagegen in zahlreichen Ufercariceta auf." (Kontkanen, 1949b). In fens (Vilbaste, 1974). Adults in June-August.

363. *Limotettix sphagneticus* Emeljanov, 1964
Text-figs. 2494-2498.

Limotettix sphagneticus Emeljanov, 1964: 418.

Resembling *L. atricapillus,* differing by characters given in the key. Male genital valve and plates as in Text-fig. 2494, style as in Text-fig. 2495, aedeagus as in Text-figs. 2496, 2497, 1st abdominal sternum in male in Text-fig. 2498. 7th abdominal sternum in female caudally less deeply concave than in *atricapillus*. Overall length of males 3.9-4.2 mm, of females 3.9-4.5 mm.

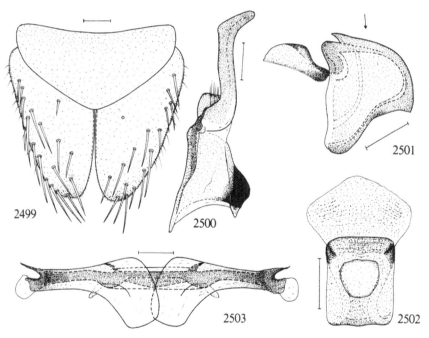

Text-figs. 2499-2503. *Limotettix ochrifrons* Vilbaste (paratype). – 2499: male genital plates and valve from below; 2500: right genital style from above; 2501: aedeagus and laminate appendage from the left; 2502: same from above (in morph. ventral aspect, in direction of arrow in 2501); 2503: 1st abdominal sternum in male from above. Scale: 0.1 mm.

Distribution. Not found in Denmark, Norway and Sweden. – East Fennoscandia: rare, found in N: vicinity of Täcktom (Håkan Lindberg); Kb: Kontiolathi 15.VIII.1942 (Lindberg); Kr: Kumsjärvi (Thuneberg). – Latvia, Lithuania, Estonia, n. and m. Russia, Tuva, w. Siberia.

Biology. In fens (Vilbaste, 1974).

364. *Limotettix ochrifrons* Vilbaste, 1973
Text-figs. 2499-2503.

Limotettix atricapilla Emeljanov, 1964: 418 (nec Boheman, 1845).
Limotettix ochrifrons Vilbaste, 1973: 200.

Resembling *L. atricapillus*, differing by characters given in the key. Male genital plates and valve as in Text-fig. 2499, style as in Text-fig. 2500, aedeagus as in Text-figs. 2501, 2502, 1st abdominal sternum in male as in Text-fig. 2503, 7th abdominal sternum in female more or less as in *atricapillus*. Overall length of males 4.1-4.5 mm, of females 4.6-4.8 mm. – Nymphs as in *striola* (Vilbaste, 1982).

Distribution. Not in Denmark, Sweden and Norway. – East Fennoscandia: Kb: Kontiolahti 15-25.VIII.1942 (Håkan Lindberg). – Estonia, Latvia, Lithuania, n. Russia, Altai Mts., Tuva, Mongolia.

Biology. In fens (Vilbaste, 1974).

Genus *Laburrus* Ribaut, 1942

Laburrus Ribaut, 1942: 268.
Type-species: *Athysanus limbatus* Ferrari, 1882, by original designation.

Body broad, robust. Head wider than pronotum, frontally rounded obtusely angular, in dorsal aspect little longer medially than near eyes. Frontoclypeus partly visible from above, face part wider than long. Anteclypeus rectangular or slightly narrower towards apex. Anterior tibiae with 3-4 setae in the inner, 4 in the outer dorsal row. Anal tube in male short, well sclerotized. Male pygofer lobes without appendages. Macrosetae of genital plates arranged in a row along lateral border. Connective Y-shaped, short and broad. Style robust, median apophyse slender, finger-shaped. Aedeagus long and slender, socle well developed, phallotreme ventral, subterminal. In Denmark and Fennoscandia one species.

365. *Laburrus impictifrons* (Boheman, 1852)
Text-figs. 2504-2510.

Deltocephalus impictifrons Boheman, 1852: 119.

Greenish yellow, shining. Frontoclypeus with brownish yellow or indistinct transverse streaks. Veins of fore wings yellow, distinct. Fore wings about as long as abdomen, apically fumose. Fore and middle femora usually with dark longitudinal stripes and a subapical dark transverse band. Posterior tibiae black dotted. Venter largely sordid yellow, abdominal dorsum black with yellow or whitish yellow lateral and segmental borders. Male pygofer lobes as in Text-fig. 2504, genital plates and valve as in Text-fig. 2505 (but inner margins of plates not diverging in specimens in repose). Male genital style as in Text-fig. 2506, connective as in Text-fig. 2507, aedeagus as in Text-figs. 2508-2509, 7th abdominal sternum in female as in Text-fig. 2510. Overall length of males 3.9-4.5 mm, of females 4.7-5.2 mm. – Nymphs with 4 complete rows of hairs on abdomen,

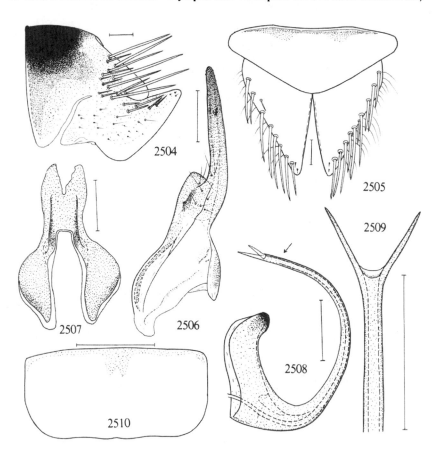

Text-figs. 2504-2510. *Laburrus impictifrons* (Boheman). – 2504: left pygofer lobe in male from the left; 2505: genital plates and valve from below; 2506: right genital style from above; 2507: connective from above; 2508: aedeagus from the left; 2509: apex of aedeagus in ventral aspect (in direction of arrow in 2508); 2510: 7th abdominal sternum in female from below (depressed under coverglass). Scale: 0.5 mm for 2510, 0.1 mm for the rest.

vertex distinctly wider than long, without black spots, body with two black longitudinal bands (Vilbaste, 1982).

Distribution. Rare in Denmark, found in LFM: Bøtø Strand 9.VIII.1914 (O. Jacobsen), and B: Boderne 14.VIII.1979 (Trolle). – Sweden: not uncommon in Scania, not found in the rest of the country. – Not found in Norway. – East Fennoscandia: found only in Vib: Valkjärvi 8.VIII.1866 (J. Sahlberg). – France, German D.R. and F.R., Austria, Bohemia, Moravia, Slovakia, Hungary, Bulgaria, Yugoslavia, Italy, Poland, Estonia, Latvia, Lithuania, n., m., and s. Russia, Ukraine, Moldavia, Georgia, Altai Mts., Kazakhstan, Manchuria, Tuva, Mongolia, m. and w. Siberia, Maritime Territory, Korean Peninsula, Japan.

Biology. In "Stranddünen, Binnendünen, Sandfeldern, besonnten Hängen" (Kuntze, 1937). On *Artemisia campestris* (Ribaut, 1952; Wagner & Franz, 1961; Schiemenz, 1969b). Hibernation takes place in the egg stage, one generation (Schiemenz, l.c.). In dry meadows on sandy soils (Vilbaste, 1974). Adults in July-September.

Genus *Euscelidius* Ribaut, 1942

Euscelidius Ribaut, 1942: 268.
   Type-species: *Athysanus variegatus* Kirschbaum, 1858, by original designation.

Body broad, robust. General aspect suggestive of *Euscelis*. Head approximately as wide as pronotum, frontally obtusely angular, as seen from above medially distinctly longer than near eyes. Frontoclypeus in frontal aspect distinctly longer than wide, anteclypeus approximately parallel-sided. Anterior tibiae with 3 setae in inner, 4 in outer dorsal row. Wings always fully developed. Male anal tube short, sclerotized dorsally and laterally. Genital plates with macrosetae arranged in a row along lateral border. Styles with median apophyse short, robust. Aedeagus slender, symmetrical, not band-shaped, phallotreme ventral. 7th abdominal sternum in female with caudal border moderately concave, medially with a short angular projection. In Denmark and Fennoscandia one species.

366. *Euscelidius schenkii* (Kirschbaum, 1868)
   Plate-fig. 188, text-figs. 2511-2517.

Jassus (Athysanus) schenkii Kirschbaum, 1868b: 111.
Athysanus zetterstedti Melichar, 1896: 275.

Resembling *Euscelis incisus* but larger. Greyish yellow, shining. Head above with some usually well-defined fuscous markings: frontally on each side three spots in a transverse row along fore border, behind these an usually triangular spot laterally often attached to the lateral spot of the frontal row; caudally of this spot there is a rounded brownish yellow more diffuse spot in front of hind border of vertex. Face with dark transverse streaks and sutures. Anteclypeus with a dark longitudinal streak. Pronotum frontally

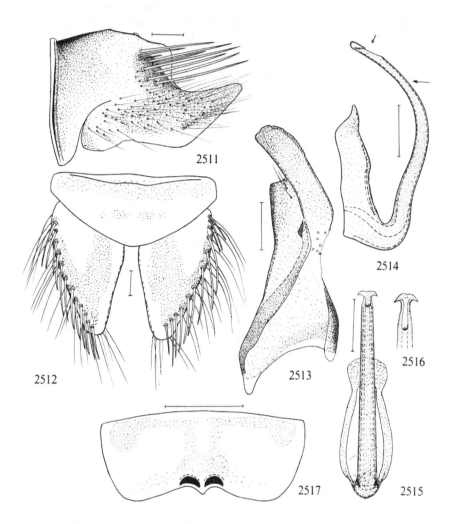

Text-figs. 2511-2517. *Euscelidius schenkii* (Kirschbaum). – 2511: left pygofer lobe in male from the left; 2512: genital plates and valve from below; 2513: right genital style from above; 2514: aedeagus from the left; 2515: aedeagus in ventral aspect (as seen in direction of longer arrow in 2514); 2516: apex of aedeagus in ventral aspect (as seen in direction of shorter arrow in 2514); 2517: 7th abdominal sternum in female from below (depressed under coverglass). Scale: 0.5 mm for 2517, 0.1 mm for the rest.

770

with a transverse row of fuscous spots, caudally with two or three pairs of longitudinal bands consisting of parallel dark transverse streaks. Fore wings longer than abdomen, densely dark mottled, mottlings missing around transverse veins and at distal ends of claval veins, resulting in whitish spots. Venter in female largely light, dark spotted; abdominal venter in male brown mottled, abdominal dorsum in both sexes black with light lateral and segmental borders. Dark markings sometimes strongly confluent; occasionally one finds almost entirely black specimens. Femora and tarsi transversely banded. Male pygofer lobes as in Text-fig. 2511, genital valve and plates as in Text-fig. 2512 (plates with median margins contiguous in repose), genital styles as in text-fig. 2513, aedeagus as in Text-figs. 2514-2516, 7th abdominal sternum in females as in Text-fig. 2517. Overall length of males 4.3-4.9 mm, of females 5.0-5.4 mm. – Nymphs with 4 complete rows of setae on abdomen; vertex without sharply delimited black spots, distinctly wider than long, abdomen with pale midline; dark points present on sides of longitudinal bands of meso- and metanotum; sides of anteclypeus roughly parallel or divergent; abdomen brown, anteriorly with a large pale patch; middle of last tergite and pygofer also pale; pale midline of vertex of uniform width, pale patch on abdomen usually extending to end of tergite IV (Vilbaste, 1982). – An excellent reproduction of the 5th instar larva is found in Walter (1975).

Distribution. Uncommon in Denmark, so far known from EJ, F, LFM, SZ and NEZ. – Widespread, not uncommon in southern and central Sweden, found Sk.-Upl., Hls. – Scarce and sporadic in Norway, found in AK, Bø, Nnv. – Scarce in southern and central East Fennoscandia, found in Al, Ab-Ta, Kb; Vib, Kr. – Widespread in Europe, also found in Canary Is., Tunisia, Cyprus, Iraq, Armenia, Azerbaijan, Kazakhstan, Uzbekistan, Altai Mts., w. Siberia; Nearctic region.

Biology. In "besonnten Hängen, Wiesen" (Kuntze, 1937). In cultivated fields (Linnavuori, 1952a). "Auf Trockenrasen" (Wagner & Franz, 1961). In xeric and, to a less extent, mesic biotopes (Schiemenz, 1969b).

Genus *Conosanus* Osborn & Ball, 1902

Athysanus (Conosanus) Osborn & Ball, 1902: 232.
Type-species: *Athysanus obsoletus* Kirschbaum, 1858, by original designation.

Closely related to *Euscelis*. Body broad, robust. Head a little wider than pronotum, frontally obtusely angular, frontoclypeus approximately as long as broad; anteclypeus broad, slightly tapering towards apex. Anterior tibiae with 4 setae in the inner, 4 in the outer dorsal row. Wing-dimorphous in both sexes. Male pygofer dorsally with a shallow incision. Genital plates with lateral margin strongly convex (Text-fig. 2518), macrosetae situated near side margins. Connective broad, H-shaped. Aedeagus with socle well developed, shaft symmetrical, dorsoventrally compressed, band-like, phallotreme ventral. Caudal margin of 7th abdominal sternum in females broadly concave, with a short angular median projection. A monotypic genus.

367. *Conosanus obsoletus* (Kirschbaum, 1858)
Text-figs. 2518-2523.

Athysanus obsoletus Kirschbaum, 1858a: 7.
Jassus (Athysanus) convexus Kirschbaum, 1868b: 109.
Jassus pauperculus Thomson, 1869: 54 (nec Flor, 1861).
Athysanus sexpunctatus J. Sahlberg, 1871: 271.

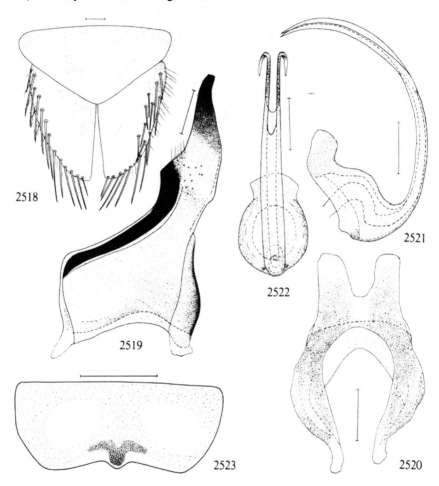

Text-figs. 2518-2523. *Conosanus obsoletus* (Kirschbaum). – 2518: male genital plates and valve from below; 2519: right genital style from above; 2520: connective in ventral aspect; 2521: aedeagus from the left; 2522: aedeagus in ventral aspect; 2523: 7th abdominal sternum in female from below. Scale: 0.5 mm for 2523, 0.1 mm for the rest.

Greyish yellow, shining. Head above with 3 pairs of black spots being more or less distinct: one median pair near apex, caudo-laterally of this another somewhat larger pair, and a third smaller pair in the caudo-lateral corner. In addition there may be one or two transverse streaks near each ocellus, one medially, another caudally. Transverse streaks and sutures of face black; anteclypeus often with a spot. Pronotum unicolorous light or with some small dark spots near anterior border, more caudally with 6 longitudinal bands consisting of short parallel dark transverse streaks. Fore wings in sub-brachypters a little shorter than abdomen, in macropters about as long as abdomen, with or without black mottlings between the light veins; in dark specimens, the cells may be entirely fuscous. Colour of venter strongly varying, largely greyish yellow to nearly black. Femora longitudinally and transversely banded. Male genital valve and plates as in Text-fig. 2518 (median margins of plates contiguous in repose); styles as in Text-fig. 2519, connective as in Text-fig. 2520, aedeagus as in Text-figs. 2521, 2522, 7th abdominal sternum in female as in Text-fig. 2523. Overall length of males 4.2-5.7 mm, of females 5.0-6.2 mm. – Nymphs with 4 complete longitudinal rows of hairs on abdomen; vertex distinctly wider than long; midline of abdomen pale; dark point present on sides of longitudinal bands of meso- and metathorax; sides of anteclypeus roughly parallel or divergent; two dark longitudinal bands usually present on dorsal surface of abdomen; distance between dark median longitudinal bands somewhat greater on pro- and mesothorax than on abdomen (Vilbaste, 1982). For a good description and reproduction of instar V, see also Walter (1975).

Distribution. Common in Denmark. – Common in southern Sweden, found Sk.-Sdm. – Norway: found only in VAy: Kristiansand (Warloe). – Not found in East Fennoscandia. – Widespread in western, central and southern Europe, also in Azores, Algeria, Cyprus, Anatolia, Moldavia, Ukraine, Iran; Nearctic region.

Biology. Mainly in "zeitweilig nassen Bentgraswiesen, ständig feuchten *Caltha*-Wiesen, extremnassen *Carex*-Wiesen" (Marchand, 1953). Hibernates in larval stage (Müller, 1957), in egg stage; 2 generations (in Central Europe) (Remane, 1958). "Scheint wegen der Eiablage an *Juncus effusus* gebunden zu sein" (Wagner & Franz, 1961).

Genus *Euscelis* Brullé, 1832

Euscelis Brullé, 1832: 109.
  Type-species: *Euscelis lineolata* Brullé, 1832, by monotypy.

Head a little wider than pronotum, in dorsal aspect obtusely angular or rounded obtusely angular. Frontoclypeus in frontal aspect a little longer than wide. Anteclypeus roughly parallel-sided. Fore wings with apical membrane narrow even in macropters. Anterior tibiae with 3 or 4 setae in inner, 4 outer dorsal row. Male pygofer only shallowly incised dorsally. Genital plates triangular, lateral margins almost straight in middle, their macrosetae arranged in a row along lateral margin. Styles well developed, median apophyse finger-like, lateral heel well marked. Connective short and broad, Y-shaped. Aedeagus with socle well developed, shaft dorsoventrally strongly compressed, band-like, phallotreme ventral. Caudal margin of 7th abdominal sternum in female straight, without processes or incisions. In Denmark and Fennoscandia three species.

Key to species of *Euscelis*

1   Aedeagus without appendages (Text-fig. 2532) Dark
     markings extensive .... 368. *incisus* (Kirschbaum), spring generation
–   Aedeagus on each side with an apical recurrent appendage
     (Text-figs. 2530, 2531, 2534, 2542) 2
2   Appendages of aedeagus long, apices reaching to level or
     proximally of level of phallotreme (Text-fig. 2542) .... 370. *ohausi* Wagner
–   Appendages of aedeagus shorter 3
3 (2) Apical incision of aedeagus deep, reaching far proximally of
     apices of appendages (Text-fig. 2534) .... 369. *distinguendus* (Kirschbaum)
–   Apical incision of aedeagus shallow, reaching to level of api-
     ces of appendages (Text-figs. 2530, 2531) ....
     .... 368. *incisus* (Kirschbaum), summer generation

368. *Euscelis incisus* (Kirschbaum, 1858)
   Text-figs. 2524-2532.

Cicada plebeja Fallén, 1806: 24 (nec Scopoli, 1763).
Athysanus incisus Kirschbaum, 1858a: 10.
Athysanus obscurellus Kirschbaum, 1858a: 10.
Athysanus communis Edwards, 1888c: 39.
Euscelis plebejus ssp. *albingensis* W. Wagner, 1939: 179.

Greyish yellow with brownish black markings, shining. Head above with the following pairs of black markings: one spot near apex of head, between this spot and ocellus a transverse streak, behind these markings a transverse band or a transverse row of dots reaching eye but not median line, and behind this band near caudal border of vertex a roundish spot with light center, or a hook-shaped spot. Transverse streaks on frontoclypeus black, between their median ends a light median line. Anteclypeus with a black median longitudinal patch. Genae and lora black spotted. Pronotum along fore border with or without a transverse row of dark spots, caudally of these 6 longitudinal bands consisting of parallel dark transverse streaks. Fore wings about as long as abdomen or somewhat longer, more or less densely black mottled along the light veins and in the cells. In light specimens, transverse veins and distal ends of claval veins each usually surrounded by a whitish spot. Venter varying in colour, greyish yellow to black. Femora transversely black banded. – The above description refers to the summer generation (= *plebejus* Fall., *communis* Edw.). In the spring generation (= *incisus* s.str., *obscurellus* Kbm.), the black markings are more or less strongly confluent, whitish spots on fore wings usually missing. The seasonal dimorphism in this species affects also the size of body and the structure of male genitalia. Observe that Text-figs. 2529-2532 are at the same scale! Male pygofer as in Text-figs. 2524, 2525, genital plates and valve as in Text-fig. 2526, connective as in Text-fig. 2527, style as in Text-fig. 2528, aedeagus of summer generation as in Text-figs. 2529-2531, aedeagus of spring generation as in Text-

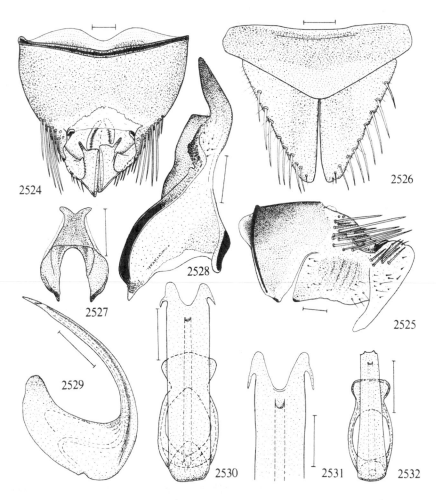

Text-figs. 2524-2532. *Euscelis incisus* (Kirschbaum). – 2524: male pygofer from above (anal apparatus removed); 2525: left pygofer lobe in male from the left; 2526: genital plates and valve from below; 2527: connective from above; 2528: right genital style from above; 2529: aedeagus from the left; 2530: aedeagus in ventral aspect; 2531: apex of aedeagus in ventral aspect (another specimen), 2532: aedeagus in ventral aspect. Text-figs. 2524, 2525, 2528-2531 refer to summer generation, 2526, 2527, and 2532 to spring generation. Scale: 0.1 mm.

fig. 2532. Overall length of males (spring generation) 3.0-3.6 mm, of females (spring generation) 3.8-4.1 mm, of males (summer generation) 3.7-4-4 mm, of females (summer generations) 3.9-4.4 mm. – Nymph with 4 rows of hairs on abdominal dorsum; vertex distinctly wider than long; abdomen with pale midline; margins of longitudinal bands of meso- and metathorax without dark spots; lateral hairs present only on posterior angles of tergite VIII (Vilbaste, 1982). Abdominal venter light in summer generation, dark in spring generation (Walter, 1975).

Distribution. Very common in Denmark. – Common in southern Sweden, found Sk.-Upl. – So far not recorded from Norway, nor from East Fennoscandia. – Widespread in the Palaearctic region except its northern parts.

Biology. In "Stranddünen, Binnendünen, Sandfeldern, besonnten Hängen, Wiesen" (Kuntze, 1937). In "trockenen Glatthaferwiesen, einzeln (zufällig) in wechselfeuchten Bentgraswiesen und spärlich in ständig feuchten *Caltha*-Wiesen" (Marchand, 1953). "Hibernation als Larve" (Schiemenz, 1964). – The biology of this species and its seasonal polymorphism has been thoroughly studied by Müller (1954 and later). Already Wagner (1939) showed that there exists a gradual variation in shape of aedeagus between the extremes illustrated in our Text-figs. 2531 and 2532. This was explained as the existence of a "Rassenkreis", but Müller (1954, 1955, 1957b) experimentally proved that this variation is related to different photoperiodic influences. "In optimal habitats . . . *Euscelis incisus* may form two generations a year, the first one (*incisus*) in the beginning of May after hibernation with older larvae by quiescence, and the second one (*plebejus*) in August. . . . But in suboptimal situations . . . the larval population in autumn is often . . . too small and also too young as not to be extirpated during the wintery quiescence. Then . . . only the little percentage of diapausing eggs – existing in all populations – is able to guarantee the further existence of the population" (Müller, 1981). – In the Swedish provinces Sk. and Bl. both the *incisus*-form and the *plebejus*-form have been found, in the rest of the Swedish area of distribution only the *plebejus*-form. – Adults in April-May and again in June-September.

369. *Euscelis distinguendus* (Kirschbaum, 1858)
Plate-fig. 189, text-figs. 2533-2537.

Athysanus distinguendus Kirschbaum, 1858a: 8.

Resembling the summer generation of *E. incisus*. Aedeagus as in Text-figs. 2533, 2534, 1st and 2nd abdominal terga in male as in Text-fig. 2535, 1st-3rd abdominal sterna in male as in Text-figs. 2536, 2537. Overall length of males 3.1-4.6 mm, of females 3.3-4.6 mm. – Nymphs as those of *incisus* but lateral hairs present on posterior angles of tergites VII and VIII; with faint pattern, at least 2 median longitudinal bands usually present (Vilbaste, 1982). For accurate descriptions and excellent reproductions of 1st-5th instars, see Walter (1975).

Distribution. Not found in Denmark. – Scarce in southern Sweden, found in Sk., Bl., Hall., Gtl., Ög., Nrk. – Scarce in Norway, found in AK: Tøyen, Oslo (Siebke), Aarnes,

Nes (Holgersen); HEs: Helgøya (Esmark); Bø: Drammen (Warloe). – East Fennoscandia: rare and local, found in Ab, N, Ta; Kr. – France, Portugal, Netherlands, German D.R. and F.R., Austria, Italy, Albania, Bohemia, Moravia, Slovakia, Romania, Bulgaria, Yugoslavia, Greece, European Turkey and Anatolia, Poland, Estonia, Latvia, Lithuania, n., m., and s. Russia, Ukraine, Tunisia, Iraq, Mongolia.

Biology. In clover fields (Kuntze, 1937). Hibernation in egg-stage, 1 generation (Schiemenz, 1969b; Müller, 1981). In dry meadows (Vilbaste, 1974). Adults in July and August.

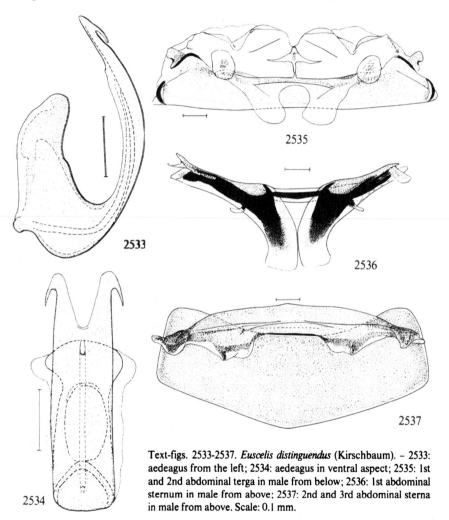

Text-figs. 2533-2537. *Euscelis distinguendus* (Kirschbaum). – 2533: aedeagus from the left; 2534: aedeagus in ventral aspect; 2535: 1st and 2nd abdominal terga in male from below; 2536: 1st abdominal sternum in male from above; 2537: 2nd and 3rd abdominal sterna in male from above. Scale: 0.1 mm.

370. *Euscelis ohausi* W. Wagner, 1939
Text-figs. 2538-2542.

Euscelis ohausi W. Wagner, 1939: 177.
Euscelis singeri W. Wagner, 1951: 55.

Dorsal black markings unequally distributed. Vertex centrally with two strong triangular black spots tending to extend laterally and medially, almost forming a band. On apex of vertex there are two small black spots, laterally of these two additional small spots often confluent with the large spot. Mottlings on fore wings in *f. typica* arranged in longitudinal rows especially distinct along commissural border, along Cu,

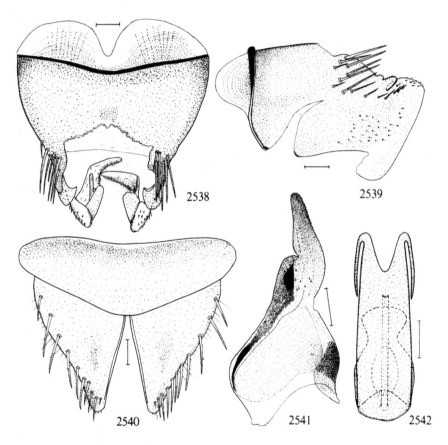

Text-figs. 2538-2542. *Euscelis ohausi* Wagner. – 2538: male pygofer from above, anal apparatus removed; 2539: left pygofer lobe in male from outside; 2540: genital plates and valve from below; 2541: right genital style from above; 2542: aedeagus in ventral aspect. Scale: 0.1 mm.

778

and in apical cells, weakly marked on both sides of claval suture. In *f. singeri* the mottlings are more evenly distributed on the wing surface. Fore wings in sub-brachypterous males as long as, in sub-brachypterous females shorter than abdomen, in macropterous males longer, in macropterous females as long as abdomen. Male pygofer as in Text-figs. 2538, 2539, genital valve and plates as in Text-fig. 2540 (but inner margins not diverging in repose); style as in Text-fig. 2541, aedeagus as in Text-fig. 2542. Overall length of males of *ohausi* s. str. 3.4-4.7 mm, of females 4.1-5.5 mm; *f. singeri* is a little longer. – 7th abdominal tergum in nymph with only 2 pairs of setae; fore wing buds with usually 4 indistinctly limited longitudinal streaks (Walter, 1975).

Distribution. Very rare in Denmark, found only in WJ: Skallingen 24.VII.1966 (Trolle). – Not recorded from Sweden, Norway, and East Fennoscandia. – France, Portugal, England, Scotland, Wales, German D.R. and F.R., Netherlands, Poland.

Biology. On *Genista anglica (ohausi)* and *Sarothamnus (singeri)* (Wagner, 1951).

Genus *Ederranus* Ribaut, 1942

Ederranus Ribaut, 1942: 267.
Type-species: *Athysanus discolor* J. Sahlberg, 1871, by original designation.

Head wider than pronotum. Transition between vertex and frontoclypeus fairly abrupt. Shaft of aedeagus dorsoventrally compressed, band-shaped, arising from ventral end of socle. Phallotreme ventral, subapical. Frontoclypeus elongate, much longer than wide. Anterior tibiae with only one seta in inner dorsal row. Anal tube in male dorsally well developed, short. Genital plates triangular, lateral margins convex, their macrosetae arranged along lateral margin. Connective Y-shaped. 7th abdominal sternum in female with caudal margin concave, medially with a small projection. Two species.

Key to species of *Ederranus*

1   Yellowish, vertex and pronotum each with a pair of black
    spots (Text-fig. 2543). Male genital plates broad (Text-fig.
    2544), aedeagus as in Text-figs. 2547, 2548 .... 371. *sachalinensis* (Matsumura)
–   Light yellow or greenish, usually with extensive dark mark-
    ings. Face usually black. Male genital plates more elongate
    (Text-fig. 2553), with long fine pilosity, aedeagus as in Text-
    figs. 2555, 2556 .... 372. *discolor* (J. Sahlberg)

371. *Ederranus sachalinensis* (Matsumura, 1911)
   Text-figs. 2543-2550.

Cicada lutea C. Sahlberg, 1842: 88 (nec Olivier, 1790).
Athysanus sachalinensis Matsumura, 1911: 26.

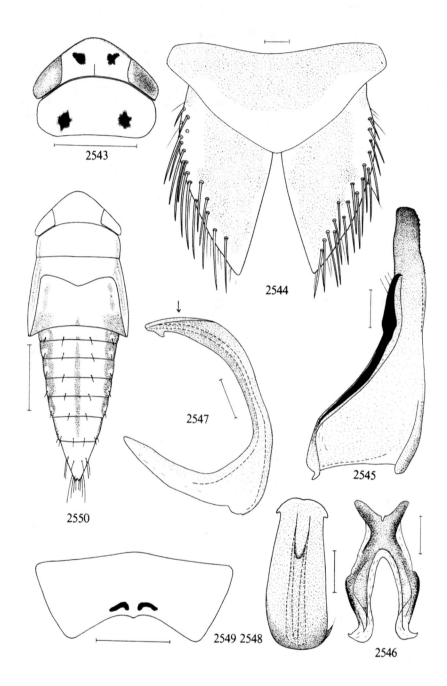

2543

2544

2547

2545

2550

2549 2548

2546

Elongate, yellow, shining. Head as seen from above obtusely angular, medially slightly longer than half width of vertex, 2/3 longer than length near eyes, with two black spots (Text-fig. 2543); frontoclypeus 1/4 longer than maximal width; pronotum with two black spots. Abdominal tergum and venter black banded. Hind tibiae black spotted, tarsi apically fuscous. Wing dimorphous. Fore wings in macropterous males distinctly, in macropterous females slightly longer, than abdomen; in sub-brachypterous specimens about as long as abdomen. Male genital plates and valve as in Text-fig. 2544 (median margins of plates not diverging in repose), styles as in Text-fig. 2545, connective as in Text-fig. 2546, aedeagus as in Text-figs. 2547, 2548. Apodemes of 1st and 2nd abdominal sterna in male small, inconspicuous. 7th abdominal sternum in female as in Text-fig. 2549. – Nymphs with 4 rows of setae on abdominal tergum; vertex distinctly wider than long; midline of abdomen dark; only a single wider longitudinal line running along abdomen and thorax; this may have a dark border (Vilbaste, 1982). See also Text-fig. 2550.

Distribution. Not found in Denmark, Sweden and Norway. – East Fennoscandia: only found in St: Yläne (C. Sahlberg, J. Sahlberg). – N. Russia, Altai Mts., w. Siberia, Tuva, Korean Peninsula, Sakhalin, Maritime Territory.

Biology. On *Calamagrostis lanceolata* in marshy forests (Emeljanov, 1964).

372. *Ederranus discolor* (J. Sahlberg, 1871)
Text-figs. 2551-2558.

Athysanus discolor J. Sahlberg, 1871: 277.
Athysanus nauta J. Sahlberg, 1871: 280.

Elongate, light yellow, shining, especially males with dark markings strongly varying in extension. Head as seen from above frontally obtusely angular, medially 1/4 shorter than maximal width of vertex, 1/3 longer than length near eyes. Frontoclypeus 1/3 longer than maximal width. Anteclypeus 1.5 times as long as basal width, slightly wider towards apex. Wing dimorphous, fore wings in macropters somewhat longer, in sub-brachypters 1/4 shorter than abdomen. Face usually largely black, in females sometimes only with a median spot varying in size, or unspotted. In males, vertex with or without a pair of black spots, pronotum caudally with or without a pair of black spots varying in size, scutellum with a black longitudinal band varying in width, thoracic venter and abdomen largely black, fore wings with cells more or less completely filled up with dark pigment. Hind tibiae black spotted. Females with black

Text-figs. 2543-2550. *Ederranus sachalinensis* (Matsumura). – 2543: head and pronotum of male from above; 2544: male genital plates and valve from below; 2545: right genital style from above; 2546: connective in ventral aspect; 2547: aedeagus from the left; 2548: apical part of aedeagus in ventral aspect (as seen in direction of arrow in 2547); 2549: 7th abdominal sternum in female from below; 2550: 5th instar nymph (♀). Scale: 0.5 mm for 2549, 1 mm for 2543, 2550, 0.1 mm for the rest.

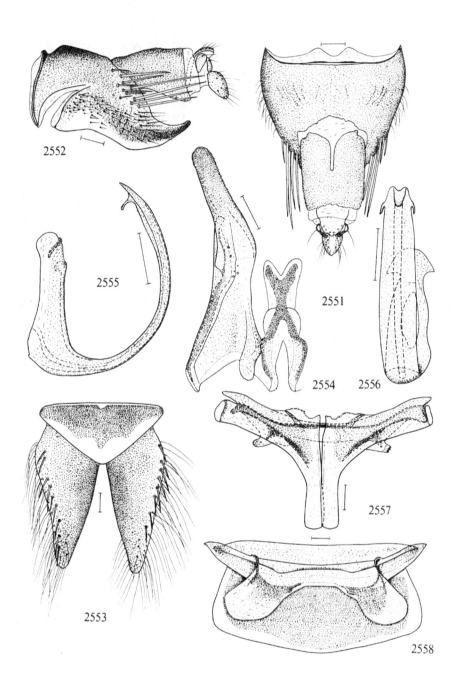

2552

2555

2551

2554

2556

2553

2557

2558

782

markings much reduced, sometimes entirely light (*f. nauta*). Male pygofer and anal apparatus as in Text-figs. 2551, 2552, genital plates and valve as in Text-fig. 2553 (but plates not diverging in repose). Styles and connective as in Text-fig. 2554, aedeagus as in Text-figs. 2555, 2556, 1st abdominal sternum in male as in Text-fig. 2557, 2nd abdominal sternum in male as in Text-fig. 2558. Overall length ($\vec{\circ}\circ$) 5-6 mm.

Distribution. Not found in Denmark, Sweden and Norway. – Very rare in East Fennoscandia, found in Ka: Räisälä (J. Sahlberg); Ta: Teisko (J. Sahlberg); Vib: Pyhäjärvi (Håkan Lindberg); Kr: Kexholm; Jaakima; Kirjavalahti; Solomina (J. Sahlberg); Sortavala (Lindberg). – N. Russia.

Biology. On *Phragmites communis* (Sahlberg, 1871); on *Glyceria spectabilis* (Lindberg, 1947); on *Glyceria aquatica* (Emeljanov, 1964). Adults in July-October.

Genus *Streptanus* Ribaut, 1942

Streptanus Ribaut, 1942: 261.
  Type-species: *Cicada sordida* Zetterstedt, 1828, by original designation.

Related to *Euscelis*. Body broad, robust, Head wider than pronotum. Anterior tibiae with 3 setae in the inner, 4 in the outer dorsal row. Wing-polymorphous leafhoppers, brachypterism predominating in most species. Male anal tube well sclerotized dorsally and laterally. Genital plates with lateral margins convex, their macrosetae arranged along lateral margin, usually in a row. Connective Y- or X-shaped. Styles robust. Shaft of aedeagus dorsoventrally flattened, band-shaped, asymmetrical, apex of aedeagus laterally dilated, more or less spatulate. Phallotreme ventral, situated far proximally of apex of aedeagus. Caudal border of 7th abdominal sternum in female medially with an angular projection. In Denmark and Fennoscandia five species.

Key to species of *Streptanus*

1   Head frontally angular, approximately as long as pronotum
    medially. Fore wings in brachypters caudally almost squarely
    cut off, usually reaching caudal border of 6th abdominal
    tergum, leaving 7th and 8th terga uncovered. Male genital
    plates apically sharply pointed (Text-figs. 2592, 2593). Aede-
    agus without basal dilatations; apical expansion (spatula) much
    wider than long (Text-figs. 2595, 2596) .... 377. *marginatus* (Kirschbaum)

Text-figs. 2551-2558. *Ederranus discolor* (Sahlberg). – 2551: male pygofer and anal apparatus from above; 2552: same from the left; 2553: genital plates and valve from below; 2554: right genital style and connective from above; 2555: aedeagus from the left; 2556: aedeagus in ventral aspect; 2557: 1st abdominal sternum in male from above; 2558: 2nd and 3rd abdominal sterna in male from above. Scale: 0.1 mm.

- Head frontally obtusely angular or rounded angular, distinctly shorter than pronotum medially. Fore wings in brachypters longer, apically rounded. Male genital plates apically rounded obtuse. Spatula of aedeagus not or just wider than long 2
2 (1) Larger, length of brachypterous males not below 4 mm, of brachypterous females at least 5 mm, macropters about 6 mm. Dark markings on vertex usually distinct. Basal dilatations present on both sides of shaft of aedeagus, rounded, not angular (Text-figs. 2562, 2563) .... 373. *aemulans* (Kirschbaum)
- Smaller species. Dark markings on vertex diffuse or absent 3
3 (2) Shaft of aedeagus without basal projections (Text-fig. 2587). Usually pale-coloured without distinct markings .... 376. *confinis* (Reuter)
- Shaft of aedeagus with basal projection on both sides, at least the left one angular 4
4 (3) Spatula distinctly longer than wide, symmetrical. Distance between phallotreme and proximal margin of spatula about twice the length of spatula (Text-fig. 2573) .... 374. *sordidus* (Zetterstedt)
- Spatula not longer than wide, slightly asymmetrical. Distance between phallotreme and proximal margin of spatula about 4×length of spatula (Text-fig. 2581) .... 375. *okaensis* Zachvatkin.

373. *Streptanus aemulans* (Kirschbaum, 1868)
Text-figs. 2559-2567.

Jassus (Athysanus) aemulans Kirschbaum, 1868b: 107.
Jassus (Athysanus) obtusus Kirschbaum, 1868b: 108.
Athysanus sahlbergi Reuter, 1880: 209.

Greyish yellow, shining. Head frontally rounded obtusely angular, above with the following brown or fuscous markings: frontally on each side an arched transverse streak, caudally of this a transverse band usually divided into spots extending from each ocellus to median line at apex of head; behind this a fairly straight transverse band between eyes, often interrupted medially, near eye and hind border on each side one or two smaller dots. Frontoclypeus with dark transverse streaks; sutures of face dark or light; a dark spot usually present below each antenna. Anteclypeus dilated towards

Text-figs. 2559-2567. *Streptanus aemulans* (Kirschbaum). – 2559: genital plates and valve from below; 2560: right genital style from above; 2561: connective from above; 2562: aedeagus from the left; 2563: aedeagus from the right; 2564: apical part of aedeagus in ventral aspect (in direction of arrow in 2563); 2565: 1st abdominal sternum in male from above; 2566: 2nd and 3rd abdominal sterna in male from above; 2567: 7th abdominal sternum in female from below: Scale: 0.5 mm for 2567, 0.1 mm for the rest.

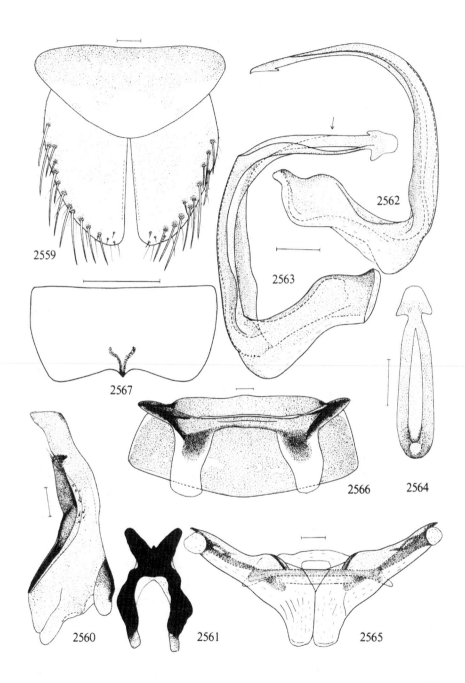

2559

2562

2563

2567

2560

2561

2566

2564

2565

apex, with or without a dark longitudinal spot. Pronotum entirely light or with a row of dark patches along anterior border; caudally of these six indistinct dark longitudinal bands may be present. Fore wings in brachypters shorter than abdomen, especially in females, in macropters slightly longer than abdomen. Veins of fore wings light, cells more or less distinctly, and more or less completely, filled up with brownish pigmentation. Abdominal tergum medially with a double dark longitudinal streaks, on each side with a dark longitudinal band. Venter of abdomen dark-bordered. Femora transversely and longitudinally dark banded, tibiae dark spotted. Genital valve and plates as in Text-fig. 2559, style as in Text-fig. 2560, connective as in Text-fig. 2561, aedeagus as in Text-figs. 2562-2564, 1st-3rd abdominal sterna in male as in Text-figs. 2565, 2566, 7th abdominal sternum in female as in Text-fig. 2567. Overall length of brachypterous males 4.1-5 mm, of brachypterous females 5.0-5.6 mm, of macropterous males 5.0-5.1 mm, of macropterous females 5.5-6.0 mm. – Nymphs with 4 longitudinal rows of hairs on abdomen; vertex distinctly wider than long; midline of abdomen pale; dark point present on sides of longitudinal bands of meso- and metathorax; 2 longitudinal bands usually present on dorsal surface; distance between median longitudinal bands more or less equal, or pattern indistinct; nymphs large, N5 longer than 4.5 mm; distance between median hairs of tergite III more or less equal to that of following tergites; pattern on vertex consisting of cross-lines (Vilbaste, 1982). – For good descriptions and illustrations of nymphs, consult also Walter (1975).

Distribution. Common in Denmark. – Common in southern and central Sweden, found Sk.-Ång., also recorded from Lu.Lpm. (Haupt, 1917). – Norway: found in Ø, AK, Bø, VE, AAy, Ry, SFi. – Fairly common in southern and central East Fennoscandia, Al, Ab-Kb, Om; ObN. – England, Scotland, Wales, Ireland, France, Belgium, Netherlands, German D.R. and F.R., Austria, Switzerland, Hungary, Romania, Bohemia, Moravia, Slovakia, Yugoslavia, Poland, Estonia, Latvia, Lithuania, Ukraine, Kazakhstan, n. and m. Siberia, Altai Mts., Kirghizia, Kamchatka.

Biology. In "Stranddünen, Flachmooren, Wäldern, Waldlichtungen" (Kuntze, 1937). In meadows, in clover fields, in forests (Vilbaste, 1974). In grass meadows, especially with *Elytrigia repens*. Adult in July-October. – On October 3, 1946, a male captured by netting was observed in a condition of "death-feigning", remaining motionless with legs contracted for about half a minute. Afterwards it behaved quite normal, producing its "common song" etc.

Text-figs. 2568-2576. *Streptanus sordidus* (Zetterstedt). – 2568: left pygofer lobe in male from the left; 2569: male genital plates and valve from below; 2570: right genital style from above; 2571: aedeagus from the left; 2572: aedeagus from the right; 2573: apical part of aedeagus in ventral aspect (as seen in direction of arrow in 2571); 2574: 1st abdominal sternum in male from above; 2575: 2nd and 3rd abdominal sterna in male from above; 2576: 7th abdominal sternum in female from below. Scale: 0.5 mm for 2576, 0.1 mm for the rest.

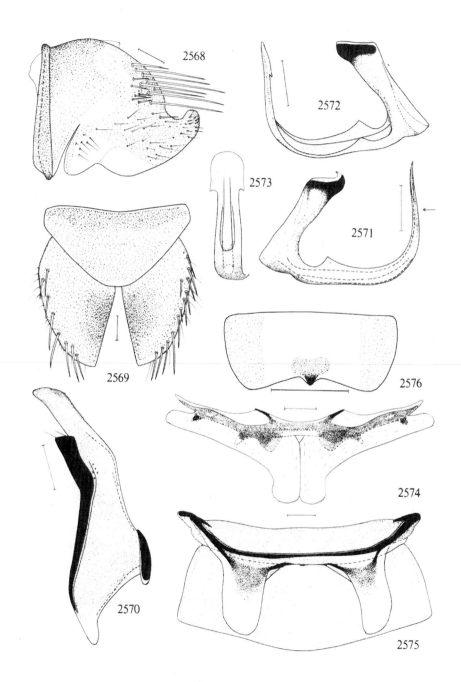

2568

2572

2573

2571

2569

2576

2574

2570

2575

787

374. **Streptanus sordidus** (Zetterstedt, 1828)
   Text-figs. 2568-2576.

Cicada sordida Zetterstedt, 1828: 531.
Jassus (Athysanus) confusus Kirschbaum, 1868b: 107.
Athysanus fraterculus Reuter, 1880: 211 (nec Berg, 1879).

Greyish yellow to pale brownish yellow, shining. Much varying in extension of dark markings. Especially females may be almost unicolorous light; on the other hand, almost entirely black specimens are not uncommon. In moderately pigmented specimens, head above with the following dark markings: frontally on each side an arched streak from ventral side of ocellus to apex of head, behind this on each side a transverse streak from ocellus usually not quite reaching a triangular spot near median line. Caudally of these markings a medially interrupted, approximately straight, transverse band from eye to eye; near each lateral corner a spot varying in size. Pronotum with or without a transverse row of dark spots along fore border, behind these 1-3 pairs of dark longitudinal bands. Wing trimorphous, females usually brachypterous, males usually sub-brachypterous, rarely brachypterous or macropterous. But in males, the difference between brachypters and sub-brachypters is unsharp. Fore wings of brachypters covering 7-8 abdominal terga, unicolorous light or with dark streaks in cells. Fore wings in macropters usually reaching considerably caudally of abdominal apex, unicolorous or with dark mottlings in cells. Face without markings or marked as in *aemulans;* colour of thoracic venter and abdomen strongly varying. Legs largely dark or transversely spotted, tibiae black spotted. Male pygofer lobes as in Text-fig. 2568, genital plates not diverging, otherwise as in Text-fig. 2569, style as in Text-fig. 2570, aedeagus as in Text-figs. 2571-2573, 1st-3rd abdominal sterna in male as in Text-figs. 2574, 2575, 7th abdominal sternum in female as in Text-fig. 2576. Overall length of sub-brachypterous males 3.5-4.2 mm, of brachypterous females 4.0-4.7 mm, of macropterous males 4.4-4.7 mm, of macropterous females 4.8-5.2 mm. – Nymphs as those of *aemulans,* smaller, N_5 than 4.5 mm, distance between median spines of tergite III clearly shorter than in the following tergites; vertex with longitudinal lines (Vilbaste, 1982). See also Walter (1975).

   Distribution. Common in Denmark, found in nearly all districts. – Common and widespread in Sweden, Sk.-Lu.Lpm. – Norway: published and unpublished records are available from most regions (Ø, AK-TRy, TRi); these have not been revised as regards the possible presence of *S. okaensis,* but a future revision will probably not change our view of the general distribution of *sordidus* in Norway very much. – East Fennoscandia:

Text-figs. 2577-2584. *Streptanus okaensis* Zachvatkin. – 2577: male pygofer and anal apparatus from the left; 2578: genital plates and valve from below; 2579: right genital style from above; 2580: aedeagus from the right; 2581: apical part of aedeagus in ventral aspect; 2582: 1st abdominal sternum in male from above; 2583: 2nd and 3rd abdominal sterna in male from above; 2584: 7th abdominal sternum in female from below (depressed under coverglass). Scale: 0.5 mm for 2584, 0.1 mm for the rest.

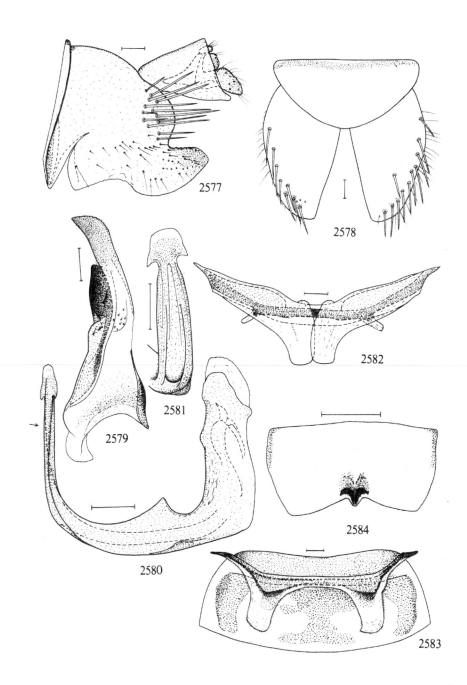

2577

2578

2579

2580

2581

2582

2583

2584

common, found in Al, Ab, N, Ka, St, Ta, Oa, Sb, Kb, Om, ObN; Vib, Kr. – Widespread in Europe, also found in Algeria.

Biology. In "Salzstellen, Hochmooren, Flachmooren, Wäldern, Waldlichtungen, Wiesen" (Kuntze, 1937). Mainly in "zeitweilig nassen Bentgraswiesen, *Caltha*-Wiesen, extremnassen *Carex*-Wiesen" (Marchand, 1953). Hibernates in the egg-stage (Remane, 1958; Törmälä & Raatikainen, 1976). In Central Europe 2 generations (Remane, 1958). Adults in July-September.

375. *Streptanus okaensis* Zachvatkin, 1948
Text-figs. 2577-2584.

Streptanus okaensis Zachvatkin, 1948b: 190.

Resembling *S. sordidus,* on an average a little larger. Wing trimorphous, males usually sub-brachypterous (hind wings as long as fore wings, these reaching hind margin of 8th abdominal tergum), females usually brachypterous (hind wings 2/3 as long as fore wings, these reaching about middle of 8th tergum). Macropters of both sexes with wings reaching considerably caudally of abdominal apex. Male pygofer and anal apparatus as in Text-fig. 2577, genital valve and plates as in Text-fig. 2578 (plates not diverging in repose), styles as in Text-fig. 2579, aedeagus as in Text-figs. 2580, 2581, 1st-3rd abdominal sterna in male as in Text-figs. 2582, 2583, 7th abdominal sternum in female as in Text-fig. 2584. Overall length of sub-brachypterous males 3.8-4.5 mm, of brachypterous females 4.5-5.5 mm, of macropters (2 ♂♂, 3 ♀♀) 4.4-5.4 mm.

Distribution. Denmark: rare, so far found only once in a forest-bog in NEZ: Mårum 17.VIII.1969 (Trolle). – Scarce in Sweden, found in Vg.: Öxnered 23.VII.1943; Upl.: Djursholm, Ösbysjön 29.VIII.1963 and 27.VII.1966, Vallentuna, vicinity of Grindstugan 20.VII.1968; Vb.: Skellefteå, Ursvik 17.VII.1933 (Ossiannilsson). – So far not recorded from Norway. – Widespread in East Fennoscandia, found in Al: Eckerö; Finström; Brändö; Geta; Jomala; Ab: Lojo; Pojo; N: Ekenäs; Ekenäs Hästö Busö; Tvärminne; Esbo; ObN: Rovaniemi Pisa; Torneå; Kemi; Li: Enare. – N. Germany (Remane, in litt.); m. and n. Russia, Kurile Isl., Tuva, Mongolia.

Biology. In tufts of *Calamagrostis canescens* and *purpurea* in marshes.

376. *Streptanus confinis* (Reuter, 1880)
Text-figs. 2585-2590.

Athysanus confinis Reuter, 1880: 211.
Euscelis wagneri Ossiannilsson, 1944: 14.
Streptanus similis Kristensen, 1965b: 291 (nec Kirschbaum, 1868).

Resembling *sordidus,* usually largely pale, dark markings obsolete. Head frontally slightly more angular. Male genital valve and plates as in Text-fig. 2585, styles as in Text-fig. 2586, aedeagus as in Text-figs. 2587, 2588, 1st-3rd abdominal sterna in male as in Text-figs. 2589-2590. Measurements as in *sordidus.* Nymphs as in *sordidus* but lateral hairs present only on posterior angle of tergite VIII, vertex usually without pattern; median hair spots on abdomen usually distinctly darker than lateral ones; distance between median hairs of tergite III not shorter than in following tergites (Vilbaste, 1982).

Distribution. Scarce in Denmark, known from a few places in NWJ, EJ and NEZ. – Scarce in Sweden, found in Sm., Öl., Gtl., Ög., Upl. – Norway: so far found only in Ø:

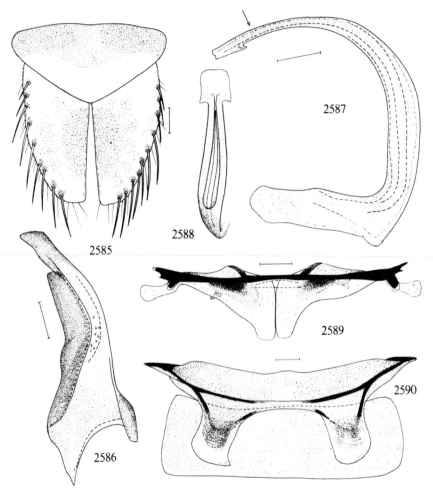

Text-figs. 2585-2590. *Streptanus confinis* (Reuter). – 2585: genital plates and valve from below; 2586: right genital style from above; 2587: aedeagus from the left; 2588: apical part of aedeagus in ventral aspect (in direction of arrow in 2587); 2589: 1st abdominal sternum in male from above; 2590: 2nd and 3rd abdominal sterna in male from above. Scale: 0.1 mm.

Mysen 18.VIII.1960 (Holgersen). – East Fennoscandia: scarce and sporadic, found in Al, Ab, N, Kb. – Latvia, Lithuania, Estonia, Bohemia, Iran.

Biology. "Auf feuchten Wiesen" (Lindberg, 1947). In Upl.: Vallentuna I found two males in tufts of *Calamagrostis canescens*, with *S. okaensis*. Adults in July and August.

377. *Streptanus marginatus* (Kirschbaum, 1858)
Plate-fig. 190, text-figs. 2591-2599.

Athysanus marginatus Kirschbaum, 1858a: 7.
Athysanus brevipennis Kirschbaum, 1858a: 9.
Jassus (Athysanus) similis Kirschbaum, 1868b: 114.
Jassus porrectus Thomson, 1869: 56.

Brownish yellow, shining. Head frontally angularly produced, above without markings or with indistinct brownish markings: a streak between each ocellus and apex of head, and an obtusely angular transverse band on disk of vertex not reaching eyes. In macropters, these markings are often more distinct, black. Pronotum unicolorous or with traces of dark spots. Scutellum usually unicolorous. Fore wings in brachypters usually leathery, ratio length: width = 1.7-1.8, hind wings little more than half length of fore wings. In some males a tendency towards sub-brachypterism can be observed, fore wings being twice as long as wide, apically rounded, length of hind wings about 3/4 of fore wing length. I have not seen macropterous males; macropterous females with fore wings an long as abdomen, hind wings little shorter. Fore wing veins in macropters often dark-bordered. Transverse streaks and sutures of face fuscous; in females, face often unicolorous light. Thoracic venter and abdomen in brachypterous females largely light or with more or less extended dark patches, in males and macropters usually largely dark. Femora transversely dark banded or largely dark, tibiae dark spotted. Male pygofer and anal apparatus as in Text-fig. 2591, genital valve and plates as in Text-figs. 2592, 2593 (latter not diverging in repose), styles as in Text-fig. 2594, aedeagus as in Text-figs. 2595, 2596, 1st-3rd abdominal sterna in male as in Text-figs. 2597, 2598, 7th abdominal sternum in female as in Text-fig. 2599. Overall length of brachypterous males 3.1-3.9 mm, of brachypterous females 3.5-4.5 mm, of macropterous females 3.5-4.5 mm. – Nymphs as in *confinis* but "vertex usually with faint pattern; median row of hairs spots on abdomen not darker than lateral ones; distance between median hairs of tergite III somewhat shorter than in the following tergites" (Vilbaste, 1982). See also Walter, 1975.

Distribution. Common in Denmark, especially in Jutland. – Common and widespread also in Sweden, Norway, and East Fennoscandia. – England, Scotland, Wales, France, Belgium, Netherlands, German D.R. and F.R., Switzerland, Austria, Bohemia, Moravia, Slovakia, Romania, Poland, Estonia, Latvia, Lithuania, n. and m. Russia, Ukraine, Tunisia.

Biology. In "Stranddünen, Heiden, Hochmooren, Wäldern, Wiesen" (Kuntze, 1937). – In drier meadow area of seashores and in rocky islets and shores, in dryish fields,

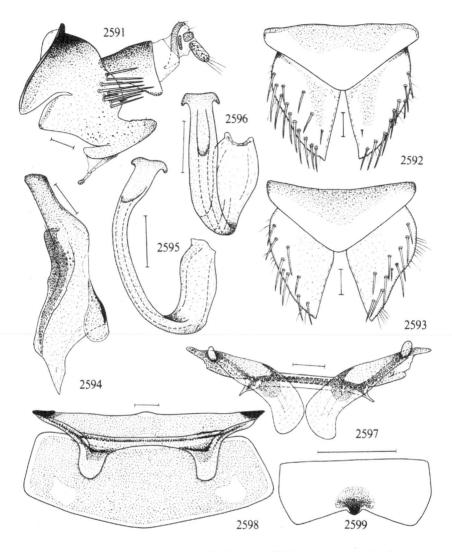

Text-figs. 2591-2599. *Streptanus marginatus* (Kirschbaum). – 2591: male pygofer and anal apparatus from the left; 2592: genital plates and valve from below; 2593: same of another specimen; 2594: right genital style from above; 2595: aedeagus from the right; 2596: aedeagus in ventral aspect; 2597: 1st abdominal sternum in male from above; 2598: 2nd and 3rd abdominal sterna in male from above; 2599: 7th abdominal sternum in female from below (depressed under coverglass). Scale: 0.5 mm for 2599, 0.1 mm for the rest.

793

moist sloping meadows, peaty meadows, cultivated fields, moist *Myrtillus* spruce woods, dry *Vaccinium* pine woods, dry *Calluna* heaths (Linnavuori, 1952a). Hibernation takes place in the larval stage (Müller, 1957; Schiemenz, 1969b). 1 generation (Schiemenz, 1969b). Adults in May-October.

Tribe Paralimnini
Genus *Paramesus* Fieber, 1866

Paramesus Fieber, 1866a: 506.
　　Type-species: *Athysanus obtusifrons* Stål, 1853, by monotypy.
(Jassus) Dochmocarus Thomson, 1869: 65.
　　Type-species: *Cicada nervosa* Fallén, 1826, by monotypy.

Body elongate, robust. Head wider than pronotum, fore margin smoothly curved or rounded obtusely angular. Transition between vertex and frontoclypeus abrupt. Vertex short, not or little longer medially than near eyes, frontally finely transversely striate. Anteclypeus dilated towards apex. Sides of pronotum not carinate. Anterior tibiae with 4 setae in each dorsal row. Male pygofer moderately excised. Genital plates triangular, their macrosetae arranged in a row along lateral border. Connective elongate, hairpin-shaped. Styles robust. Socle of aedeagus well developed, dorsally with a transverse bar-like appendage. In Denmark and Fennoscandia one species.

378. *Paramesus obtusifrons* (Stål, 1853)
　　Plate-fig. 191, text-figs. 2600-2608.

Cicada nervosa (Fallén, 1826: 39, nec Linné, 1758).
Athysanus obtusifrons Stål, 1853: 175.

Brownish yellow, shining. Vertex frontally behind the light fore border with an arched black transverse line between eyes, behind this a broader brownish transverse band. Face immediately below upper border with a black arched line; transverse streaks on frontoclypeus brown. Anteclypeus with a T-shaped brownish marking, sutures of face dark. Pronotum with 4 indistinct dark longitudinal bands. Scutellum with dark lateral corners and a dark median longitudinal band. Veins of fore wings light, dark bordered, apical part of fore wing darker in males. Apical and subapical cells of fore wings sometimes with some secondary transverse veins. Thoracic venter and abdomen in male largely black, segmental borders partly light; in female venter largely light,

Text-figs. 2600-2608. *Paramesus obtusifrons* (Stål). – 2600: left pygofer lobe and anal apparatus in male from the left; 2601: male pygofer from below, anal apparatus and genital plates removed; 2602: male genital plates and valve from below; 2603: right genital style from above; 2604: connective from above; 2605: aedeagus from the left; 2606: aedeagus in ventral aspect; 2607: 1st abdominal sternum in male from above; 2608: 2nd and 3rd abdominal sterna in male from above. Scale: 0.1 mm.

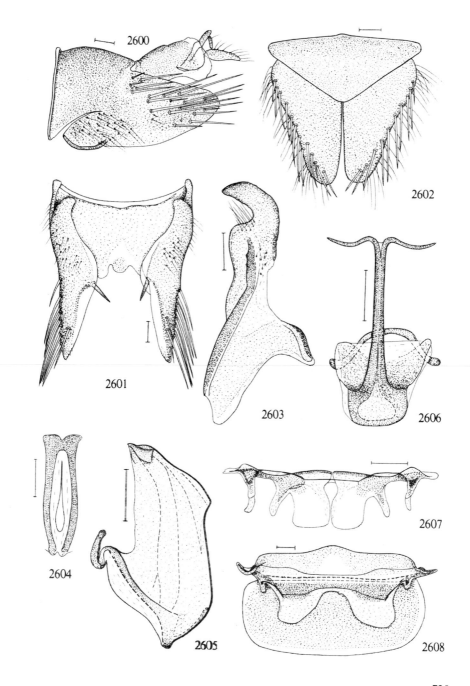

2600

2602

2601

2603

2606

2604

2605

2607

2608

abdominal dorsum with largely light lateral margins. Legs longitudinally striped, tibiae with black dots. Middle femora ventrally near apex with two dark transverse bands; femora often partly dark proximally. Male pygofer and anal apparatus as in Text-figs. 2600, 2601, genital valve and plates as in Text-fig. 2602, styles as in Text-fig. 2603, connective as in Text-fig. 2604, aedeagus as in Text-figs. 2605, 2606, 1st-3rd abdominal sterna in male as in Text-figs. 2607, 2608. Caudal border of 7th abdominal sternum in female almost straight or medially somewhat convex with a small median incision. Overall length of males 4.7-5.4 mm, of males 5.8-6.8 mm. – Nymphs with 4 complete rows of hairs on abdomen; vertex distinctly wider than long; abdomen with 4 more or less distinct longitudinal bands; pattern of dorsal surface (blackish) brown (Vilbaste, 1982).

Distribution. Not uncommon in Denmark, found in most districts. – Not uncommon in southern Sweden, found in Sk., Bl., Sm., Öl., Gtl., Boh., Upl. – Norway: recorded from VE: Berger (Ossiannilsson), and VAy: Kristiansand (Warloe). – Scarce in southern and south-western East Fennoscandia, found in Al, Ab, N, St. – England, France, Netherlands, German D.R. and F.R., Bulgaria, Romania, Albania, Hungary, Italy, Poland, Estonia, Latvia, s. Russia, Ukraine, Georgia, Anatolia, Cyprus, Moldavia, Tunisia, Afghanistan, Azerbaijan, Mongolia; Nearctic region.

Biology. On *Juncus maritima* and *Scirpus maritimus* (Kuntze, 1937; Ossiannilsson, 1947b). Halobiont, On *Scirpus* spp. (Linnavuori, 1952a). On *Scirpus maritimus* and *Phragmites communis* (Ribaut, 1952, Linnavuori, 1969a). Adults in July-September.

Genus *Parapotes* Emeljanov, 1975

Paranastus Emeljanov, 1972: 107 (nec Roewer, 1943).
    Type-species: *Paramesus reticulatus* Horváth, 1897, by original designation.
Parapotes Emeljanov, 1975: 390 (n. n.).
    Type-species: *Paramesus reticulatus* Horváth, 1897, by original designation.

In exterior structure resembling *Paramesus*. Anteclypeus parallel-sided. Male pygofer elongate, with a ventral process. Genital valve triangular, elongate. Genital plates apically broadly cut off, laterally slightly concave. Styles elongate with small apical apophyse. Aedeagus with small socle, apically with a pair of slender pointed processes, phallotreme terminal. In front of socle of aedeagus there is a large phragmoid body with a complicated structure (Text-figs. 2610, 2611). Caudal margin of 7th abdominal sternum in female with a short, sharply angular median projection. In Northern Europe one species.

Text-figs. 2609-2618. *Parapotes reticulata* (Horváth). – 2609: male pygofer and anal apparatus from the left (outline of phragma structure dotted); 2610: phragma structure inside male pygofer from the left; 2611: same from below; 2612: male genital plates and valve from below; 2613: right genital style from above; 2614: connective; 2615: aedeagus from the left; 2616: aedeagus in ventral aspect; 2617: 1st abdominal sternum in male from above; 2618: 2nd and 3rd abdominal sterna in male from above. Scale: 0.1 mm.

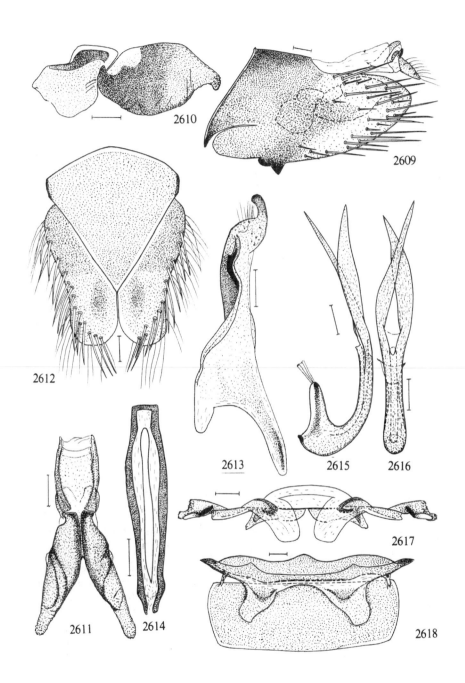

2610

2609

2612

2613

2615

2616

2611

2614

2617

2618

797

379. **Parapotes reticulata** (Horváth, 1897)
Text-figs. 2609-2618.

Paramesus reticulatus Horváth, 1897: 628.

Head anteriorly smoothly arched. Frontoclypeus blackish brown with narrow dirty yellowish transverse streaks, upper margin jet black; a lighter yellow transverse band near lower margin. Anteclypeus dark brownish, proximally with a diffusely delimited lighter transverse band. Genae and lora largely dirty yellow with more or less extended fuscous surface. Vertex frontally with a transverse depression, sordid yellow with two more or less distinct fuscous transverse bands. Pronotum and scutellum sordid yellow with indistinct whitish and fuscous markings. Fore wings in clavus and apical half of corium with some secondary cross-veins, brownish yellow, veins light, dark-bordered, light spots present around distal ends of claval veins and around some transverse veins. In male, thoracic venter and abdomen largely black, genital valve and genital plates with light margins, apical 3/5 rusty yellowish. Venter of female light, extensively dark banded and spotted, abdominal dorsum largely black. Femora dark banded, setae of tibiae arising from dark spots. Male pygofer and anal apparatus as in Text-fig. 2609, phragmoid body as in Text-figs. 2610, 2611, male genital plates and valve as in Text-fig. 2612, styles as in Text-fig. 2613, connective as in Text-fig. 2614, aedeagus as in Text-figs. 2615, 2616, 1st-3rd abdominal sterna in male as in Text-figs. 2617, 2618. Overall length of one male 5.04 mm, of one female 6.41 mm. – The males here described (one dissected, one measured) were borrowed from the Hungarian Natural History Museum, by courtesy of Dr. A. Soós.

Distribution. Not found in Denmark, Norway and East Fennoscandia. – Very rare in Sweden, only one female found in Bl.: Förkärla, Vambåsa 14.IX.1958, by N. Gyllensvärd. – Hungary, Slovakia, Romania, German D.R., Poland, s. Russia, Iran.

Biology. "Auf sumpfigem Gelände" (Haupt, 1935).

Genus *Paralimnus* Matsumura, 1902

Paralimnus Matsumura, 1902: 386.
Type-species: *Paralimnus fallaciosus* Matsumura, 1902, by subsequent designation.

Body elongate. Head wider than pronotum. Transition between vertex and face fairly abrupt. Anteclypeus parallel-sided. Frontal margin of head rounded angular. Ocelli close to eyes. Pronotum laterally not carinate. Anterior tibiae with one seta in the inner, 4 in the outer dorsal row. Apical part of fore wing symmetrically rounded. Anal tube of male dorsally sclerotized. Male pygofer lobes ventrally with a process. Genital plates elongate, medially mutually contiguous, lateral margins straight or concave, their macrosetae arranged in a row along side margin. Styles short. Connective tuning-fork-shaped, stem long. In Northern Europe two species.

Key to species of *Paralimnus*

1   Face above with a narrow arched black transverse line parallel with upper margin. In male, no black spots present

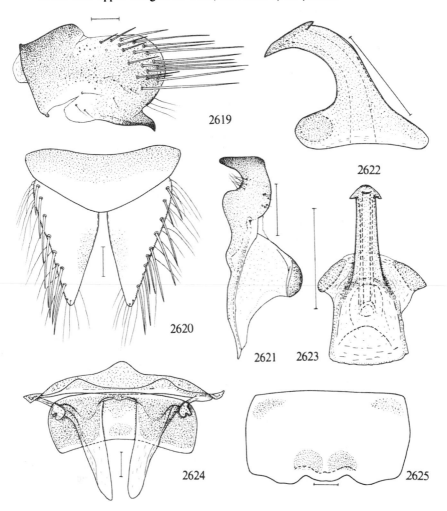

Text-figs. 2619-2625. *Paralimnus phragmitis* (Boheman). – 2619: left pygofer lobe in male from the
left; 2620: male genital plates and valve from below; 2621: right genital style from above; 2622:
aedeagus from the left; 2623: aedeagus in ventral aspect; 2624: 2nd and 3rd abdominal sterna in
male from above; 2625: 7th abdominal sternum in female from below (depressed under
coverglass). Scale: 0.1 mm.

near costal border. Aspect as in Plate-fig. 192............:... 380. *phragmitis* (Boheman)
–   Face above with a broad black transverse band extending
       from eye to eye. Fore wing of male with two black spots at
       costal border, viz., one near middle and one half-way be-
       tween middle and apex (Plate-fig. 193) .... 381. *rotundiceps* (Lethierry).

380. *Paralimnus phragmitis* (Boheman, 1847)
   Plate-fig. 192, text-figs. 2619-2625.

Thamnotettix phragmitis Boheman, 1847b: 265.

Straw-coloured, dull shining. Head frontally obtusely angular, above with an unin-
terrupted black arched line near and parallel with the light fore border, caudally of this
with two large orange-coloured patches or an indistinctly delimited orange-coloured
transverse band from eye to eye. Face along upper border with a narrow black arched
transverse line. Frontoclypeus with brownish yellow transverse streaks. Pronotum with
two orange-coloured patches at fore border, and four smaller spots in a transverse row
caudally of middle. Scutellum with the usual spots on lateral corners and a median double spot orange-coloured. Veins of fore wings pale except on a few dark stretches, dark-
bordered. Three oblique dark streaks between costal border and the nearest
longitudinal vein in apical half of fore wing are especially conspicuous. Apical part of
fore wing largely fumose, dark-bordered and with a pair of small dark spots (Plate-fig.
192). Legs longitudinally striped, tibiae fuscous spotted, posterior tarsi transversely banded, claws brownish. Thoracic venter in males usually largely dark, in females lighter;
abdominal tergum black with light lateral and segmental margins, venter more or less
extensively transversely banded or spotted. Pygofer lobes in male as in Text-fig. 2619,
genital plates and valve as in Text-fig. 2620 (but plates with inner margins contiguous in
repose), styles as in Text-fig. 2621, aedeagus as in Text-figs. 2622, 2623, 2nd and 3rd
abdominal sterna in male as in Text-fig. 2624, 7th abdominal sternum in female as in
Text-fig. 2625. Overall length of males 3.8-4.6 mm, of females 4.2-5.3 mm. – In nymphs,
frons immediately beneath vertex with one broad brownish cross-band, pattern of
dorsal surface also brownish; longitudinal bands of dorsal surface irregular, with
toothed margins (Vilbaste, 1982).

   Distribution. Common in Denmark. – Fairly common in southern and central
Sweden, Sk.-Hls. – So far not found in Norway. – Fairly common in southern and cen-
tral East Fennoscandia, Al, Ab, N-Om; Kr. – England, France, Netherlands, German
D.R. and F.R., Austria, Bohemia, Moravia, Slovakia, Hungary, Italy, Romania,
Yugoslavia, Poland, Estonia, Latvia, n. and s. Russia, Ukraine, Armenia, Moldavia,
Georgia, Israel, Iran, Altai Mts., m. Siberia, Azerbaijan.

   Biology. In "Hochmooren, Uferzone; an Schilf" (Kuntze, 1937).On *Phragmites communis* (Sahlberg, 1871; Ossiannilsson, 1947b; Ribaut, 1952; Wagner & Franz, 1961;
Schiemenz, 1976). In the *Phragmites* zone and the *Scirpus Tabernaemontani – S.
maritimus* zones of seashores, and in quagmire marshes (Linnavuori, 1952a). Adults in
June-September.

381. *Paralimnus rotundiceps* (Lethierry, 1885)
Plate-fig. 193, text-figs. 2626-2631.

Deltocephalus rotundiceps Lethierry, 1885: 111.

Strawcoloured, dull shining. Head frontally rounded, along the light fore border with a black arcuate line often interrupted medially. Frontoclypeus above with a broad black transverse band. Laterally of the frontal suture this band is continuing in shape of two narrow parallel streaks reaching eye. Anteclypeus with a black spot. Markings on ver-

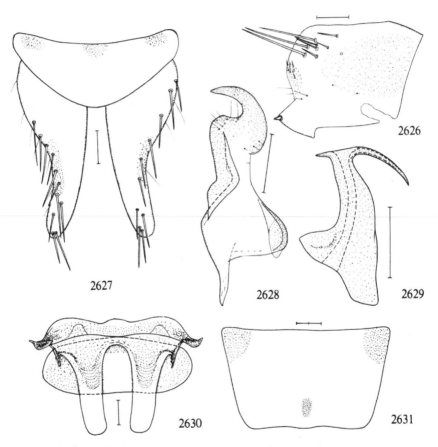

Text-figs. 2626-2631. *Paralimnus rotundiceps* (Lethierry). – 2626: right pygofer lobe in male from outside (seen as depressed under coverglass); 2627: male genital plates and valve from below; 2628: right genital style from above; 2629: aedeagus from the left; 2630: 2nd and 3rd abdominal sterna in male from above; 2631: 7th abdominal sternum in female from below (depressed under coverglass). Scale: 0.1 mm.

tex, pronotum and scutellum less clearly orange-coloured than in *phragmitis*, partly brownish, partly indistinct. Veins of fore wings only partly dark-bordered. Fore wing in male (Plate-fig. 193) with the following black markings missing in female: one large patch occupying wing apex, two smaller spots at costal border, the first approximately half-way between base and apex of the latter, the second half-way between that spot and apex of wing. Venter dark spotted, abdominal dorsum as in *phragmitis*. Tibiae dark spotted, apices of tibiae, basal segments of fore and middle tarsi and apices of 2nd segment of hind tarsi and 3rd segment of all tarsi, usually black or at least dark. Male pygofer lobes as in Text-fig. 2626, genital valve and plates as in Text-fig. 2627 (but median borders of plates contiguous in repose). Style as in Text-fig. 2628, aedeagus as in Text-fig. 2629, 2nd and 3rd abdominal sterna in male as in Text-fig. 2630, 7th abdominal sternum in female as in Text-fig. 2631. Overall length ($\male\female$) 4.0-5.5 mm.

Distribution. Not found in Denmark. – Rare in Sweden, only found in Dlsl.: Holm, Vita Sannar 18.VIII.1942 (Ander). – Norway: AK: Oslo (Siebke); VAy: Kristiansand (Warloe). – East Fennoscandia: only found in Om: Jakobstad 11.VII.1940 (Håkan Lindberg). – France, Belgium, Switzerland, Austria, Italy, Romania, Yugoslavia, Hungary, Georgia.

Biology. On *Phragmites* (Haupt, 1935, Lindberg, 1947); "sur les *Carex*" (Ribaut, 1952). Adults in July-September.

Genus *Metalimnus* Ribaut, 1948

Metalimnus Ribaut, 1948: 59.
Type-species: *Deltocephalus formosus* Boheman, 1845, by original designation.

Related to *Paralimnus*. Fore wings apically truncate or asymmetrically rounded. Vertex almost as long as pronotum or even longer, frontally angular or rounded angular. Transition between vertex and face abrupt. Male pygofer caudally on each side with a row of short and stout macrosetae. Connective elongate, stem long, branches apically fused. Anal tube dorsally well sclerotized. In Fennoscandia two species.

Key to species of *Metalimnus*

1 Head frontally rounded angular. Vertex with two rounded orange-coloured or brownish spots, also pronotum with rounded spots. Male genital plates apically rounded. 7th abdominal sternum in female as in Text-fig. 2637 .... 382. *formosus* (Boheman)
– Head frontally angular. Vertex, pronotum and scutellum longitudinally orange-banded. Male genital plates apically acutely angular. 7th abdominal sternum in female as in Text-fig. 2644 .... 383. *marmoratus* (Flor).

382. *Metalimnus formosus* (Boheman, 1845)
Plate-figs. 194, 215, text-figs. 2632-2637.

Deltocephalus formosus Boheman, 1845b: 155.
Scaphoideus formosus confluens Lindberg, 1924a: 18.

Sordid whitish yellow, above with orange-coloured, milky-white and brown markings, shining. Head as seen from above almost rectangularly produced, vertex fairly plain with two orange-coloured or brownish spots. Near fore border of head on each side often two brownish streaks parallel with same. Face above with two transverse brownish arched lines; transverse streaks on frontoclypeus more or less dark brown.

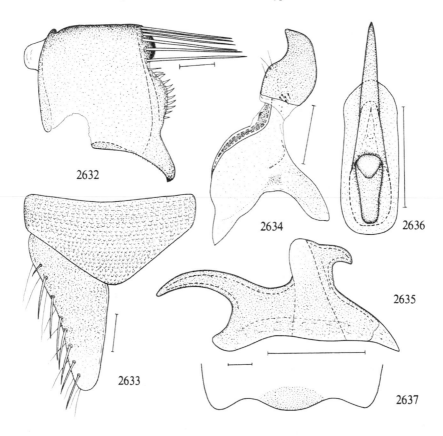

Text-figs. 2632-2637. *Metalimnus formosus* (Boheman). – 2632: left pygofer lobe in male from the left; 2633: genital valve and right genital plate from below; 2634: right genital style from above; 2635: aedeagus from the left; 2636: aedeagus in ventral aspect; 2637: 7th abdominal sternum in female from below (depressed under coverglass). Scale: 0.1 mm.

Pronotum near fore border with two orange-coloured or black-brown spots, caudally with 4 similar spots arranged in a transverse row. In strongly pigmented specimens, these may be enclosed in a fuscous transverse band. Caudal border of pronotum often milky white. Scutellum largely orange-coloured. In *f. confluens* (Lindb.), spots on vertex and scutellum and median spots on pronotum fusing, forming a longitudinal band. – Fore wing veins light, margined with fuscous or reddish. In strongly coloured specimens (males), dark markings partly extended forming large black-brown spots, interjacent surface partly milky white. Thus, a large irregular zigzag-shaped patch is often present in clavus, a somewhat smaller spot in corium extending obliquely from claval suture to costal border, and one spot at costal border distally of middle; apical part of fore wing sometimes largely dark. Our Plate-fig. 215 represents a moderately pigmented female. Femora transversely banded, middle and hind tibiae with large black spots. Abdomen usually largely black. Male pygofer lobes as in Text-fig. 2632, genital valve and plates as in Text-fig. 2633, styles as in Text-fig. 2634, aedeagus as in Text-figs. 2635, 2636, caudal border of 7th abdominal sternum in female as in Text-fig. 2637. Overall length of males 3.0-3.6 mm, of females 4.2-4.5 mm. – Nymphs with frons, immediately beneath vertex, with one broad orange cross-band, vertex and thorax with orange longitudinal bands, vertex approximately as long as wide between eyes, or somewhat shorter (Vilbaste, 1982).

Distribution. No reliable records from Denmark. – Widespread but scarce in Sweden, Sk.-Lu.Lpm. – So far not found in Norway. – Common in southern and central East Fennoscandia, found in Ab, N-ObN; Vib, Kr. – England, France, Belgium, Netherlands, German D.R. and F.R., Austria, Switzerland, Italy, Moravia, Slovakia, Bulgaria, Yugoslavia, Hungary, Romania, Poland, Estonia, Latvia, Lithuania, m. and s. Russia, Ukraine, Moldavia, Altai Mts., m. Siberia, Azerbaijan, Tuva, Mongolia, Manchuria.

Biology. On *Phragmites communis* and *Carex* ssp. (Sahlberg, 1871). In moist meadows (Lindberg, 1924). On *Glyceria* and *Pseudacorus* (Haupt, 1935). In "Flachmooren" (Kuntze, 1937). "Auf *Carex*-Weissmooren und gewissen Weissmoor-Reisermoor-komplexen" (Kontkanen, 1938). Tyrphobiont, in tall-sedge and short-sedge bogs, wet "rimpi" bogs and quagmire marshes (Linnavuori, 1952a). Adults in July-October.

383. *Metalimnus marmoratus* (Flor, 1861)
Plate-fig. 216, text-figs. 2638-2644.

Jassus (Deltocephalus) formosus var. *marmorata* Flor, 1861a: 236.
Deltocephalus stdli J. Sahlberg, 1871: 301.

Vertex medially slightly longer than pronotum. Transition between vertex and face acute-angled. Vertex, pronotum and scutellum whitish, vertex with 2, pronotum with 4 orange longitudinal bands, scutellum with 3 orange-coloured spots (Plate-fig. 216). Pigmentation of face more or less as in *formosus*. Markings of fore wings, thorax, abdomen and legs resembling those of *formosus*. Male pygofer lobes as in Text-fig.

2638, genital valve and plates as in Text-fig. 2639 (plates contiguous in repose), styles as in Text-fig. 2640, aedeagus as in Text-figs. 2641, 2642, 2nd and 3rd abdominal sterna in male as in Text-fig. 2643, 7th abdominal sternum in female as in Text-fig. 2644. Overall length (♂♀) 3.0-3.8 mm. – Nymphs as in *formosus* but vertex clearly longer (1.1-1.2 ×) than wide between eyes (Vilbaste, 1982).

Distribution. Not in Denmark, Sweden and Norway. – Fairly scarce in southern and central East Fennoscandia, found in Al, Ab, N-Kb; Vib, Kr. – German F.R., Poland,

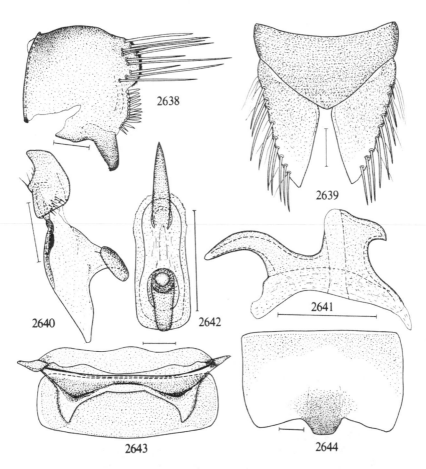

Text-figs. 2638-2644. *Metalimnus marmoratus* (Flor). – 2638: left pygofer lobe in male from the left; 2639: genital plates and valve from below; 2640: right genital style in male from above; 2641: aedeagus from the left; 2642: aedeagus in ventral aspect; 2643: 2nd and 3rd abdominal sterna in male from above; 2644: 7th abdominal sternum in female from below. Scale: 0.1 mm.

Estonia, Latvia, Lithuania, n. Russia, Ukraine, Altai Mts., Tuva, Mongolia, Maritime Territory.

Biology. On *Carex limosa* in bogs (Sahlberg, 1871). Stenotopic species, "ausschliesslich auf *Carex limosa*-Weissmooren" (Kontkanen, 1938). In tall-sedge bogs, wet "rimpi" bogs, quagmire marshes, cloudberry-*Sphagnum fuscum* bogs (Linnavuori, 1952a).

Genus *Arocephalus* Ribaut, 1946

Arocephalus Ribaut, 1946: 85.
Type-species: *Jassus (Deltocephalus) longiceps* Kirschbaum, 1868, by original designation.
Ariellus Ribaut, 1952: 272 (subgenus).
Type-species: *Jassus (Deltocephalus) punctum* Flor, 1861, by original designation.

Small leafhoppers. Head frontally angular, wider than pronotum. Frontoclypeus longer than wide, anteclypeus broad, tapering towards apex. Anterior tibiae with 1 seta in inner, 4 in outer dorsal row. Anal tube in male well sclerotized dorsally. Male pygofer without processes. Genital plates short, their macrosetae arranged in a row along lateral margin. Connective long, hairpin-shaped, stem short. Aedeagus with appendages near apex symmetrical. Phallotreme ventral. In Denmark and Fennoscandia three species.

Key to subgenera of *Arocephalus*

1   Phallotreme situated much closer to apex than to base of aedeagus. Male pygofer dorsally excised .... *Arocephalus* s. str.
–   Phallotreme situated much closer to base than to apex of aedeagus. Male pygofer not excised .... *Ariellus* Ribaut.

Key to species of *Arocephalus*

1   Larger, overall length > 3.4 mm. Male pygofer dorsally black. Shaft of aedeagus strongly compressed laterally, with two pairs of appendages (Text-figs. 2662, 2663) .... 386. *longiceps* (Kirschbaum)
–   Smaller, overall length < 3.4 mm. Male pygofer dorsally yellowish. Shaft of aedeagus not compressed laterally, with one pair of appendages 2
2 (1) Fore wing with a black spot at apex of 3rd (inner) subapical cell (Plate-fig. 195). Anal tube in male dorsally black. Aedeagus as in Text-figs. 2648, 2649. 7th abdominal sternum in female with caudal border medially slightly concave (Text-fig. 2652) .... 384. *punctum* (Flor)

– Fore wing without a black spot at apex of 3rd subapical cell.
   Anal tube in male dorsally light. Aedeagus as in Text-figs. 2656,
   2657. Caudal border of 7th abdominal sternum in female me-
   dially slightly convex (Text-fig. 2658) .... 385. *languidus* (Flor).

384. *Arocephalus (Ariellus) punctum* (Flor, 1861)
   Plate-fig. 195, text-figs. 2645-2652.

Cicada costalis Wallengren, 1851: 254 (nec Fallén, 1826).
Jassus (Deltocephalus) punctum Flor, 1861a: 247.

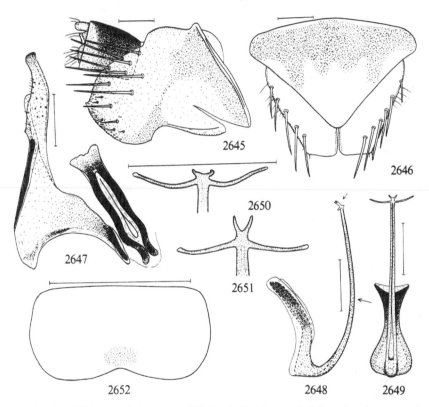

Text-figs. 2645-2652. *Arocephalus punctum* (Flor). – 2645: right pygofer lobe and anal apparatus in male from the right; 2646: genital plates and valve from below; 2647: right genital style and connective from above; 2648: aedeagus from the left; 2649: aedeagus in ventral aspect (as seen in direction of longer arrow in 2648); 2650: apex of aedeagus in ventral aspect (in direction of longer arrow in 2648); 2651: same in direction of shorter arrow in 2648; 2652: 7th abdominal sternum in female from below (depressed under coverglass). Scale: 0.5 mm for 2652, 0.1 mm for the rest.

Deltocephalus costalis Fieber, 1869: 204 (nec Fallén, 1826).
Deltocephalus paucinervis J. Sahlberg, 1871: 318.

Ivory-white, pale light yellow, shining. Head above near apex with a pair of more or less distinct brownish oblique streaks, behind these with two indistinct brownish yellow longitudinal bands. Pronotum with two or three pairs of brownish yellow longitudinal bands. Wing dimorphous in both sexes, prevailingly sub-brachypterous; macropters with fore wings about as long as abdomen, hind wings slightly shorter; sub-brachypters with fore wings usually reaching caudal margin of 8th abdominal tergum, hind wings only 1/2-2/3 of fore wings. Apical cells and distal ends of subapical cells usually dark-edged, more extensively in macropters. Thoracic venter largely light, black pigmentation of abdomen much varying in extension. Posterior tibiae black spotted. Male anal tube dorsally black, female pygofer with a large black spot on each side. Male pygofer and anal apparatus as in Text-fig. 2645, genital valve and plates as in Text-fig. 2646, connective and style as in Text-fig. 2647, aedeagus as in Text-figs. 2648-2651, 7th abdominal sternum in female as in Text-fig. 2652. Overall length of sub-brachypterous males 2.5-3.1 mm, of sub-brachypterous females 2.8-3.2 mm, of macropters 2.4-2.9 mm. – Nymphs with 4 complete rows of hairs on abdomen, pattern of frons brownish or indistinct, vertex yellowish green; thorax brownish; abdomen more or less uniformly brown (Vilbaste, 1982).

Distribution. Widespread and common in Denmark, also in Sweden up to Vb. – Norway: found Ø: Rakkestad (Holgersen); AK: Oslo (Reuter); On: Dovre, Toftemo (Siebke); VAy: Lyngdal (Holgersen); TRy: Skjervøy (Zetterstedt). – Fairly scarce in southern East Fennoscandia, found in Al, Ab, N, Ta; Kr. – England, Scotland, Wales, France, Spain, Belgium, Netherlands, German D.R. and F.R., Austria, Italy, Bohemia, Moravia, Slovakia, Bulgaria, Poland, Estonia, Latvia, Lithuania, n. Russia, Tunisia.

Biology. Especially among *Nardus stricta* (Sahlberg, 1871). In "Salzstellen, Sandfeldern" (Kuntze, 1937). In dry meadows but also in fens (Vilbaste, 1974). Hibernates in egg stage, 2 generations (Remane, 1958). One generation (Schiemenz, 1969a and b).

385. *Arocephalus (Arocephalus) languidus* (Flor, 1861)
Text-figs. 2653-2658.

Jassus (Deltocephalus) languidus Flor, 1861a: 246.
Jassus (Deltocephalus) pusillus Kirschbaum, 1868b: 136.
Jassus productus Thomson, 1869: 72.
Deltocephalus haupti Lindberg, 1924b: 40.

Fairly like *punctum*, still smaller, head relatively slightly shorter. Yellowish white to light yellow, shining. Head without distinct markings. Wing dimorphous in both sexes, usually sub-brachypterous. Fore wings in macropters (one female only available) longer than abdomen, hind wings slightly shorter than fore wings. Fore wings in sub-brachypterous males about as long as abdomen, in sub-brachypterous females usually reaching caudal margin of 8th abdominal tergum, hind wings in both sexes usually less than half as long as fore wings. Fore wings unicolorous, or cells more or less distinctly dark-edged, especially in males and macropters. Thoracic venter in males and macropters largely black, in sub-brachypterous females light, legs entirely light or dark striped and transversely banded. Hind tibiae dark spotted. Abdomen in males black with light lateral and segmental borders, in sub-brachypterous females usually entirely light. Male pygofer as in Text-fig. 2653, genital plates and valve as in Text-fig. 2654, style as in Text-fig. 2655, aedeagus as in Text-figs. 2656, 2657, 7th abdominal sternum in female as in Text-fig. 2658. Overall length of sub-brachypterous males 2.2-2.6 mm, of sub-brachypterous females 2.4-2.9 mm, of one macropterous female 2.7 mm. – Nymphs pale grey (brownish yellow in alcohol) with pale olive pattern, thorax longitudinally banded; abdomen speckled dark; for the rest as in *punctum* (Vilbaste, 1982).

Distribution. So far not found in Denmark, nor in Norway. – Scarce, locally abundant in Sweden, found in Sk., Öl., Gtl., Ög., Vg., Nrk., Sdm., Upl., Vstm. – Rare in East Fennoscandia, found in Ab: Karislojo (J. Sahlberg); Kb: Hammaslahti (Kontkanen);

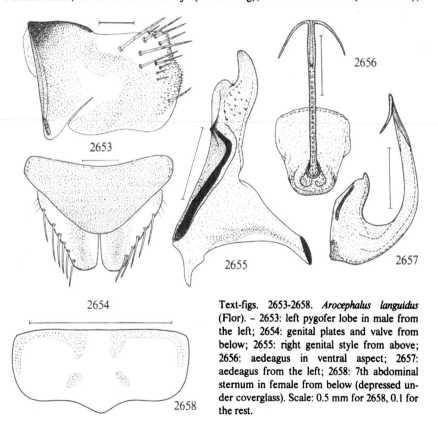

Text-figs. 2653-2658. *Arocephalus languidus* (Flor). – 2653: left pygofer lobe in male from the left; 2654: genital plates and valve from below; 2655: right genital style from above; 2656: aedeagus in ventral aspect; 2657: aedeagus from the left; 2658: 7th abdominal sternum in female from below (depressed under coverglass). Scale: 0.5 mm for 2658, 0.1 for the rest.

Kr, Lr. – France, Belgium, German D.R. and F.R., Switzerland, Austria, Hungary, Bohemia, Moravia, Slovakia, Italy, Bulgaria, Romania, Poland, Estonia, Latvia, Lithuania. m. Russia, Ukraine, Moldavia, Georgia, Kazakhstan, Altai Mts., m. and w. Siberia, Tuva, Mongolia, Tunisia.

Biology. In "besonnten Hängen" (Kuntze, 1937). "Hauptart der Haargras-Hügelsteppen" (Schiemenz, 1969a). Hibernates in egg-stage, in Central Europe 2 generations (Schiemenz, 1969b). In dry meadows (Vilbaste, 1974). Monovoltine dominant in the Seslerietum, hibernates in egg-stage (Müller, 1978). Adults in June-August.

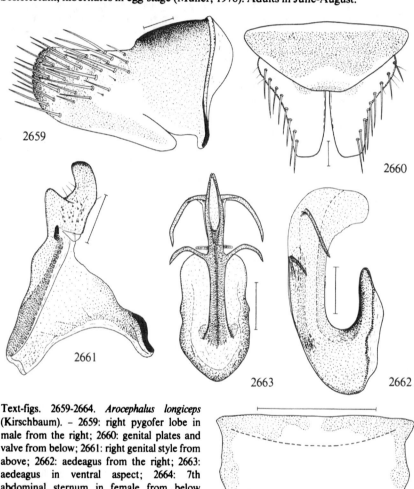

Text-figs. 2659-2664. *Arocephalus longiceps* (Kirschbaum). – 2659: right pygofer lobe in male from the right; 2660: genital plates and valve from below; 2661: right genital style from above; 2662: aedeagus from the right; 2663: aedeagus in ventral aspect; 2664: 7th abdominal sternum in female from below (depressed under coverglass). Scale: 0.5 mm for 2664, 0.1 mm for the rest.

386. *Arocephalus (Arocephalus) longiceps* (Kirschbaum, 1868)
    Text-figs. 2659-2664.

Jassus (Deltocephalus) longiceps Kirschbaum, 1868b: 135.
Deltocephalus linnei Fieber, 1869: 214.

Resembling certain *Psammotettix* spp. Vertex slightly shorter than pronotum, frontal margin almost rectangular. Both sexes macropterous, wings considerably longer than abdomen. Frontoclypeus brown with pale yellow transverse streaks, in lower part usually also a pale median line, on upper apex with a small black V-shaped marking. Anteclypeus with a T-shaped black marking. Vertex ivory-white with two broad ochraceous or pale brownish longitudinal bands, frontally with a pair of obliquely arched black streaks sometimes interrupted near middle; coronal suture black. Pronotum whitish with four ochraceous or brownish longitudinal bands. Scutellum orange spotted. Fore wings greyish with or without a reddish tinge, veins light, dark-bordered. Femora transversely and longitudinally black banded and spotted, tibiae brownish spotted. Male pygofer lobes as in Text-fig. 2659, male genital plates and valve as in Text-fig. 2660, styles as in Text-fig. 2661, aedeagus as in Text-figs. 2662, 2663, 7th abdominal sternum in female as in Text-fig. 2664. Overall length (♂♀) 3.45-4.2 mm.

Distribution. Very rare in Denmark, found only in B: Almindingen 1.VII.1974 (Trolle). – Not found in Sweden, Norway, and East Fennoscandia. – France, Spain, Belgium, Netherlands, German D.R. and F.R., Switzerland, Austria, Italy, Bohemia, Moravia, Slovakia, Bulgaria, Romania, Hungary, Greece, Poland, Georgia, Moldavia, Fergana Valley, Mongolia.

Biology. "An Waldgräsern" (Wagner & Franz, 1961). Hibernation takes place in the egg-stage, 2 generations. "In xerophilen bis hygrophilen Biotopen, wobei erstere bevorzugt werden". Adults (in Central Europe) end of May to end of July; early August to early November (Schiemenz, 1969b).

Genus *Psammotettix* Haupt, 1929

Psammotettix Haupt, 1929b: 262.
    Type-species: *Athysanus maritimus* Perris, 1857, by original designation.

Head wider than pronotum. Frontoclypeus elongate. Anteclypeus narrowing in lower half. Transition between frontoclypeus and vertex rounded angular. Sides of pronotum short, not carinate. Wings well developed in most species but some are wing-dimorphous. Fore tibiae with 1 seta in inner, 4 in outer dorsal row. Male pygofer dorsally deeply incised, without processes. Male anal tube sclerotized only laterally. Genital valve trapezoidal (Text-figs. 2672, 2697). Genital plates very short, their macrosetae arranged in a row along lateral border. Styles usually slightly dilated apically. Connective long and narrow, stem long, branches approximately parallel (Text-fig. 2676). Aedeagus without appendages, shaft more or less spatulate. Socle of aedeagus connected with a transverse laminate phragmoid structure. Phallotreme large, ventral. Caudal margin of 7th abdominal sternum of female in most (all our) species straight without processes or incisions. In many species only males can be identified with certainty. Nymphs with 4 complete rows of hairs on abdominal dorsum. In Denmark and Fennoscandia 13 species.

Key to species of *Psammotettix*

1 Dorsum of fore body and fore wings greenish ....
.... 388. *cephalotes* (Herrich-Schäffer)
– Dorsum and fore wings not greenish 2
2 (1) Dark markings of fore wings extended, more or less distinctly arranged in two or three irregular transverse bands. Distal half of aedeagus with lateral brim very wide, in lateral aspect strongly convex (Text-fig. 2713) .... 399. *poecilus* (Flor)
– Dark markings of fore wings, if present, not arranged as transverse bands. Shape of aedeagus different 3
3 (2) Large species, overall length 3.9-4.4 mm. Macropterous. Vertex comparatively short, anterior margin of head obtuse-angled (Text-fig. 2665). Aedeagus as in Text-figs. 2667, 2668 .... 387. *alienus* (Dahlbom)
– Smaller species. Vertex usually longer, anterior margin more or less rectangular (Text-figs. 2686, 2693, 2695) 4
4 (3) Fore wings very pale-coloured with a few strongly marked black spots not tending to form streaks. Aedeagus as in Text-figs. 2678, 2679. On coastal sand-hills .... 390. *sabulicola* (Curtis)
– Black markings of fore wings, if present, tending to form streaks along cells (Plate-fig. 196, text-figs. 2694, 2695) 5
5 (4) Shaft of aedeagus dorsoventrally compressed, in ventral aspect very broad (Text-fig. 2675) .... 389. *confinis* (Dahlbom)
– Shape of aedeagus different 6
6 (5) Vertex without dark markings. Small species, length 2.5-2.8 mm. Aedeagus as in Text-figs. 2710, 2711 .... 248. *pallidinervis* (Dahlbom)
– Vertex normally with dark markings 7
7 (6) Shaft of aedeagus in lateral aspect with a conspicuous bulge dorsally (Text-figs. 2681, 2684, 2688, 2691, 2700) *(putoni* group) 8
– Shaft of aedeagus in lateral aspect with dorsal outline almost straight or faintly curved 13
8 (7) Larger species, length of posterior tibiae 1.6-1.7 mm 9
– Smaller, length of posterior tibiae 1.3-1.5 mm 11
9 (8) Vertex convex, especially in female. Interocellar line of vertex (see description of *alienus*) not interrupted, caudally coalescing with dark surface of disk of vertex. Usually subbrachypterous, fore wings without or with sparse mark-

ings, veins concolorous. Dorsal bulge on aedeagal shaft
somewhat abrupt (Text-fig. 2684) .... 392. *putoni* (Then)
–   Vertex flat. Dorsal bulge on shaft of aedeagus smoothly
marked (Text-figs. 2681, 2700) 10
10 (9) Macropterous. Dark markings on vertex usually not
coalescing. Interocellar line interrupted. Fore wings usually with strong markings, veins lighter than cell mem-
brane, transverse veins whitish .... 391. *nodosus* (Ribaut)
–   Usually sub-brachypterous. In macropters, fore wings only
slightly longer than abdomen. Interocellar line interrupted
or continuous (Text-figs. 2693, 2695). Markings of vertex
often confluent or indistinct. Pigmentation of fore wings
usually moderately developed .... 395. *dubius* Ossiannilsson
11 (8) Dorsal bulge on shaft of aedeagus abrupt (Text-fig. 2688).
Interocellar lines not interrupted, coalescing with nearest
dark patch on vertex. Dark colour on vertex usually
divided into patches (Text-fig. 2686). Usually sub-brachypterous, fore wings often (not always) strongly black marked,
subcostal cell always light .... 393. *albomarginatus* W. Wagner
–   Dorsal bulge of shaft of aedeagus smoothly marked 12
12 (11) Macropterous. Body structure delicate. Interocellar li-
nes interrupted, vertex for the rest usually without mar-
kings. Length of aedeagus about 0.25 mm .... 394. *excisus* (Matsumura)
–   Usually sub-brachypterous. Body robust. Markings of
head and fore wings varying. Length of aedeagus about
0.29 mm .... 395. *dubius* Ossiannilsson
13 (7) Smaller, length 2.5-3.0 mm. Ratio "bowl" of aedeagus (part
situated distally of proximal margin of phallotreme): total
length of aedeagus (as seen from below: Text-fig. 2705) =
about 0.33 .... 396. *frigidus* (Boheman)
–   Larger, length 3.0-3.4 mm. Ratio "bowl": total length of
aedeagus (seen as in Text-fig. 2708) = about 0.39 ....
.... 397. *lapponicus* (Ossiannilsson).

Note. The insect described by Ribaut (1952) as *Psammotettix striatus* (L.) has not
been found in Sweden so far.

387. *Psammotettix alienus* (Dahlbom, 1851)
Plate-fig. 196, text-figs. 2665-2668.

Thamnotettix aliena Dahlbom, 1851: 187.
Jassus (Deltocephalus) breviceps Kirschbaum, 1868b: 132.
Psammotettix striatus Razvyazkina & Pridantzeva, 1968: 690, p.p. (an Linné, 1758?).
Swedish: randig dvärgstrit.

Head obtuse-angular, near fore border with a pair of fine evenly curved transverse lines extending between apex of head and a point below each ocellus. Caudally of these there is a pair of "interocellar lines" between a point near apex of head and each ocellus; in the present species this line is interrupted near its middle (Text-fig. 2665). Vertex caudally of interocellar lines with usually distinctly separate spots; these are often quite small and dark. Pronotum usually with four (or six) distinct dark longitudinal bands. Macropterous, fore wings considerably longer than abdomen, veins usually partly dark-bordered. In dark specimens, the cells may be largely brownish or fuscous. Abdomen black, laterally sordid yellow, or largely light, with more or less extended black markings. Legs greyish yellow, with black spots and streaks. Male genital style as in Text-fig. 2666, aedeagus as in Text-figs. 2667, 2668. – Nymphs with body pattern brown; vertex shorter than wide; fore body with longitudinal bands, abdomen distinctly darker than thorax (Vilbaste, 1982); for illustrations, see Tullgren (1925).

Distribution. Widespread and common in Denmark, as well as in southern and central Sweden (Sk.-Dlr., also found in Ås.Lpm. and T.Lpm.). – Distribution in Norway so

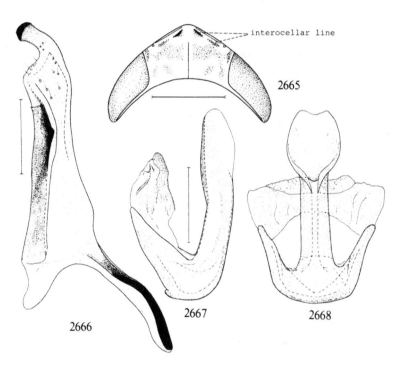

Text-figs. 2665-2668. *Psammotettix alienus* (Dahlbom). – 2665: head of female from above; 2666: right genital style in male from above; 2667: aedeagus from the left; 2668: aedeagus in ventral aspect. Scale: 0.5 mm for 2665, 0.1 mm for the rest.

814

far not investigated; I have seen specimens from AK: Asker (Taksdal) and Ås (Rygg), and from Bø: Ringerike (Warloe). – Fairly common in southern and central East Fennoscandia, found in Al, Ab-Om; Kr. – Widespread in the Palaearctic region, also present in the Nearctic.

Biology. In cultivated fields (Linnavuori, 1952a). Hibernation in the egg-stage (Müller, 1957; Raatikainen, 1971). "An Gräsern und Kräutern; Kulturfolger" (Wagner & Franz, 1961). In central Europe 2 generations a year (Schiemenz, 1969b). "In xerophilen, mesophilen und hygrophilen Biotopen, ohne eine derselben zu bevorzugen" (Schiemenz, 1969b). In dry meadows, in fields, dry wood margins, clear-cut areas, etc. (Vilbaste, 1974). "Very frequent and abundant in leys and pastures; from these it migrated in large numbers to cereal crops. . . . The first adults were caught in leys on 14.VI.60 and flight occurred between 15.VI. and 18.VII. Flight was observed when the daily maxima reached 18°C, and 81% of the specimens were caught in days warmer than the average" (Raatikainen & Vasarainen, 1973). "Nowadays, almost exclusively in fields and meadows and on wasteland. Its host plants are grasses and cereals, e.g. oats" (Raatikainen & Vasarainen, 1976). Adults in June-October.

Economic importance. *Psammotettix alienus* is a vector of wheat dwarf virus (Lindsten & al., 1970).

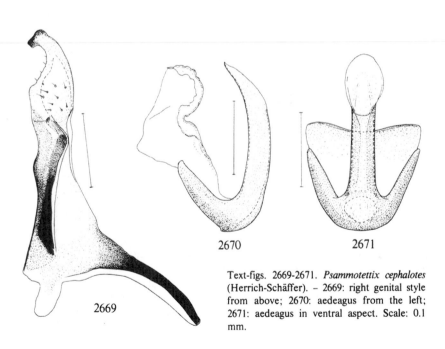

2670   2671

2669

Text-figs. 2669-2671. *Psammotettix cephalotes* (Herrich-Schäffer). – 2669: right genital style from above; 2670: aedeagus from the left; 2671: aedeagus in ventral aspect. Scale: 0.1 mm.

815

388. *Psammotettix cephalotes* (Herrich-Schäffer, 1834)
Text-figs. 2669-2671.

Jassus cephalotes Herrich-Schäffer, 1834c: 6.
Jassus (Deltocephalus) citrinellus Kirschbaum, 1838b: 134.

Upper side unicolorous light green or yellowish green, only vertex sometimes with indistinct traces of markings. Fore margin of head almost rectangular. Frontoclypeus with brownish transverse streaks. Thoracic venter and abdomen largely black, lateral borders of abdomen and genital segments (♂♀) light, also 7th abdominal sternum of female often light. Fore wings largely semi-transparent, veins yellow or green. Macropterous, fore wings distinctly (1/5-1/4) longer than abdomen. Male genital styles as in Text-fig. 2669, aedeagus as in Text-figs. 2670, 2671. Overall length (♂♀) 2.7-3.7 mm. – Nymphs with vertex distinctly longer than wide; entirely pale, brownish yellow, usually with a slightly paler middle line which may have darker margins; eyes grey or greenish yellow; body sticky (Vilbaste, 1982).

Distribution. Widespread but not very common in Denmark. – Moderately common in southern and central Sweden, found Sk.-Dlr., Jmt. – Apparently rare in Norway, found in TEy: Heistad (Holgersen). – East Fennoscandia: fairly common in Al, for the rest scarce and sporadic, found in Ab, Kb; Kr. – England, Scotland, Ireland, France, Spain, Belgium, Netherlands, German D.R. and F.R., Austria, Switzerland, Italy, Albania, Bohemia, Moravia, Slovakia, Bulgaria, Hungary, Yugoslavia, Poland, Estonia, Latvia, Lithuania, n. and m. Russia, Ukraine, Anatolia, Moldavia, Algeria.

Biology. In "Salzstellen" (Kuntze, 1937). Hibernates in the egg-stage (Müller, 1957; Schiemenz, 1969b). "Auf Weiden und Wiesen besonders im Gebirge, in der Ebene seltener und dort nur an xerothermen Stellen. Nach Steiner in Nordtirol auf Trockenweiden, auf mehr oder weniger feuchten Wiesen und im Molienetum. Auch im Gebiete sowohl auf trockenen als auch auf sehr nassen Grünlandflächen" (Wagner & Franz, 1961). In central Europe 2 generations (Schiemenz, 1969b). In various meadows, fens, margins of woods, etc. (Vilbaste, 1974). Adults in June-September.

389. *Psammotettix confinis* (Dahlbom, 1851)
Text-figs. 2672-2676.

Deltocephalus confinis Dahlbom, 1851: 193.
Deltocephalus thenii Edwards, 1915: 208.
Deltocephalus spathifer Ribaut, 1925: 19.

Fore margin of head almost rectangular. Interocellar lines interrupted. Vertex caudally of these usually with 3 pairs of distinct dark spots. Extension and intensity of dark pattern much varying. In dark specimens there are four distinct longitudinal bands on pronotum, and well developed dark streaks along veins of fore wings. Macropterous, fore wings usually considerably longer than abdomen.Male genital valve and plates as in Text-fig. 2672, styles as in Text-fig. 2673, aedeagus as in Text-figs. 2674, 2675, connective as in Text-fig. 2676. Overall length ($\male\female$) 3.2-4.0 mm. – Nymphs as in *alienus*, abdomen approximately coloured as thorax (Vilbaste, 1982).

Distribution. Common in Denmark. Very common in Sweden, Sk.-Lu.Lpm. – Common and widespread in Norway, found in most regions up to TRi. – Very common in southern and central East Fennoscandia, found in Al, Ab, N-ObN; Ks; Vib. – Widespread in Europe and palaearctic part of Asia, also in the Nearctic region.

Biology. In "Stranddünen, besonnten Hängen, Wiesen" (Kuntze, 1937). "Auf recht verschiedenartigen ± trocknen Standorten mit Graswuchs" (Kontkanen, 1938). In dryish fields, moist sloping meadows, peaty meadows, cultivated fields, also in the drier meadow area of seashores (Linnavuori, 1952a). In the "Arrhenatheretum elatioris" (Marchand, 1953). Hibernates in egg-stage (Remane, 1958; Schiemenz, 1969b). In central Europe 2 generations (Schiemenz, 1969b). In various meadows, fens, high peat bogs, on river banks, lake and sea shores, in fields, on wood margins, woodland clearings, etc. (Vilbaste, 1974). Adults in April-October.

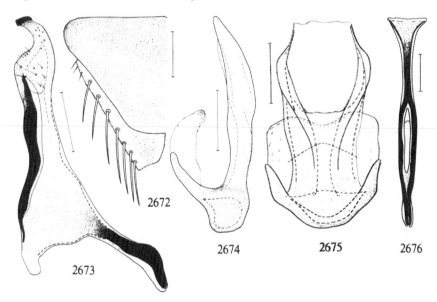

Text-figs. 2672-2676. *Psammotettix confinis* (Dahlbom). – 2672: genital plates and valve from below; 2673: right genital style from above; 2674: aedeagus from the left; 2675: aedeagus in ventral aspect; 2676: connective. Scale: 0.1 mm.

390. *Psammotettix sabulicola* (Curtis, 1837)
Text-figs. 2677-2679.

Aphrodes sabulicola Curtis, 1837a: pl. 633.

Deltocephalus arenicola J. Sahlberg, 1871: 343.

Head fairly obtuse-angular. Pale whitish yellow, shining. Interocellar lines of vertex each divided into a larger median and a smaller lateral spot; behind these, as usual, 3 pairs of brownish yellow spots. Pronotum with four (six) somewhat darker longitudinal bands. Fore wings usually with some very distinct blackish brown markings, the following being comparatively constant: 2 longitudinal streaks along claval commissure, one median and one distal; one spot in sutural clavus cell proximally of middle; one spot immediately distally of basal transverse vein connecting M with Cu. In addition there are a few spots in apical part of fore wing. Macropterous, but fore wings not or little longer than abdomen. Abdomen ventrally light with a median row of black spots more extended in males than in females. Male genital style as in Text-fig. 2677, aedeagus as in Text-figs. 2678, 2679. Overall length of males 3.3-3.9 mm, of females 3.6-4.1 mm.

Distribution. Widespread and common in Denmark. – Scarce, locally common in southern Sweden, found in Sk., Hall., Bl., Gtl., G. Sand. – Recorded from southern Norway by Strand (1906) but this statement has not been confirmed. Found in Fø: Sør-Varanger, Småstrammen and Grense Jakobselv in July, 1937, by Soot-Ryen. – Fairly scarce and sporadic in East Fennoscandia, along the coasts of the Baltic and the Gulf of Finland; found in Ka, St, Om, ObS; Vib, Kr. – England, Scotland, Wales, Ireland, France, Belgium, Netherlands, German D.R. and F.R., Italy, Bulgaria, Poland, Estonia, Latvia, Lithuania, n. Russia, Kazakhstan, Tunisia.

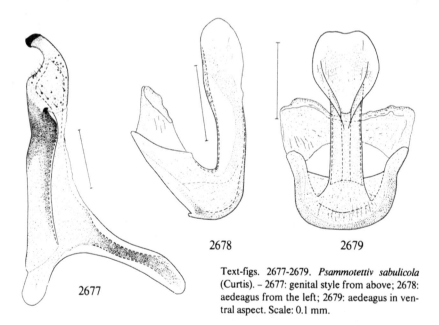

2678 2679 2677

Text-figs. 2677-2679. *Psammotettiv sabulicola* (Curtis). – 2677: genital style from above; 2678: aedeagus from the left; 2679: aedeagus in ventral aspect. Scale: 0.1 mm.

Biology. "Auf offenen Sandflächen (weissen Dünen) besonders an *Ammophila*" (Wagner, 1941). On sand-dunes with *Ammophila* and *Elymus* (Ossiannilsson, 1947b). In *Carex arenaria*-associations in "Binnendünen", hibernation takes place in the egg-stage; 2 generations (Remane, 1958). On coastal sand dunes, on *Elymus* (Vilbaste, 1974). Adults in June-August.

391. *Psammotettix nodosus* (Ribaut, 1925)
Text-figs. 2680-2682.

Deltocephalus nodosus Ribaut, 1925: 17.

Resembling *P. confinis*. Vertex flat. Dorsal pigmentation usually well developed. Markings of vertex usually distinct, not fused. Interocellar lines interrupted near middle, median part often enlarged into a triangular spot. Macropterous. Veins of fore wings whitish, dark-bordered. Length of posterior tibiae 1.6-1.7 mm. Male genital style as in Text-fig. 2680, aedeagus as in Text-figs. 2681, 2682. Overall length (♂♀) 3.2-3.6 mm. – Nymphs almost uniformly brown, abdomen with rows of pale spots; vertex distinctly longer than wide (Vilbaste, 1982).

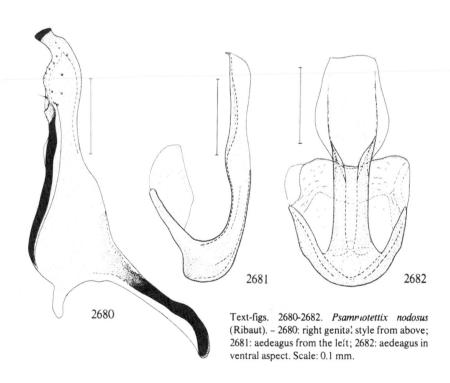

2680

2681

2682

Text-figs. 2680-2682. *Psammotettix nodosus* (Ribaut). – 2680: right genital style from above; 2681: aedeagus from the left; 2682: aedeagus in ventral aspect. Scale: 0.1 mm.

Distribution. Common in Denmark. – Common and widespread in Sweden, Sk.-T.Lpm. – Distribution in Norway imperfectly known; I have seen specimens from VAy, Ry, Ri, TRi; Huldén (1982) recorded *P. nodosus* from Fi. – Scarce and sporadic in East Fennoscandia, found in N, St, Oa, ObN, Ks, Li; Lr. – Widespread in Europe, also in Morocco and Moldavia.

Biology. On sandy soils, hibernation in egg-stage, 2 generations (Remane, 1958; Schiemenz, 1969b). "In Moorwiesen, auf *Eriophorum* und *Molinia* . . . im *Nardeto-Molinietum*" (Wagner & Franz, 1961). "Im Bereich der grauen Düne" (Remane, 1965). "In xerophilen und mesophilen Biotopen, wobei der Schwerpunkt des Auftretens im xerophilen Bereich liegt; auch in Hochmooren" (Schiemenz, 1969b). On *Festuca rubra* and *Agrostis* (Schaefer, 1973). Adults in June-October.

392. *Psammotettix putoni* (Then, 1898)
Text-figs. 2683-2685.

Deltocephalus putoni Then, 1898: 49 (nec *Psammotettix putoni* Ribaut, 1952: 244).
Deltocephalus halophilus Edwards, 1924: 53.

Vertex convex. Transition between vertex and frontoclypeus comparatively rounded.

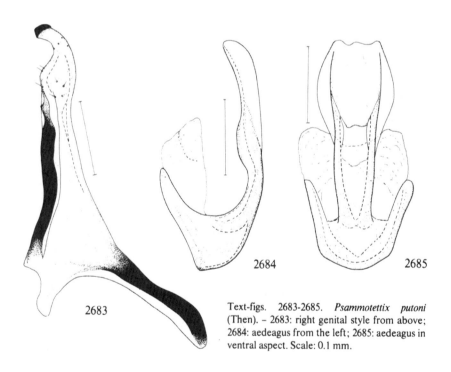

2684

2685

2683

Text-figs. 2683-2685. *Psammotettix putoni* (Then). – 2683: right genital style from above; 2684: aedeagus from the left; 2685: aedeagus in ventral aspect. Scale: 0.1 mm.

Usually sub-brachypterous with fore wings little longer (♂) than or almost as long (♀) as abdomen. Fore wings of macropters distinctly longer than abdomen. Light ochraceous to dark brown. Pigmentation usually faintly developed. Interocellar lines of vertex continuous. Vertex largely unicolorous brownish, markings indistinct, interocellar lines confluent with and being the anterior delimitation of the brown colour, a narrow median line and a spot around each ocellus remaining light. Fore wings often without markings, veins concolorous; if present, dark colour usually arranged in longitudinal streaks along veins, or only a dark spot present in outer claval cell 1/3 from base (var. *halophilus* Edwards). Apical cells often with a few dark markings. Male genital style as in Text-fig. 2683, aedeagus as in Text-figs. 2684, 2685. Length of posterior tibiae 1.6-1.75 mm. Overall length of sub-brachypterous males about 2.7 mm, of sub-brachypterous females about 3 mm, of macropterous males 3.2-3.3 mm.

Distribution. Very rare in Denmark, known from NEZ: Amager Fælled 7.IX.1919 (Oluf Jacobsen). – Very rare in southern Sweden, so far only found in Bl.: Mjällby, Nogersund 6.VII.1967 (Ossiannilsson). – Norway: only found in Ry: Solbakk, Strand 9.VIII.1952 (Holgersen). – Not recorded from East Fennoscandia. – England, Scotland, Wales, Ireland, France, Netherlands, German F.R.

Biology. Litoral – halophilous (Wagner, 1941). "An Salzwiesen mit dem Andelgras *Puccinellia maritima* (Huds.) gebunden" (Remane, 1965). Adults in June-October (Le Quesne, 1969).

393. *Psammotettix albomarginatus* W. Wagner, 1941
Text-figs. 2686-2689.

Psammotettix albomarginatus W. Wagner, 1941: 128.

Vertex acute-angled, flat. Sub-brachypterous, fore wings in male slightly longer, in female from somewhat shorter to longer than abdomen. Usually with strong markings. Vertex unicolorous dark brown except along median line and in a light spot around each ocellus, sometimes also with a transverse light band between eyes (Text-fig. 2686). Interocellar lines completely confluent with the dark colour of vertex. Pronotum often with four brown longitudinal bands. Fore wings with strong whitish veins, cells of inner half usually completely filled with dark colour, cells of outer half whitish. In weakly pigmented specimens, inner cells with brown spots, claval suture always dark-bordered on outside. Genital style in male as in Text-fig. 2687, aedeagus as in Text-figs. 2688, 2689. Length of posterior tibiae 1.3-1.5 mm. Overall length of males 2.5-2.8 mm, of females 2.8-2.9 mm.

Distribution. Denmark: so far known only from EJ: Kalø Vig 31.VIII.1966 (Trolle). – Sweden: found in Sk.: Maglehem, Möllegården and Skogadal 1.-6.VIII.1982 (Ossiannilsson), and in Öl.: Karlevi 6.VIII.1981 (Ossiannilsson). – So far not found in Norway. – East Fennoscandia: recorded from Ab: vicinity of Åbo (Linnavuori, 1969a), and from Ks: Kuusamo, July, 1949 (Linnavuori, 1950). – England, Ireland, Netherlands, German D.R. and F.R., Poland.

Biology. "Auf Dünen, die mit Flechten und *Agrostis* bewachsen sind" (Wagner, 1941). In the "Corynephoretum agrostidetosum aridae" (Marchand, 1953). Hibernates in egg-stage, 2 generations (Remane, 1958; Schiemenz, 1969b). Only on *Corynephorus canescens* (Remane, 1965). Stenotopic species, on *Corynephorus;* adults in May-October (in Central Europe) (Schiemenz, 1969b).

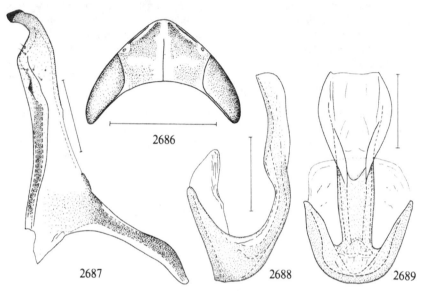

Text-figs. 2686-2689. *Psammotettix albomarginatus* Wagner. – 2686: head of male from above; 2687: right genital style from above (specimen from Hamburg, W. Wagner leg.); 2688: aedeagus from the left (same specimen); 2689: same in ventral aspect. Scale: 0.5 mm for 2686, 0.1 mm for the rest.

394. *Psammotettix excisus* (Matsumura, 1906)
Text-figs. 2690-2692.

Deltocephalus excisus Matsumura, 1906: 79.
Psammotettix exilis W. Wagner, 1941: 127.

"Very small, delicate species. Vertex acute, flat. Fore wings in ♂ and ♀ much longer than abdomen. Ground colour of upper side whitish yellow, usually sparsely marked. Vertex usually unicolorous pale yellowish white, with fine black median streak almost reaching apex. Interocellar lines distinct, each interrupted at middle. Additional markings of vertex consisting of separate spots, if present. Cells of clavus, median cell and apical cells entirely or partly finely dark-bordered. Even in sparsely pigmented specimens, the markings are linear, not spotty. Length of posterior tibiae 1.3-1.4 mm. Overall length in macropters (♂♀) 2.7 mm" (After Wagner, 1941). – Male genital style

822

as in Text-fig. 2690, aedeagus as in Text-figs. 2691, 2692. – Vertex of nymph distinctly longer than wide; abdomen dark, with rows of pale spots, ventrally also dark; anterior body without longitudinal bands; thorax distinctly paler than abdomen; frons with distinct arch-lines; legs pale with darker rings and spots (Vilbaste, 1982).

Distribution. Denmark: so far known from NWJ: Hansted, and NEZ: Tibirke Bakker. – Not found in Sweden, Norway, and East Fennoscandia. – France, Portugal, Netherlands, German D.R. and F.R., Hungary, Bohemia, Moravia, Poland, Estonia, Lithuania.

Biology. "Auf Sandboden an Gramineen zwischen *Artemisia campestris*" (Wagner, 1941). In the Corynephoretum cladonietosum (Marchand, 1953). Hibernation in egg stage, 2 generations (Remane, 1958; Schiemenz, 1969b). On various grasses "im Bereich der grauen Düne" (Remane, 1965). Stenotopic xerophilous species of dry grassy plains (Schiemenz, 1969b). In dry meadows, sand areas, dry forest margins, etc. (Vilbaste, 1974).

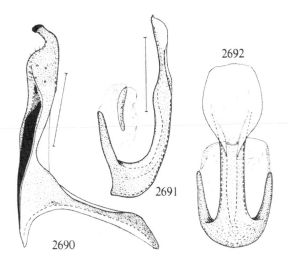

Text-figs. 2690-2692. *Psammotettix excisus* (Matsumura). – 2690: right genital style from above (paratype of *P. exilis* Wagner, from Pommern: Geesover Hügel 12.VIII.1937, W. Wagner); 2691: aedeagus from the left (same specimen); 2692: same in ventral aspect. Scale: 0.1 mm.

395. **Psammotettix dubius** Ossiannilsson, 1974
    Text-figs. 2693-2702.

Psammotettix albomarginatus Ossiannilsson, 1943b: 15 (nec Wagner, 1941).
Psammotettix exilis Ossiannilsson, 1953b: 108 (nec Wagner, 1941).
Psammotettix dubius Ossiannilsson, 1974: 24.

Body comparatively robust, head angular. Usually sub-brachypterous; fore wings in macropterous specimens only slightly longer than in sub-brachypters (Text-figs. 2694, 2695). Upper side pale dingy yellow with brownish spots and stripes, as reproduced in

Text-figs. 2693-2695. Face with the usual brownish transverse streaks. Dorsal pigmentation of head more or less distinct, interocellar lines sometimes indistinctly interrupted but often confluent with pigmented areas of vertex, these greatly varying in size and intensity of the dark colour. In strongly pigmented specimens there is a bottle-shaped pale median band on caudal 2/3 of vertex, bordering the black coronal suture (Text-fig. 2693). Pronotum brownish with 5 milk-coloured longitudinal bands varying in width. In strongly pigmented specimens, the dark colour of the fore wings is concentrated in clavus and corial cells near clavus, a band along costal border being pale, more or less as in *albomarginatus*. In most specimens, pigmentation is moderately developed, as in Text-figs. 2693-2695, or even less. Scutellum pale with or without diffuse fuscous spots. Extent of pigmentation of thoracic venter and abdomen varying. Legs as usual in *Psammotettix*. Male pygofer lobes as in Text-fig. 2697, genital valve and plates as in Text-fig. 2698, styles as in Text-fig. 2699. Aedeagus varying especially in shape of the terminal margin in ventral aspect (Text-figs. 2696, 2700-2702). Length of posterior tibia 1.44-1.76 mm. Overall length of sub-brachypterous males 2.5-3.0 mm, of sub-brachypterous females 2.7-3.3 mm, of macropters (♂♀) 2.95-3.4 mm.

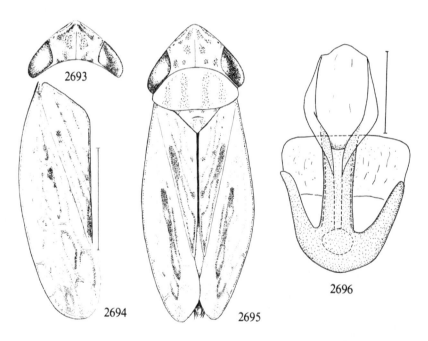

Text-figs. 2693-2696. *Psammotettix dubius* Ossiannilsson. – 2693: head of male (holotype); 2694: left fore wing of macropterous male; 2695: sub-brachypterous female; 2696: aedeagus of paratype in ventral aspect. Scale: 0.1 mm for 2696, 1 mm for the rest.

Distribution. Not recorded from Denmark. – Widespread, often abundant in suitable localities in Sweden, found in Sk., Hall., Sm., Öl., Gtl., Hls., P.Lpm. – Widespread in Norway, found in HEn, Bø, Bv, AAi, HOi, STi, Nsi. – Not recorded from East Fennoscandia, further distribution unknown.

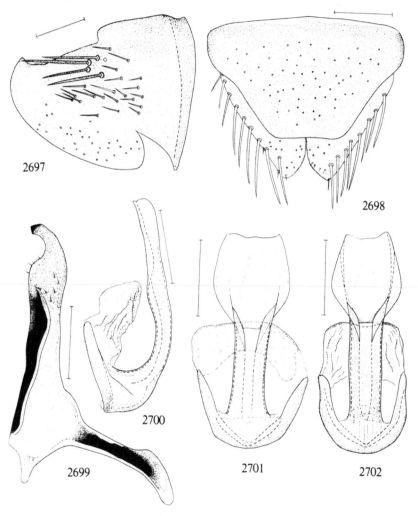

Text-figs. 2697-2702. *Psammotettix dubius* Ossiannilsson. – 2697: right pygofer lobe in male from the right; 2698: male genital plates and valve from below; 2699: right genital style from above; 2700: aedeagus from the left; 2701: aedeagus in ventral aspect; 2702: aedeagus in ventral aspect (another specimen). (2697-2701 after a specimen from Norway, HEn: Nybergsund, Holgersen). Scale: 0.1 mm.

825

Biology. On grasses on dry, sandy soils like the "alvar" of Öland. Also in heather marshes and dry moors. Adults in June-August.

396. *Psammotettix frigidus* (Boheman, 1845)
Text-figs. 2703-2705.

Deltocephalus frigidus Boheman, 1845b: 156.
Deltocephalus aquilonis Ossiannilsson, 1938a: 6.

Body robust. Upper side greyish yellow or brownish yellow, shining, with fuscous markings. Head medially usually distinctly longer than pronotum. Fore margin of head almost rectangular; interocellar lines usually continuous, sometimes interrupted. Markings on disk of vertex diffuse. Pronotum with indistinct longitudinal bands. Subbrachypterous, fore wings shorter to just longer than abdomen, hind wings roughly 3/4 of length of fore wings. Veins of fore wings prominent, light, partly dark-bordered in clavus and apical part of wing. Thoracic venter and abdomen largely black, lateral margins of abdomen and female pygofer light. Middle femora more or less distinctly transverse dark banded, tibiae dark spotted. Male genital styles as in Text-fig. 2703, aedeagus as in Text-figs. 2704, 2705. Length of posterior tibiae 1.47-1.65 mm. Overall length (♂♀) 2.5-3.0 mm.

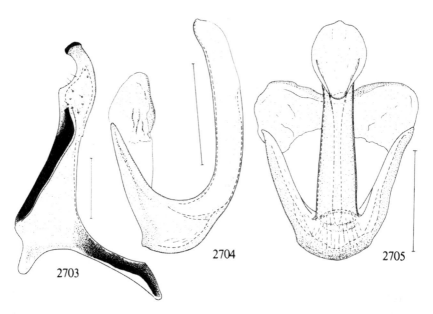

Text-figs. 2703-2705. *Psammottettix frigidus* (Boheman). – 2703: right genital style from above; 2704: aedeagus from the left; 2705: aedeagus in ventral aspect. Scale: 0.1 mm.

826

Distribution. Not found in Denmark. – Scarce in northern Sweden, found in Jmt.: Skurdalshöjden (Tullgren); Ås.Lpm.: Dikanäs, Kittelfjäll 29.VII.1934 (E. Runquist); Lu.Lpm.: Kvickjokk (Boheman). – Norway: found in several localities in the Hardangervidda: HOi: Ulvik and Eidsfjord, Bv: Uvdal and Hol (cf. Ossiannilsson, 1974). Also found in TEi: Rjukan, altitude 1017 m (Holgersen), and in Ri: Sandvatn, Suldal (Holgersen). – Very rare in East Fennoscandia, found in Le: Kilpisjärvi 27.VII.1924 (Håkan Lindberg); Kr, Lr. – Scotland, Latvia, n. Russia.

Biology. On dry meadows, heather moors and dry moraine, also in bogs. Adults in July-September.

397. *Psammotettix lapponicus* (Ossiannilsson, 1938)
   Text-figs. 2706-2708.

Deltocephalus lapponicus Ossiannilsson, 1938a: 4.

Resembling *frigidus,* somewhat larger and broader. Head frontally more obtuse-angular, about as long as pronotum. Interocellar lines divided into two parts, the median one dilated into a triangular spot. Wing dimorphous, usually sub-brachypterous, fore wings in sub-brachypterous males distinctly longer than, in sub-brachypterous females about as long as abdomen, hind wings as in *frigidus.* Markings of fore wings less distinct and less extended than in *frigidus,* sometimes absent. I have seen only one

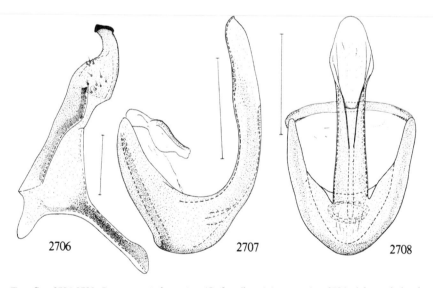

Text-figs. 2706-2708. *Psammotettix lapponicus* (Ossiannilsson) (paratype). – 2706: right genital style from above; 2707: aedeagus from the left; 2708: aedeagus in ventral aspect. Scale: 0.1 mm.

827

macropterous specimen (♀) of the present species. In this specimen, face and vertex are almost entirely brown without markings, pronotum brown with 5 indistinctly lighter longitudinal stripes, fore wings brown with light veins. Fore wings as long as abdomen, hind wings little shorter. Male genital styles as in Text-fig. 2706, aedeagus as in Text-figs. 2707, 2708. Length of posterior tibiae 1.70-1.93 mm. Overall length (♂♀) 3.0-3.4 mm.

Distribution. Not found in Denmark. – Rare in northern Sweden, only found in P.Lpm.: Arvidsjaur near the community 26.VII.1933 (Ossiannilsson). – Norway: found in Bv: Uvdal and Hol; HOi: Kinsarvik, Stavali; Ullensvang, Bersavikvann; Eidfjord, Rjota and Sysenvann, 870-1300 m altitude. – Not recorded from East Fennoscandia. – N. Siberia, Mongolia; Nearctic region.

Biology. In grass meadows, on heather moors, both dry and moist, and in marshes. Adults in July and August.

398. *Psammotettix pallidinervis* (Dahlbom, 1851)
   Text-figs. 2709-2711.

Deltocephalus pallidinervis Dahlbom, 1851: 190.
Deltocephalus hannoveranus W. Wagner, 1937a: 70.

A small, delicate species. Upper side shining greyish yellow, almost without dark markings. Head medially distinctly longer than pronotum. Transverse streaks on

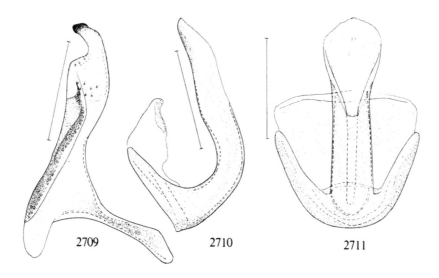

Text-figs. 2709-2711. *Psammotettix pallidinervis* (Dahlbom). – 2709: right genital style from above; 2710: aedeagus from the left; 2711: aedeagus in ventral aspect. Scale: 0.1 mm.

frontoclypeus light brownish. Vertex, pronotum and scutellum pale yellow without distinct markings. Wing dimorphous, fore wings in macropters as long as abdomen, hind wings slightly shorter; fore wings in sub-brachypterous males a little shorter than abdomen, in sub-brachypterous females reaching to caudal margin of 8th abdominal tergum, hind wings considerably shorter. Fore wings usually pale yellow or brownish yellow, veins paler, subcostal cell whitish. Exceptionally the fore wings are entirely fuscous with veins and subcostal cell light. Dark markings (streaks along claval veins and bordering apical cells) absent or vestigial. Thoracic venter and abdomen black, the latter with narrow light segmental borders. Male genital styles as in Text-fig. 2709, aedeagus as in Text-figs. 2710, 2711. Length of posterior tibiae 1.3-1.6 mm., overall length of males 2.5-2.7 mm, of females 2.9-3.2 mm. – Nymphs with vertex distinctly longer than wide, abdomen dark with rows of pale spots, ventrally also dark; thorax distinctly paler than abdomen; arch-lines of frons usually indistinct; legs almost entirely blackish brown (Vilbaste, 1982).

Distribution. So far not found in Denmark. – Sweden: only found in Öl. and Gtl.; Öl.: Alböke; Resmo; Borgholm; Karlevi; Ekelunda; Vickleby (Wahlgren, Bornfeldt, Ossiannilsson). Dahlbom found the types in Gtl.: Västerby. – Norway: recorded from Fi: Bassevuovde 10.VIII.1977 (Huldén). – East Fennoscandia: very rare, found in Ab: Laitila (J. Sahlberg); St: Yläne (J. Sahlberg); Kr, Lr. – Netherlands, German D.R. and F.R., Austria, Hungary, Moravia, Bulgaria, Romania, Poland, Estonia, Latvia, Lithuania, n. and m. Russia, m. Siberia.

Biology. "Bewohnt Sandfelder und Dünen (Wagner & Franz, 1961). Stenotopic, xerophilous; hibernates in the egg-stage, in Central Europe 2 generations (Schiemenz, 1969b). "In dry meadows, on sand dunes, in dry pine forests on sandy soil, clear cut areas" (Vilbaste, 1974). Adults in June-August.

399. *Psammotettix poecilus* (Flor, 1861)
Text-figs. 2712-2714.

Jassus (Deltocephalus) striatus var. 1. *poecilus* Flor, 1861a: 261.
Psammotettix scutuliferus W. Wagner, 1939: 160.

"Upper side with very strong markings. Vertex with two small triangles at apex; behind these on each side two dark brown spots occupying almost entire vertex, the following details remaining whitish: a spot bordering each ocellus, a patch around apical dark triangles, a median band, and an arched line on each side extending from caudal margin to eye. Pronotum dark mottled, with five light longitudinal bands. Cells of fore wings strongly brown or brownish black edged. The brown pattern is interrupted by two indistinct light transverse bands, one in the basal 1/4, another distally of middle. Within these bands, veins, especially transverse veins, are very strong, bright white. In some specimens, the light colour is dominating, transverse veins strongly prominent. Wings in females as long as abdomen, in males only slightly longer. Vertex in male shorter than basal width, ratio length: width = 0.73-0.88, in female as long as basal width or a little shorter, ratio = 0.94-1.00. Pronotum medially as long as vertex or

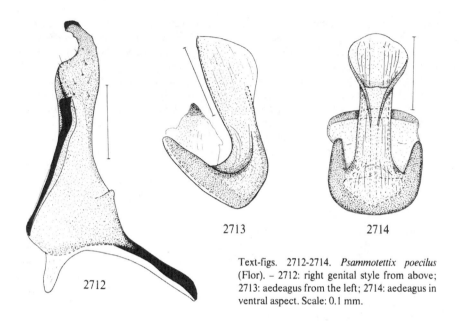

Text-figs. 2712-2714. *Psammotettix poecilus* (Flor). – 2712: right genital style from above; 2713: aedeagus from the left; 2714: aedeagus in ventral aspect. Scale: 0.1 mm.

somewhat longer, in female as long as vertex". (After Wagner, 1939). Male genital style as in Text-fig. 2712, aedeagus as in text-figs. 2713, 2714, – Nymphs with vertex distinctly longer than wide, abdomen dark, with rows of pale spots, ventrally also dark; anterior body with dark brown longitudinal bands (Vilbaste, 1982).

Distribution. Not found in Denmark, nor in Norway. – Very rare in Sweden, only found in Jmt.: Bispgården 6.VIII.1964 by A. Sundholm. – Rare and sporadic in East Fennoscandia, found in N, Ka, St, Ta, Sa, Kb; Vib, Kr. – German D.R. and F.R., Austria, Poland, Estonia, Latvia, Lithuania, n. and m. Russia, Ukraine, Moldavia, Altai Mts., Tuva, Mongolia.

Biology. "Auf Dünen und heideartigen Biotopen auf *Calamagrostis*" (Lindberg, 1947). "In xerophilen und mesophilen Biotopen, wobei der Schwerpunkt des Auftretens im xerophilen Bereich liegt. Ei-Überwinterer, 2 Generationen [in Central Europe]" (Schiemenz, 1969b). "On sandy areas: dunes, dry meadows, woodland clearings" (Vilbaste, 1974). Adults in June-August.

Genus *Ebarrius* Ribaut, 1947

Ebarrius Ribaut, 1947: 82.
Type-species: *Deltocephalus cognatus* Fieber, 1869, by original designation.

830

Resembling *Psammotettix*. Body fairly elongate. Head frontally angular, wider but not longer than pronotum. Chaetation of fore tibiae as in *Psammotettix*. Male anal tube dorsally sclerotized. Male pygofer lobes without appendages. Genital plates short, more or less broad, their macrosetae arranged in a row along lateral border. Connective hairpin-shaped, branches much longer than stem. Apical half of style broad, oval. Aedeagus asymmetrical, phallotreme terminal. Caudal margin of 7th abdominal sternum in female with 3 projections. In Denmark and Fennoscandia one species.

400. *Ebarrius cognatus* (Fieber, 1869)
Plate-fig. 197, text-figs. 2715-2723.

Deltocephalus cognatus Fieber, 1869: 214.
Deltocephalus interstinctus J. Sahlberg, 1871: 341 (nec Fieber, 1869).

Dirty yellow, shining. Vertex frontally with 2 pairs of black-brown streaks or spots (= interrupted interocellar lines, cf. *Psammotettix alienus*), more caudally usually with some yellow brown or black-brown spots: one immediately medially of the adocellar black-brown spot, and two longitudinal streaks on each side near caudal border. Dark transverse streaks of frontoclypeus medially partly confluent, a narrow median longitudinal stripe usually remaining light. Sutures of face and two parallel longitudinal streaks on anteclypeus dark. Pronotum with six more or less distinct dark longitudinal bands, frontally often with a few spots. Macropterous, fore wings longer than abdomen, pigmentation resembling that of certain *Psammotettix* species, veins partly dark bordered. Thoracic venter largely black, fore and middle femora dark spotted and transversely banded, hind femora with longitudinal bands, tibiae dark spotted. Abdomen black with light segmental borders and partly light lateral margins and genitalia. Male pygofer as in Text-fig. 2715, genital plates and valve as in Text-fig. 2716, style as in Text-fig. 2717, connective as in Text-fig. 2718, aedeagus as in Text-figs. 2719, 2720, 1st abdominal sternum in male as in Text-fig. 2721, 2nd abdominal sternum in male as in Text-fig. 2722, 7th abdominal sternum in female as in Text-fig. 2723. Overall length of males 3.0-3.6 mm, of females 3.35-4.15 mm.

Distribution. Not found in Denmark. – Scarce and sporadic in southern Sweden, commoner in the north, found in Bl.-T.Lpm. – Widespread in Norway, found in Ø, AK, On, Bv, TEi, HOi, MRy, MRi, STi, Fi. – Scarce and sporadic in East Fennoscandia, found in Ab, St, Kb, Ks; Lr. – England, Scotland, France, German F.R., Austria, Italy, Albania, Bohemia, Moravia, Slovakia, Hungary, Bulgaria, Yugoslavia, Romania, Poland, Anatolia, Armenia, Georgia.

Biology. On grassland, also arable land (Ossiannilsson, 1947b). "Scheint im Gebiete [Nordost-Alpen] nur höhere Gebirgslagen zu besiedeln und aus geprägt heliophil zu sein" (Wagner & Franz, 1961). On dry meadows and wood edges (Linnavuori, 1969a). On more or less dry slopes, hills and moors. Altitude up to 1150 m. (Ossiannilsson, 1974). Adults in July-September.

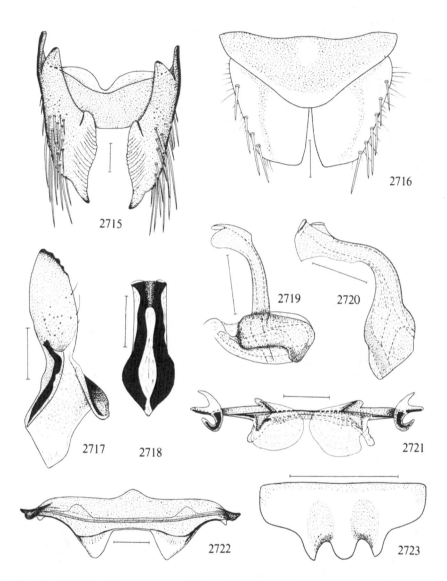

Text-figs. 2715-2723. *Ebarrius cognatus* (Fieber). – 2715: male pygofer from below (genital plates, valve and anal apparatus removed); 2716: genital plates and valve from below; 2717: right genital style from above; 2718: connective from above; 2719: aedeagus from the left; 2720: aedeagus in ventral aspect; 2721: 1st abdominal sternum in male from above; 2722: 2nd abdominal sternum in male from above; 2723: 7th abdominal sternum in female from below (depressed under coverglass). Scale: 0.5 mm for 2723, 0.1 mm for the rest.

Genus *Adarrus* Ribaut, 1947

Adarrus Ribaut, 1947: 83.

Type-species: *Deltocephalus multinotatus* Boheman, 1847, by original designation.

General structure of body and chaetation of fore tibiae as in *Psammotettix*. Fore wings about as long as abdomen. Anal tube in male dorsally sclerotized. Male pygofer lobes without appendages. Macrosetae of genital plates in disorder. Apical part of style elongate. Socle of aedeagus projecting beyond distal end of connective. Aedeagus apically with appendages, phallotreme apical or subapical. In Denmark and Fennoscandia one species.

401. *Adarrus multinotatus* (Boheman, 1847)
Plate-fig. 198, text. figs. 2724-2729.

Deltocephalus multinotatus Boheman, 1847b: 264.

Light yellow, shining. Head as seen from above with frontal border rectangular, with 3 pairs of black-brown spots (Plate-fig. 198). In strongly pigmented specimens these spots may transversely coalesce to form cross-bands. They may also be obsolescent. Face with or without dark transverse streaks. Pronotum frontally often with a pair of black spots behind the caudal pair of spots on vertex. In addition there is often a transverse row of spots across the surface of pronotum. Scutellum usually with brownish basal spots. Fore wings sometimes entirely pale but usually with the pale veins bordered with fuscous, the dark streaks often forming closed cells especially in apical part of wing. Transverse veins each enclosed in a milky spot. Thoracic venter and abdomen often entirely light; these parts may also be largely black-brown. In strongly marked specimens, femora are transversely dark banded, tarsi with terminal segments fuscous. Hind tibiae dark spotted. Male pygofer and anal apparatus as in Text-fig. 2724, genital valve and plates as in Text-fig. 2725, styles as in Text-fig. 2726, aedeagus as in Text-figs. 2727, 2728, 7th abdominal sternum in female as in Text-fig. 2729. Overall length of males 2.65-3.3 mm, of females 2.8-3.6 mm. – Nymphs with 4 complete rows of hairs on abdomen; vertex more than 1.5 × as long as wide between eyes, its anterior margin more or less convex (Vilbaste, 1982).

Distribution. Not found in Denmark, Norway and East Fennoscandia. – Rare in Sweden, found in Öl.: St. Rör 9.VIII. 1939 (Ossiannilsson), Halltorps hage in July and August 1976 (H. Anderson and R. Danielsson); Gtl.: Lummelund 13.VIII.1847 (Boheman), Roma 1.VIII.1935, Etelhem 2.VIII.1935, Tingstäde and Martebo 3.VIII.-1935 (Ossiannilsson); Upl.: Börje, Hässelby hage 20.IX.1972 (Sten Jonsson). – England, France, Belgium, Netherlands, German D.R. and F.R., Austria, Switzerland, Italy, Hungary, Bohemia, Moravia, Slovakia, Romania, Yugoslavia, Poland, Latvia, Estonia, Ukraine, Anatolia, Altai Mts., Kazakhstan, m. Siberia, Korean Peninsula, Algeria, Tunisia.

Biology. In "Binnendünen, besonnten Hängen, Hochmooren, Wäldern" (Kuntze,

1937). Ubiquitous (Schwoerbel, 1957). "Bewohner von Trockenrasen, xerotherm" (Wagner & Franz, 1961). On *Brachypodium;* hibernation takes place in egg stage; in Central Europe 2 generations (Schiemenz, 1969b).

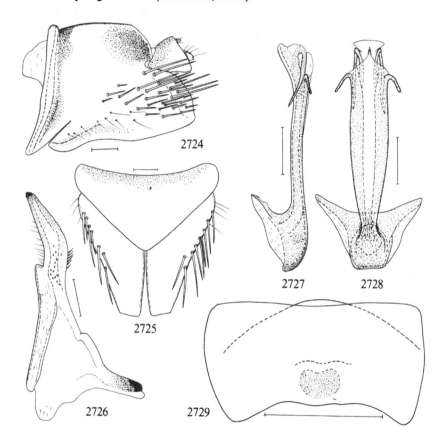

Text-figs. 2724-2729. *Adarrus multinotatus* (Boheman). – 2724: male pygofer and anal apparatus from the left; 2725: genital plates and valve from below; 2726: genital style from above; 2727: aedeagus from the left; 2728: aedeagus in ventral aspect; 2729: 7th abdominal sternum in female from below (depressed under coverglass). Scale: 0.5 mm for 2729, 0.1 mm for the rest.

Genus *Errastunus* Ribaut, 1947

Errastunus Ribaut, 1947: 83.
    Type-species: *Cicada ocellaris* Fallén, 1806, by original designation.

834

As *Adarrus*. Outer claval cell with some secondary cross-veins. Socle of aedeagus not projecting beyond distal end of connective. Phallotreme apical. Genital plates elongate, apices in repose crossing each other. Style with apical apophyse elongate. Caudal margin of 7th abdominal sternum in female with median two-pointed projection. In the Palaearctic region one species.

402. *Errastunus ocellaris* (Fallén, 1806)
Plate-fig. 199, text-figs 2730-2735.

Cicada ocellaris Fallén, 1806: 20.

Upper side ivory-white to brownish yellow, with orange-red and black-brown markings, shining. Head above near fore border with two pairs of black-brown oblique streaks, one near apex and another medially of ocellus. Caudally of the median streaks on each side a triangular orange spot, behind which is a large trapezoidal spot frontolaterally reaching the adocellar oblique streak. Dark transverse streaks of face often confluent. Pronotum with 6 orange longitudinal bands. Scutellum with reddish basal triangles, between these two narrow orange longitudinal stripes. Fore wings as long as abdomen or a little shorter, veins light, dark bordered, the dark colour often forming closed cells. Primary and secondary transverse veins milk-white, enclosed in white spots. Thoracic venter largely black-brown. Legs light, or femora proximally black, tibiae black dotted. Abdomen in male black with light segmental borders and lateral patches, in female usually with basal terga black, caudal terga light, venter largely black. Male pygofer and anal apparatus as in Text-fig. 2730, genital valve and plates as in Text-fig. 2731, styles as in Text-fig. 2732, aedeagus as in Text-figs. 2733, 2734, 7th abdominal sternum in female as in Text-fig. 2735. Overall length of males 2.7-3.4 mm, of females 3.2-4.0 mm. – Nymphs with abdominal chaetation as in *Adarrus*, vertex brownish yellow or brownish, thorax usually more or less unicolorous, tips of wing pads whitish, lower parts of frons with large central whitish spot (Vilbaste, 1982).

Distribution. Common in Denmark. – Widespread and moderately common in Sweden, found Sk.-Lu.Lpm. – Widespread, apparently scarce in Norway, found in HEn: Tynset 25.VII.1900 (Warloe); On: Dovre, Fokstua 1933 (Håkan Lindberg); Nsi: Salten (Sahlberg, 1880); Fi: Karasjok 1907 (spec. in Mus. Oslo), Bassevuovdde 10.VIII.-1977 (Huldén). – East Fennoscandia: not uncommon in the north, scarce in the south, found in Ab, St, ObN, Ks, Li; Kr, Lr. – Widespread in the Palaearctic region, also present in the Nearctic.

Biology. In "Binnendünen, besonnten Hängen, Waldlichtungen, Wiesen" (Kuntze, 1937). In the "Arrhenatheretum elatioris" (Marchand, 1953). Hibernation in egg-stage, 2 generations (Remane, 1958; Schiemenz, 1969b). "Nicht nur auf feuchten, sondern auch auf sommertrockenen Rasenflächen" (Wagner & Franz, 1961). Mainly in mesophilic but also in xerophilic and hygrophilic biotopes (Schiemenz, 1969b). Adults in May-September.

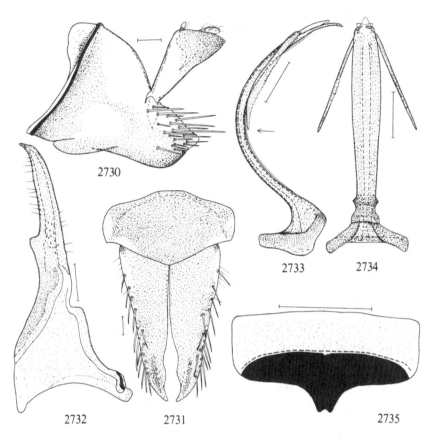

Text-figs. 2730-2735. *Errastunus ocellaris* (Fallén). – 2730: male pygofer and anal apparatus from the left; 2731: genital plates and valve from below; 2732: right genital style from above; 2733: aedeagus from the left; 2734: aedeagus in ventral aspect (in direction of arrow in 2733); 2735: 7th abdominal sternum in female from below (depressed under coverglass). Scale: 0.5 mm for 2735, 0.1 mm for the rest.

Genus *Turrutus* Ribaut, 1947

Turrutus Ribaut, 1947: 83-84.
Type-species: *Jassus (Deltocephalus) socialis* Flor, 1861, by original designation.

General body structure and chaetation of fore tibiae as in *Adarrus*. Fore wings apically not symmetrically rounded. Male pygofer lobes shorter than genital plates, with a caudoventral process. Genital plates short and broad, macrosetae arranged in a lateral

836

row. Styles moderately elongate. Aedeagus short and broad, phallotreme terminal, surrounded by a hairy collar. 7th abdominal sternum in female with a median incision. One species.

403. *Turrutus socialis* (Flor, 1861)
Plate-fig. 200, text-figs. 2736-2741.

Jassus (Deltocephalus) socialis Flor, 1861a: 242.

Above dirty yellow, shining. Fore border of head in dorsal aspect approximately rectangular. Head longer than pronotum, above near apex on each side with a triangular or clavate black-brown spot continuing on frontoclypeus, there touching its counterpart on the opposite side. A smaller black-brown streak present medially of each ocellus. Caudally of these spots there are some yellowish brown spots. Frontoclypeus with dark transverse streaks often confluent medially. Sutures on face dark. Pronotum with six yellowish brown longitudinal bands, in dark specimens also with some dark spots along frontal border. Usually sub-brachypterous, fore wings somewhat shorter than abdomen, apically obliquely rounded, hind wings little longer than half length of fore wings. Veins of fore wings usually partly dark-bordered. Thoracic venter dark spotted, fore and middle femora transversely banded, hind femora longitudinally striped. Tibiae dark spotted. Colour of abdomen varying, brownish yellow to almost entirely black. Male pygofer (Text-fig. 2736) always with a pair of large black patches dorsally. Male genital valve and plates as in Text-fig. 2737, styles as in Text-fig. 2738, aedeagus as in Text-figs. 2739, 2740, 7th abdominal sternum in female as in Text-fig. 2741. Overall length of males 2.5-3.0 mm, of females 2.9-4.2 mm. – Nymphs with 4 complete rows of hairs on abdominal dorsum; abdomen with four more or less distinct longitudinal bands; vertex usually longer than wide; a large triangular pale spot present on lower part of face (extending also to thorax) (Vilbaste, 1982).

Distribution. Uncommon in Denmark, found mainly on the coasts of Jutland, but also in a few inland localities near Silkeborg (EJ). – Fairly common in southern Sweden, Sk.-Ög., Vg.: once found in the vicinity of Stockholm (Boheman), and in Vstm.: Horn, Strömsholm (Gyllensvärd). – Not recorded from Norway. – East Fennoscandia: only found in Al: Eckerö (Håkan Lindberg). – Widespread in the Palaearctic region.

Biology. In "Stranddünen, Binnendünen, Sandfeldern, besonnten Hängen, Heiden, Wäldern, Wiesen" (Kuntze, 1937). "Auf Trockenrasen, jedoch im Gebiete wiederholt auch in Sumpfwiesen und Mooren gesammelt" (Wagner & Franz, 1961). "Hauptart der submediterranen Halbtrockenrasen" (Schiemenz, 1969a). "In xerophilen bis hygrophilen Biotopen, wobei erstere bevorzugt werden; Ei-Überwinterer, 2 Generationen" (Schiemenz, 1969b). Adults in July and August.

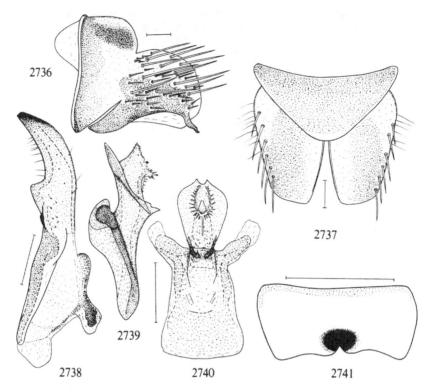

Text-figs. 2736-2741. *Turrutus socialis* (Flor). – 2736: left pygofer lobe in male from outside; 2737: genital plates and valve from below; 2738: right genital style from above; 2739: aedeagus from the left; 2740: aedeagus in ventral aspect; 2741: 7th abdominal sternum in female from below (depressed under coverglass). Scale: 0.5 mm for 2741, 0.1 mm for the rest.

Genus *Mongolojassus* Zachvatkin, 1953

Mongolojassus Zachvatkin, 1953d: 247-250.

Type-species: *Deltocephalus sibiricus* Horváth, 1901, by original designation.

General structure of body as in *Arocephalus*. Anterior tibiae with 3 setae in the inner, 4 in the outer dorsal row. Male pygofer dorsally not incised, enclosing basal part of anal tube, pygofer lobes without processes and denticles. Genital plates apically obliquely truncate. Connective short and broad, closely attached to socle of aedeagus. Shaft of aedeagus slender, apically with a pair of appendages. Phallotreme subapical, ventral. In Fennoscandia one species.

404. *Mongolojassus bicuspidatus* (J. Sahlberg, 1871)
Text-figs. 2742-2748.

Deltocephalus bicuspidatus J. Sahlberg, 1871: 307.

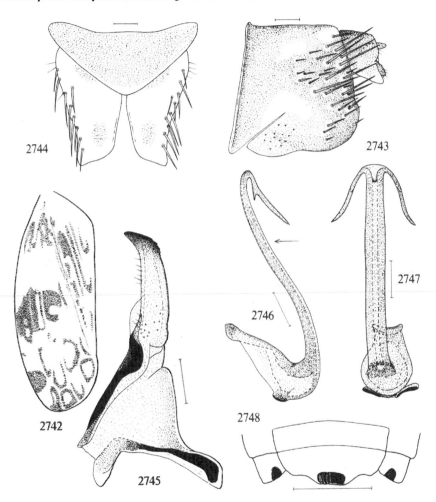

Text-figs. 2742-2748. *Mongolojassus bicuspidatus* (Sahlberg). – 2742: pattern of pigmentation of left fore wing of female; 2743: left pygofer lobe and anal appatus of male from the left; 2744: genital plates and valve from below; 2745: right genital style from above; 2746: aedeagus from the left; 2747: aedeagus in ventral aspect (in direction of arrow in 2746); 2748: 6th and 7th abdominal sterna and pleura in female from below. Scale: 0.5 mm for 2748, 0.1 mm for the rest.

Elongate, whitish with black and fuscous markings. Head in dorsal aspect obtuse-angular, medially a little shorter than wide basally, 1/3 longer than pronotum, frontally with four small, roughly triangular dark spots, caudally with two spots close together. Frontoclypeus 1/4 longer than basal width, black-brown, with a whitish longitudinal band strongly narrowing towards apex, and whitish lateral streaks; anteclypeus 1/4 longer than basal width, tapering towards apex. Pronotum short. Scutellum with fuscous basal triangles. Fore wings whitish, markings as in Text-fig. 2742. Abdominal dorsum in male blackish, with a median line widening on caudal tergum and lateral margins narrowly light, venter largely light. Legs pale yellow, tibiae black dotted, tarsi with fuscous rings. Male pygofer and anal apparatus as in Text-fig. 2743, genital valve and plates as in Text-fig. 2744, styles as in Text-fig. 2745, aedeagus as in Text-figs. 2746, 2747, 7th abdominal sternum in female as in Text-fig. 2748. Overall length about 3 mm.

Distribution. Not found in Denmark, Sweden, and Norway. – Very rare in East Fennoscandia, found in Vib: Kexholm, Ampiala 9.VII. and 11.VII.1923 (Karvonen); Kr: Valamo in Ladoga 10.VII.1866 (J. Sahlberg). – N. Russia.

Biology. On a dry sandy slope with grass and *Thymus* (Lindberg, 1924a).

Genus *Jassargus* Zachvatkin, 1933

Jassargus Zachvatkin, 1933: 268.
    Type-species: *Jassus (Deltocephalus) distinguendus* Flor, 1861, by original designation.
Lausulus Ribaut, 1947: 84.
    Type-species: *Deltocephalus pseudocellaris* Flor, 1861, by original designation.
Arrailus Ribaut, 1952: 251 (subgenus).
    Type-species: *Deltocephalus flori* Fieber, 1869, by original designation.
Sayetus Ribaut, 1952: 251 (subgenus).
    Type-species: *Deltocephalus sursumflexus* Then, 1902, by original designation.

General body structure more or less as in *Errastunus*. Vertex as long as or longer than pronotum, apically roughly rectangular. Anterior tibiae with one seta in the inner, 4 in the outer dorsal row. Fore wings not or little longer than abdomen. A hyaline triangle present on each of the two lateral apical veinlets. Male anal tube short, cordate. Male pygofer lobes ventrally each with a tooth usually directed upwards. Apical margin of style curled. Caudal margin of 7th abdominal sternum in female with 3 shallow depressions. Identification of females of this genus is often difficult (but consult Schulz, 1976). In Denmark and Fennoscandia 5 species.

Key to species of *Jassargus*

1   Phallotreme dorsal (Subgenus *Jassargus* s.str.). Male pygofer
    lobes without dorsal tooth. Caudal margin of 7th abdominal
    sternum in female shallowly sinuous (Text-fig. 2754) ...... 405. *distinguendus* (Flor)
–   Phallotreme not dorsal. Male pygofer lobes each with a dor-

405. *Jassargus (Jassargus) distinguendus* (Flor, 1861)
Text-figs. 2749-2754.

Jassus (Deltocephalus) distinguendus Flor, 1861a: 240.
Deltocephalus pseudocellaris Flor, 1861a: 547 (n.n.).
Deltocephalus falleni Fieber, 1869: 210.
Deltocephalus paleaceus J. Sahlberg, 1871: 316.
Deltocephalus kemneri Ossiannilsson, 1935b: 130 (pipunculized specimens).

Dirty yellow, shining. Markings of dorsal side much varying; in strongly pigmented specimens, vertex on each side near fore border with two oblique streaks, one longer near apex and another shorter medially of ocellus. Behind these streaks there are two or three pairs of indistinctly delimited orange-coloured or brownish yellow patches, one smaller touching the apical black streaks, and another larger on disk of vertex; a smaller pair may be present at caudal border of the head. The space between these spots may be apprehended as a pale median longitudinal band and a cross-band between eyes. Face with dark, often confluent, transverse streaks. Pronotum with six usually indistinctly brownish yellow longitudinal bands. Fore wings apically rounded obtuse, from a little longer to a little shorter than abdomen; veins light, dark bordered, the dark streaks often forming closed cells. Such specimens are suggestive of *Errastunus ocellaris* but secondary transverse veins in clavus are absent. Distal ends of claval veins and some transverse veins enclosed in milky white spots. All the dark markings may also be completely absent. Colour of thoracic venter, abdomen and legs also strongly varying; at least apices of femora always light. Tibiae black spotted. Male pygofer as in Text-fig. 2749, genital plates and valve as in Text-fig. 2750, style and connective as in Text-fig. 2751, aedeagus as in Text-figs. 2752, 2753. 7th abdominal sternum in female as in Text-fig. 2754. Overall length (♂♀) 2.7-3.4 mm. – Nymphs with 4 complete rows of hairs on abdominal dorsum; vertex less than 1.5 × as long as wide, brownish yellow or brownish; abdomen with 2 wide brownish longitudinal bands, each containing two rows of whitish spots; thorax with more or less distinct longitudinal bands; nymphs

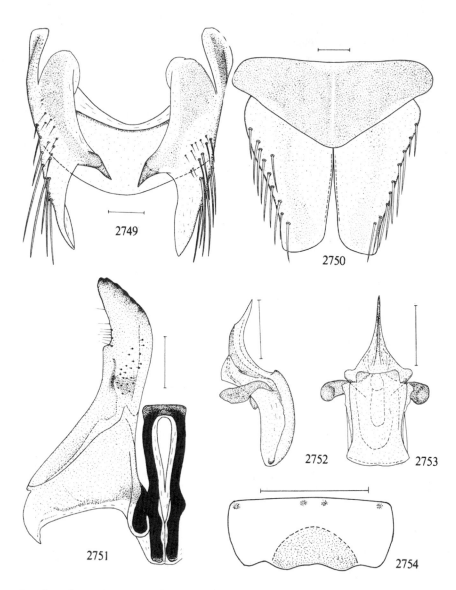

Text-figs. 2749-2754. *Jassargus distinguendus* (Flor). – 2749: male pygofer from below (anal apparatus and genital plates removed); 2750: genital plates and valve from below; 2751: right genital style and connective from above; 2752: aedeagus from the left; 2753: aedeagus in ventral aspect; 2754: 7th abdominal sternum in female from below (depressed under coverglass). Scale: 0.5 mm for 2754, 0.1 mm for the rest.

pale-coloured, pattern often indistinct, ventral surface not darker than dorsum; frontoclypeus, above lower margin, with rather narrow pale transverse band and narrow longitudinal band (Vilbaste, 1982).

Distribution. Common and widespread in Denmark as well as in Sweden up to Vb. (not in Gtl. and G. Sand.). – Distribution in Norway imperfectly known; the records given by Siebke (1874) have not been revised. Sahlberg (1871) recorded *paleaceus* from AK: vicinity of Oslo. Holgersen found the present species in AK: N. Eidsvold, Bø: Y. Sandsvaer, and VAy: Sutevik, Vanse. – Not recorded from East Fennoscandia. – Widespread in Europe (also Iceland); also found in Armenia and w. Siberia.

Biology. In "Salzstellen, Stranddünen, Binnendünen, Sandfeldern, besonnten Hängen, Wäldern, Waldlichtungen, Wiesen" (Kuntze, 1937). In all types of grassland (Marchand, 1953). "Auf feuchten Wiesen" (Wagner & Franz, 1961). Mainly in mesophilous and hygrophilous biotopes (Schiemenz, 1969b). In dry meadows, on field margins, sandy river banks, etc. (Vilbaste, 1974). Hibernation takes place in the egg-stage, in Central Europe 2 generations (Remane, 1958; Schiemenz, 1969b). Adults in July-October.

406. *Jassargus (Arrailus) flori* (Fieber, 1869)
Plate-fig. 201, text-figs. 2755-2761.

Deltocephalus flori Fieber, 1869: 210.
Deltocephalus oculatus J. Sahlberg, 1871: 308.
Deltocephalus pseudocellaris J. Sahlberg, 1871: 314, p.p. (nec Flor, 1861).
Deltocephalus falleni J. Sahlberg, 1871: 315, p.p. (nec Fieber, 1869).

Resembling *J. distinguendus*, entirely light-coloured specimens less common that in that species. Apodemes of 1st abdominal sternum in male inconspicuous, those of 2nd abdominal sternum short, semilunar. Male pygofer and anal apparatus as in Text-figs. 2755, 2756, genital valve and plates as in Text-fig. 2757 (but plates not diverging in repose), styles as in Text-fig. 2758, aedeagus as in Text-figs. 2759, 2760, 7th abdominal sternum in female as in Text-fig. 2761. Overall length ($\male\female$) 2.7-3.3 mm. – Nymphs as in *distinguendus*, pale brownish with darker brownish yellow pattern, ventral surface often darker than dorsum; frontoclypeus with a pale irregular patch not extending to an-teclypeal suture but sometimes connected with it by a narrow longitudinal line; middle bands of mesonotum broader posteriorly; abdominal sternites usually dark with pale middle line and lateral margins (Vilbaste, 1982).

Distribution. Fairly common in Denmark. – Common and widespread in Sweden, Sk.-Lu.Lpm. – Widespread in Norway; I have examined specimens from Ø, AK, HEs, HEn, Os, Bø, TEi, AAy, VAy, Ry, HOy, HOi, MRy, STi, Nsy, Nsi. Also recorded from TRi and TRy. – Common and widespread om East Fennoscandia. – Widespread in Europe, also found in Moldavia.

Biology. In "Wäldern, Waldlichtungen" (Kuntze, 1937). On dry meadows (Kontkanen, 1938). "Vor allem auf Trockenrasen, aber auch auf feuchten Wiesen und selbst im Sumpfland" (Wagner & Franz, 1961). In xerophilous, mesophilous and

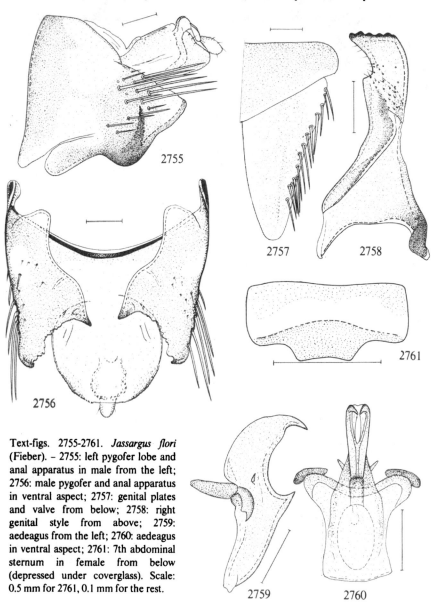

Text-figs. 2755-2761. *Jassargus flori* (Fieber). – 2755: left pygofer lobe and anal apparatus in male from the left; 2756: male pygofer and anal apparatus in ventral aspect; 2757: genital plates and valve from below; 2758: right genital style from above; 2759: aedeagus from the left; 2760: aedeagus in ventral aspect; 2761: 7th abdominal sternum in female from below (depressed under coverglass). Scale: 0.5 mm for 2761, 0.1 mm for the rest.

hygrophilous biotopes without discrimination; hibernates in egg-stage, probably 2 generations (Schiemenz, 1969b). "A species of swampy forests, spruce-birch swamps and swampy meadows which also thrives in peatland meadows used for grazing, and even in dry pastures and on arable land, where, however, it is rare" (Raatikainen & Vasarainen, 1976). On *Festuca ovina, Festuca rubra* and *Deschampsia flexuosa* (Trümbach, Waloff, according to Schulz, 1976). Adults in July-September.

407. *Jassargus (Arrailus) alpinus* (Then, 1896), ssp. *neglectus* (Then, 1896)
Text-figs. 2762-2767.

Deltocephalus alpinus Then, 1896: 175.

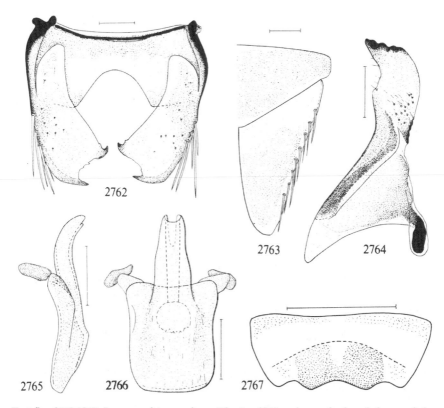

Text-figs. 2762-2767. *Jassargus alpinus neglectus* (Then). – 2762: male pygofer from below, genital plates and anal apparatus removed; 2763: genital plates and valve from below; 2764: right genital style from above; 2765: aedeagus from the left; 2766: aedeagus in ventral aspect; 2767: 7th abdominal sternum in female from below (depressed under coverglass). Scale: 0.5 mm for 2767, 0.1 mm for the rest.

Deltocephalus neglectus Then, 1896: 173.
Jassargus alpinus neglectus W. Wagner, 1958: 438.

Resembling *J. flori*. Male pygofer as in Text-fig. 2762, genital plates and valve as in Text-fig. 2763, styles as in Text-fig. 2764, aedeagus as in Text-figs. 2765, 2766, 7th abdominal sternum in female as in Text-fig. 2767. Overall length (♂♀) 2.7-3.0 mm. – Nymphs as in *flori* but frontoclypeus with broad pale middle band or large pale patch occupying the whole ventral part of frontoclypeus (Vilbaste, 1982).

Distribution. Not recorded from Denmark. – Not uncommon in northern Sweden, found in Nb.: Edefors (Tullgren), Pajala (Ehnbom, Ardö), Övertorneå (Ehnbom), Luleå, Notviken (Ossiannilsson); Lu.Lpm.: Kvickjokk, Njunjes (Gyllensvärd); T.Lpm.: Abisko (Lindberg). – Not uncommon in northern Norway, found in Nsy: Saltvik (Holgersen); Nsi: Stormo (Holgersen): Nnö: Bonnå (Holgersen), and in several localities in TRi by Holgersen and Ardö; Fi: Alta (Munster). – East Fennoscandia: Li: Enare (Huldén, in litt.). – "Species" *alpinus* has been recorded from Austria, Bulgaria, Bohemia, Moravia, Slovakia, France, German D.R. and F.R., n. Italy, Romania, Poland, Ukraine, Moldavia, Altai Mts.; ssp. *neglectus* from Moravia, Latvia, Lithuania, Estonia, Tuva, Kamchatka.

Biology. "Bewohnt die höheren Gebirge von etwa 800 m aufwärts" (Wagner & Franz, 1961). "In woods, especially in felled areas" (Vilbaste, 1974). In *Calamagrostis* stands, also in sparse larch forest, tundra, and willow-alder thicket (Vilbaste, 1980).

Note. Concerning the status of this taxon, see Wagner (1958) and Schulz (1976).

408. *Jassargus (Sayetus) sursumflexus* (Then, 1902)
Text-figs. 2768-2774.

Deltocephalus sursumflexus Then, 1902: 189.

Resembling *J. flori*, general colour usually a little paler. Male pygofer as in Text-fig. 2768, genital valve and plates as in Text-fig. 2769 (but plates not diverging in repose), styles as in Text-fig. 2770, connective as in Text-fig. 2771, aedeagus as in Text-figs. 2772, 2773, 7th abdominal sternum in female as in Text-fig. 2774. Overall length (♂♀) 2.4-3.4 mm. – Nymphs as those of *J. distinguendus* but pale transverse band on lower part of frontoclypeus quite broad (approximately 1/4-1/5 width of frontoclypeus beneath), often with a pale irregularly shaped spot in the middle (Vilbaste, 1982).

Distribution. Rare in Denmark, found only on the west-coast of Jutland – NWJ: Hansted 23.VII.1962 (N. P. Kristensen) and WJ: Skallingen 17.VII.1980 (Trolle). – Not uncommon in southern Sweden, scarce and sporadic in the north, found in Sk., Bl., Hall., Sm., Öl., Nrk., Vrm., Dlr., Hls., P.Lpm. – Norway: found in Ø: Holen, Degernes 25.VIII.1960, and in AAy: Birkenes, Åmli by Holgersen. – Very rare in East Fennoscandia, found in Kb: Pielisjärvi (Kontkanen); ObN: Pisavaara Nature Reserve (Lindberg). – England, Belgium, German D.R. and F.R., Austria, Switzerland, Italy, Bohemia, Hungary, Romania, Poland, Estonia, Latvia, Ukraine.

Biology. In "Heiden, Hochmooren, Wäldern" (Kuntze, 1937). "Ausschliesslich auf der Randzone des *Molinia*-Bewuchses gefunden. Ich nehme an, dass *Molinia coerulea* die Nähr- und Brutpflanze darstellt" (Strübing, 1955). Hibernation in egg-stage

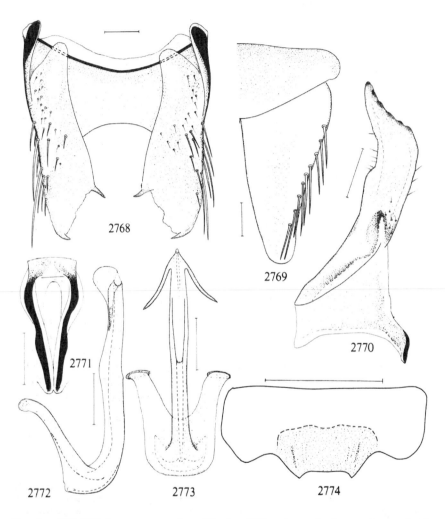

Text-figs. 2768-2774. *Jassargus sursumflexus* (Then). – 2768: male pygofer in ventral aspect (anal apparatus and genital plates and valve removed); 2769:genital plates and valve from below; 2770: right genital style from above; 2771: connective in dorsal aspect; 2772: aedeagus from the left; 2773: aedeagus in ventral aspect; 2774: 7th abdominal sternum in female from below (depressed under coverglass). Scale: 0.5 mm for 2774, 0.1 mm for the rest.

(Remane, 1958; Schiemenz, 1976). "Existenz-Optimum im hygrophilen Bereich" (Schiemenz, 1969b). One generation (Schiemenz, 1976). Adults in July and August.

409. *Jassargus (Sayetus) allobrogicus* (Ribaut, 1936)
Text-figs. 2775-2784.

Deltocephalus pseudocellaris J. Sahlberg, 1871: 314, p.p. (nec Flor, 1861).
Deltocephalus falleni J. Sahlberg, 1871: 315, p.p. (nec Fieber, 1869).
Deltocephalus allobrogicus Ribaut, 1936: 264.
Deltocephalus allobrogicus kocoureki, Dlabola, 1944: 97.

Resembling *J. flori*. Male pygofer as in Text-figs. 2775, 2776, genital valve and plates as in Text-fig. 2777, styles as in Text-fig. 2778, aedeagus with shaft and appendages much varying in shape as shown in Text-figs. 2779-2781, 2783, 2784. 7th abdominal sternum in female as in Text-fig. 2782. – Nymphs as in *flori* but middle bands on mesonotum roughly parallel; abdomen mostly pale, genital segment and preceding segment darker, sometimes only laterally; more rarely, the sides of the remaining sternites are to a certain extent also darkened (Vilbaste, 1982).

Distribution. Rare in Denmark, found in NWJ/NEJ: Thorup Strand 4.VII.1974 (E. Bøggild), and WJ: Skallingen 17.VII.1980 (Trolle). – Rare in Sweden, only found in Nb.: Karl Gustav 23.VII.1966 (Sundholm), and Nedertorneå 22.VII.1966 (Gyllensvärd). – So far not recorded from Norway. – Sporadic in southern and central East Fennoscandia, in south-western Finland commoner than *J. flori*, for the rest less common; found in Ab, N, Ta, Sa, Oa, Sb, Om, ObN, Ks. – France, Netherlands, German D.R. and F.R., Austria, Italy, Switzerland, Bohemia, Poland, Latvia, m. Russia.

Biology. "Surtout sur *Aira flexuosa* L." (Ribaut, 1936). On "Flachmooren" (Kuntze, 1937). In dryish fields, moist sloping meadows, cultivated fields, dry *Vaccinium* pine woods, dry *Calluna* heaths (Linnavuori, 1952a). "Scheinen als Larven zu überwintern" (Schiemenz, 1964). In pine forests (Vilbaste, 1974). "Its original habitats were dry heath forests, but it has spread to natural meadows on wasteland and to arable land, perhaps more frequently than the previous species [*J. flori*]" (Raatikainen & Vasarainen, 1976). Adults in July and August.

Text-figs. 2775-2782. *Jassargus allobrogicus* (Ribaut). – 2775: left pygofer lobe in male from outside; 2776: male pygofer in ventral aspect (anal apparatus and genital plates removed); 2777: male genital plates and valve from below; 2778: right genital style from above; 2779: aedeagus from the left; 2780: aedeagus in ventral aspect; 2781 (another specimen), shaft of aedeagus in ventral aspect; 2782: 7th abdominal sternum in female from below (depressed under coverglass). Scale: 0.5 mm for 2782, 0.1 mm for the rest.
Text-figs. 2783, 2784. *Jassargus allobrogicus* (Ribaut), v. *kocoureki* (Dlabola). – 2783: aedeagus in ventral aspect; 2784: apical part of shaft from the left. (After Dlabola, 1944).

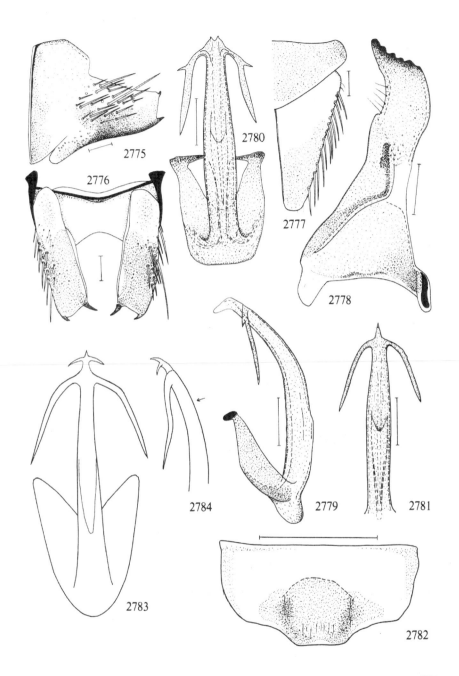

2775

2776

2780

2777

2778

2784

2779

2781

2783

2782

849

Genus *Mendrausus* Ribaut, 1947

Mendrausus Ribaut, 1947: 84.

Type-species: *Deltocephalus chyzeri* Horváth, 1897, by original designation.

As *Jassargus* but pygofer lobes without processes. 7th abdominal sternum in female as in Text-fig. 2790. No hyaline spots on lateral apical veins in fore wing. A monotypic genus.

410. *Mendrausus pauxillus* (Fieber, 1869)
Text-figs. 2785-2790.

Deltocephalus pauxillus Fieber, 1869: 217.
Deltocephalus chyzeri Horváth, 1897: 636.

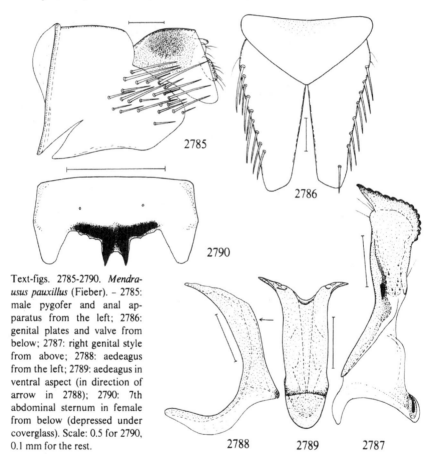

Text-figs. 2785-2790. *Mendrausus pauxillus* (Fieber). - 2785: male pygofer and anal apparatus from the left; 2786: genital plates and valve from below; 2787: right genital style from above; 2788: aedeagus from the left; 2789: aedeagus in ventral aspect (in direction of arrow in 2788); 2790: 7th abdominal sternum in female from below (depressed under coverglass). Scale: 0.5 for 2790, 0.1 mm for the rest.

850

Whitish yellow. Head wider than pronotum, vertex frontally rectangular, medially distinctly longer than pronotum, with two large, often obsolete orange patches, sometimes with six small dark streaks in a transverse row along fore border. Frontoclypeus with brownish transverse streaks. Anteclypeus broad, slightly narrower towards apex. Pronotum with four indistinct orange-coloured longitudinal bands, scutellum with basal triangles orange. Sub-brachypterous, fore wings apically rounded, shorter than abdomen, length of hind wings about 3/5 of fore wings. Fore wings pale, veins concolorous, apical cells sometimes dark spotted. Thoracic venter and abdomen entirely or largely pale, abdominal dorsum often dark banded and spotted. Anal tube in male dorsally with 2 black spots. Legs pale, fore and middle femora often transversely dark banded, tibiae dark spotted. Male pygofer and anal apparatus as in Text-fig. 2785, genital valve and plates as in Text-fig. 2786, style as in Text-fig. 2787, aedeagus as in Text-figs. 2788, 2789, 7th abdominal sternum in female as in Text-fig. 2790. Overall length (♂♀) 1.95-3.0 mm.

Distribution. Not found in Denmark, Norway and East Fennoscandia. – Very rare in Sweden, only one male found in Jtm.: Brunflo, Ope, 29.VII.1968, by Gyllensvärd. – German D.R., Bohemia, Moravia, Slovakia, Hungary, Yugoslavia, Poland, Ukraine, m. and w. Siberia, Tuva, Mongolia.

Biology. On *Festuca sulcata* (Emeljanov, 1964). Stenotopic, xerophilous; hibernates in egg-stage, one generation (Schiemenz, 1969b).

Genus *Pinumius* Ribaut, 1947

Pinumius Ribaut, 1947: 84.
  Type-species: *Deltocephalus areatus* Stål, 1858, by original designation.

Body elongate. Fore wings longer than abdomen, with a hyaline triangle around each transverse veinlet bounding 1st apical cell. Male pygofer ventro-apically with a hook-like process. Shaft of aedeagus distinct from socle, arising from its ventral part, bifurcate; phallotreme ventral, situated near base of shaft. Styles elongate. Connective short. 7th abdominal sternum in female with a broad and shallow trapezoidal incision. In Europe one species.

411. *Pinumius areatus* (Stål, 1858)
  Text-figs. 2791-2798.

Deltocephalus areatus Stål, 1858a: 193.

Frontal margin of vertex obtuse-angular in male, rectangular in female, yellowish, black spotted as in Text-fig. 2791, or with a more or less distinct brown transverse band extending between adocellar spots; frontoclypeus black with narrow light transverse streaks; anteclypeus slightly narrower towards apex, black, light-bordered; pronotum whitish to grey, along fore border dark spotted or mottled. Scutellum unspotted or with 2-6 small black spots. Fore wings whitish to grey, cells brownish edged. Venter largely dark, legs yellowish, black spotted. Male pygofer and anal apparatus as in Text-fig. 2792, genital plates and valve as in Text-fig. 2793, styles as in Text-fig. 2794, connective as in Text-fig. 2795, aedeagus as in Text-figs. 2796, 2797, caudal part of female abdominal venter as in Text-fig. 2798. Overall length (♂♀) 3.0-3.5 mm. – Abdomen in nymphs with 4 complete rows of dorsal hairs; vertex less than 1.5 × as long as wide, brownish yellow or brownish; thorax without longitudinal bands, usually variegated, only tips of wing pads whitish; frontoclypeus uniformly dark with short whitish arched lines and, occasionally, with a narrow longitudinal band (Vilbaste, 1982).

Distribution. Not found in Denmark, Sweden and Norway. – Very rare in East Fennoscandia, only found in Vib: Metsäpirtti, 3.VII.1866 (J. Sahlberg). – German D.R.

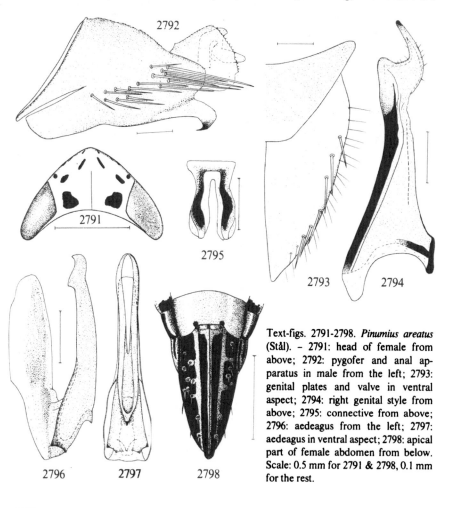

Text-figs. 2791-2798. *Pinumius areatus* (Stål). – 2791: head of female from above; 2792: pygofer and anal apparatus in male from the left; 2793: genital plates and valve in ventral aspect; 2794: right genital style from above; 2795: connective from above; 2796: aedeagus from the left; 2797: aedeagus in ventral aspect; 2798: apical part of female abdomen from below. Scale: 0.5 mm for 2791 & 2798, 0.1 mm for the rest.

and F.R., Austria, Italy, Hungary, Poland, Latvia, Lithuania, Estonia, n. Russia, Ukraine, Altai Mts., Kazakhstan, w. and m. Siberia, Tuva, Mongolia, Kirghizia; Nearctic region.

Biology. Among *Elymus arenarius* (Sahlberg, 1871). Stenotopic, xerophilous, hibernates in egg-stage, in Central Europe 2 generations (Schiemenz, 1969b). On sand dunes, dry meadows, *Koeleria*-patch, sandy river-bank (Vilbaste, 1974).

Genus *Diplocolenus* Ribaut, 1947

Diplocolenus Ribaut, 1947: 82.
Type-species: *Deltocephalus calceolatus* Boheman, 1845, by original designation.

Fore parts of body and fore wings as in *Psammotettix* but vertex longer than pronotum. Anterior tibiae with three setae in the inner, four in the outer dorsal row. Male pygofer with a spiniform process. Genital plates with a lateral incision and a dorsal tooth or tubercle; their macrosetae in disorder. Connective long, branches longer than stem, converging towards apices. Shaft of aedeagus arising from ventral end of socle, with two apical appendages. Phallotreme ventral, situated distinctly proximally of apex of shaft. Apodemes of 1st and 2nd abdominal sterna in male vestigial. In Denmark and Fennoscandia one species.

412. *Diplocolenus bohemani* (Zetterstedt, 1838)
Plate-figs. 202, 203, text-figs. 2799-2805.

Cicada bohemanni (sic) Zetterstedt, 1838: 290.
Deltocephalus calceolatus Boheman, 1845a: 23.

Straw-coloured to whitish yellow, shining, with black-brown markings varying in extension. Vertex on each side along fore border with a frontally concave arched line from apex of head to near ocellus; this line is often interrupted forming a larger median and a shorter lateral streak. In light specimens, a broad brownish yellow longitudinal band extends from each arched line to caudal margin of head. In darker individuals these bands are divided into brownish spots. Laterally and medially the bands are defined by ivory white lines. Frontoclypeus with dark transverse streaks. Pronotum with six brownish yellow longitudinal bands, sometimes also with some fuscous spots along fore border. Fore wings sometimes entirely light but usually with dark-bordered light veins, the streaks rarely forming closed cells. Usually only scattered short streaks or coils are present in distal end of cells. This description refers to the common form *calceolatus* (Boheman); in *f. typica,* the upper side is decorated with two black-brown transverse bands, the first occupying pronotum, or at least its caudal part, and basal parts of fore wings, the second extending across distal part of fore wings (Plate-fig. 203). Thoracic venter and abdomen in *f. typica* largely fuscous, in *f. calceolata* usually largely light; 7th abdominal sternum in female always with a large black patch medially (Text-fig. 2805). Legs light or with diffusely defined dark markings, at least hind tibiae black spotted.

853

Male pygofer and anal apparatus as in Text-fig. 2799, genital plates and valve as in Text-fig. 2800, styles as in Text-fig. 2801, connective as in Text-fig. 2802, aedeagus as in Text-figs. 2803, 2804. Overall length of males 3.7-4.6 mm, of females 4.0-5.1 mm. – Abdominal dorsum in nymphs with 4 complete rows of hairs, with four more or less distinct dark longitudinal bands, pattern pale brown, arched lines of frontoclypeus distinct down to its lower margin (Vilbaste, 1982).

Distribution. Scarce in Denmark, found mainly in central Jutland but also along the west-coast of the peninsula: WJ: Fjand. – Scarce in southern and northern Sweden,

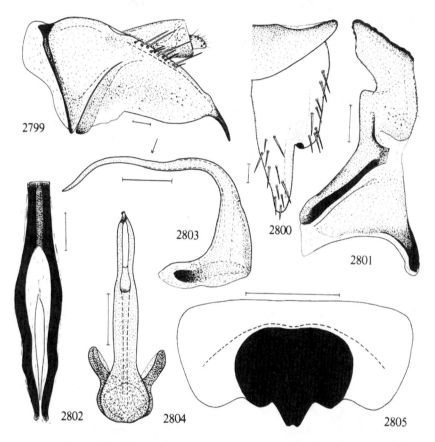

Text-figs. 2799-2805. *Diplocolenus bohemani* (Zetterstedt). – 2799: male pygofer and anal apparatus from the left; 2800: genital plates and valve from below; 2801: right genital style from above; 2802: connective from above; 2803: aedeagus from the left; 2804: aedeagus in ventral aspect; 2805: 7th abdominal sternum in female from below (depressed under coverglass). Scale: 0.5 mm for 2805, 0.1 mm for the rest.

commoner in the central part, found Sm.-Hls. – Norway: found in On: Dovre, by Boheman. – Scarce in southern and central East Fennoscandia, found in Ab, N-Kb; Vib, Kr. – Widespread in Europe, also recorded from Algeria, Tunisia, Moldavia, Altai Mts., Kazakhstan, Tuva, Mongolia, m. Siberia.

Biology. "Bewohnt Trockenrasen und Ericeten, in Mitteleuropa nur im Gebirge. . . . Sie ist ausgeprägt heliophil" (Wagner & Franz, 1961). In xerophilous, mesophilous and hygrophilous biotopes, the first-mentioned ones being preferred; heliophilous? – Hibernation takes place in the egg-stage, one generation (Schiemenz, 1969b). In dry meadows, on sand dunes (Vilbaste, 1974). – I found *D. bohemani* especially in forest glades and on sun-exposed hills etc. with *Calamagrostis arundinacea* (L.) Roth. Adults in June-August.

Genus *Verdanus* Oman, 1949

Verdanus Oman, 1949: 165.
Type-species: *Deltocephalus evansi* Ashmead, 1904, by original designation.

As *Diplocolenus*, upper side normally green. Apodemes of 1st and 2nd abdominal sterna in male vestigial. Male pygofer lobes with or without processes. Genital plates of male laterally indented but without dorsal tooth, their macrosetae usually arranged along lateral borders. Phallotreme terminal. In Denmark and Fennoscandia two species.

Key to species of *Verdanus*

1 Pygofer lobes in male in lateral aspect apically rounded (Text-fig. 2806), extending considerably beyond genital plates. Genital plates comparatively long, median border as long as basal margin (along valve). 7th abdominal sternum in female medially little projecting caudad (Text-fig. 2812) .... 413. *abdominalis* (F.)
– Pygofer lobes in male with angular ventral apex (Text-fig. 2813). Genital plates shorter, median border distinctly shorter than border along genital valve. 7th abdominal sternum in female medially strongly projecting caudad (in intact specimens more distinctly so than in Text-fig. 2819) .... 414. *limbatellus* (Zetterstedt).

413. *Verdanus abdominalis* (Fabricius, 1803)
Plate-fig. 204, text-figs. 2806-2812.

Cicada bicolor Fabricius, 1794: 40 (nec Olivier, 1790).
Cercopis abdominalis Fabricius, 1803: 98 (n.n.).
Cicada balteata Zetterstedt, 1838: 290.
Aphrodes juvenca Hardy, 1850: 425.
Deltocephalus abdominalis rufus J. Sahlberg, 1871: 329.

855

Upper side light green or yellowish green, rarely reddish (f. *rufa* Sahlb.), shining. Vertex unicolorous or sometimes with a pair of short oblique dark streaks near apex. Apical cells of fore wings often brownish bordered caudally, or entirely fumose. Wing dimorphous. In the sub-brachypterous form, the fore wings are about as long as abdomen, in macropters considerably longer. Frontoclypeus with black transverse streaks, in males often mutually confluent and with black patches on genae; sometimes the face is entirely black. Venter in females usually largely light, abdomen with a black median band; in males, thoracic venter and abdomen often largely black, only segmen-

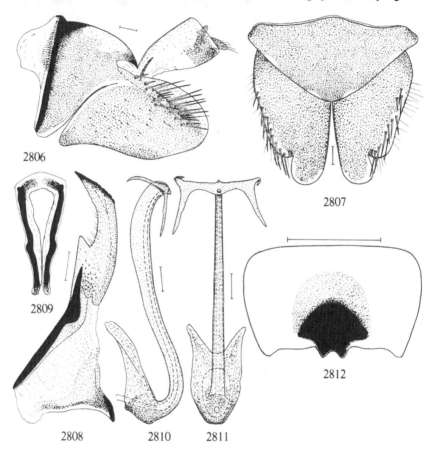

Text-figs. 2806-2812. *Verdanus abdominalis* (Fabricius). – 2806: pygofer and anal apparatus in male from the left; 2807: genital plates and valve from below; 2808: right genital style from above; 2809: connective from above; 2810: aedeagus from the left; 2811: aedeagus in ventral aspect; 2812: 7th abdominal sternum in female from below (depressed under coverglass). Scale: 0.5 mm for 2812, 0.1 mm for the rest.

tal borders narrowly light. Legs entirely light with tibiae black spotted, or partly black. Male pygofer and anal apparatus as in Text-fig. 2806, genital valve and plates as in Text-fig. 2807, genital styles as in Text-fig. 2808, connective as in Text-fig. 2809, aedeagus as in Text-figs. 2810, 2811, 7th abdominal sternum in female as in text-fig. 2812. Overall length of sub-brachypters 3.3-5.0 mm, of macropters about 4.7 mm. – Nymphs with 4 complete rows of hairs on abdominal dorsum; entire dorsal surface of anterior body more or less uniformly pale whitish green; frons without longitudinal band; ventral surface mainly black (Vilbaste, 1982).

Distribution. Denmark: very common in most districts. – Very common in Sweden, found in all provinces, Sk.-T.Lpm. – Common and widespread in Norway, AK, Ø-Fø. – East Fennoscandia: common and widespread, less common in the north. – Widespread in the Palaearctic region (in England represented by ssp. *juvencus* (Hardy)).

Biology. In "Stranddünen, Waldlichtungen, Wiesen" (Kuntze, 1937). In the drier meadow area of seashores, also in dryish fields, moist sloping meadows, peaty meadows, cultivated fields (Linnavuori, 1952a). Hibernation takes place in the egg-stage (Müller, 1957; Schiemenz, 1969b). Mainly mesophilous; 1 generation (Schiemenz, 1969b). "In cages the species fed on oats, and in fields it occurred on grasses" (Raatikainen & Vasarainen, 1976). Adults in May-October.

414. *Verdanus limbatellus* (Zetterstedt, 1828)
Text-figs. 2813-2819.

Cicada abdominalis var. *limbatella* Zetterstedt, 1828: 522.

Resembling *V. abdominalis*, on an average a little smaller, upper side often with more or less extended dark markings. In addition to the oblique streaks present on apex of head there are often a pair of large patches behind these and sometimes also two pairs of short longitudinal streaks at caudal border of vertex. Face often entirely or almost entirely black. Pronotum unicolorous green or with 6 brownish longitudinal bands. Fore wings often with dark streaks along veins, or the cells may be almost entirely filled up with brownish pigment. Wing dimorphous; fore wings in macropters slightly longer than abdomen, in sub-brachypters a little shorter. Venter usually largely black, even in females. Femora at least basally black, tibiae partly black and black dotted. Male pygofer and anal apparatus as in Text-fig. 2813, genital plates and valve as in Text-fig. 2814, styles as in Text-fig. 2815, connective as in Text-fig. 2816, aedeagus as in Text-figs. 2817, 2818, 7th abdominal sternum in female as in Text-fig. 2819. Overall length of sub-brachypters 3.0-4.5 mm, of macropters about 3.9 mm.

Distribution. Not found in Denmark. – Common in the north of Sweden, Dlr., Hls.-T.Lpm. – Widespread in Norway, AK-Fn and Fø. – Common in northern East Fennoscandia, found in N, Sb, Ks-Li; Lr. – N. Russia, Altai Mts., m. and n. Siberia, Tuva, Kamchatka.

Biology. "Auf mehr oder weniger trockenen Graswiesen, hauptsächlich innerhalb der subarktischen Region" (Lindberg, 1932). "Bis in die arktische Zone" (Lindberg, 1947). Adults in end of June-September.

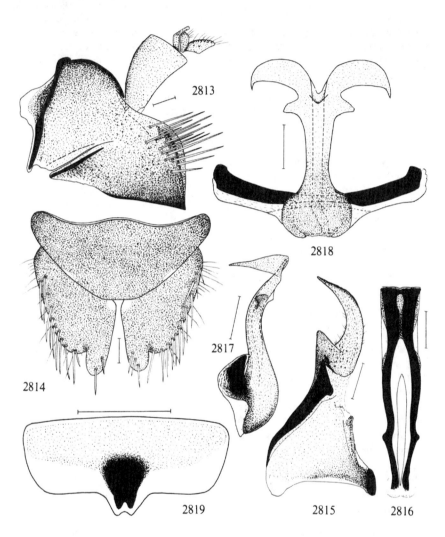

Text-figs. 2813-2819. *Verdanus limbatellus* (Zetterstedt). – 2813: male pygofer and anal apparatus from the left; 2814: genital plates and valve from below; 2815: right genital style from above; 2816: connective from above; 2817: aedeagus from the left; 2818: aedeagus in ventral aspect; 2819: 7th abdominal sternum in female from below (depressed under coverglass). Scale: 0.5 mm for 2819, 0.1 mm for the rest.

858

Genus *Arthaldeus* Ribaut, 1947

Arthaldeus Ribaut, 1947: 83.
  Type-species: *Cicada pascuella* Fallén, 1826, by original designation.

Small, moderately elongate leafhoppers. Head medially as long as pronotum or longer. Chaetation of fore tibiae as in *Psammotettix*. Upper side usually greenish or greenish yellow (colour after death often changing into yellow). Fore wings usually a little longer (♂) or a little shorter (♀) than abdomen. Male pygofer lobes each with a long thong-like appendage arising from ventral border; these appendages crossing each other. Genital plates comparatively long and narrow, each apically with a small dorsal tubercle, their macrosetae is disorder. Styles elongate, slender. Shaft of aedeagus simple, without appendages, phallotreme ventral. 7th abdominal sternum in female with a deep and narrow median incision. In Denmark and Fennoscandia two species.

Key to species of *Arthaldeus*

1 Frontoclypeus with a pale median longitudinal band widening
   towards apex (Plate-fig. 211). Appendages of male pygofer
   lobes simple, tapering towards apex (Text-fig. 2826) .. 416. *striifrons* (Kirschbaum)
– Frontoclypeus with transverse streaks but without a distinct
   light median band. Appendages of male pygofer lobes each
   with a long recurrent spine near apex (Text-fig. 2820) ..... 415. *pascuellus* (Fallén).

415. *Arthaldeus pascuellus* (Fallén, 1826)
  Text-figs. 2820-2825.

Cicada pascuella Fallén, 1826: 32.
Cicada punctipes Zetterstedt, 1828: 525.
Deltocephalus minki Fieber, 1869: 217.

Upper side in living specimens light green to yellowish green, shining, in dead specimens often yellowish. Vertex near apex with a pair of short dark oblique streaks. Frontoclypeus with brownish or black transverse streaks which often confluent medially. Genae often partly brown or fuscous. Pronotum, scutellum and fore wings unicolorous light, veins in apical part of fore wing sometimes indistinctly dark-bordered. Thoracic venter varying in colour, entirely or largely light to largely black-brown. Femora unicolorous light, or fore and middle femora transversely banded, hind femora longitudinally striped. Posterior tibiae, often also anterior and middle tibiae, fuscous spotted. Colour of abdomen also strongly varying, from largely black with narrow light segmental and lateral borders to largely light with an elongate black patch of varying size dorsally and ventrally. Lateral lobes of male pygofer (Text-fig. 2820) about as long as genital plates (Text-fig. 2821), the latter with a longitudinal dark streak not reaching apex, and with a small apical dark spot. Styles as in Text-fig. 2822, aedeagus as in Text-figs. 2823, 2824, 7th abdominal sternum in female as in Text-fig. 2825. Overall length of males 2.9-3.5 mm, of females 3.2-4.1 mm. – Thorax in nymphs dorsally with longitudinal bands or uniformly brown; vertex greenish yellow with broad whitish middle band; frontoclypeus with a wedge-shaped whitish spot lying below level of 4th-5th arch-lines; anterior margin of head forms a right angle (Vilbaste, 1982).

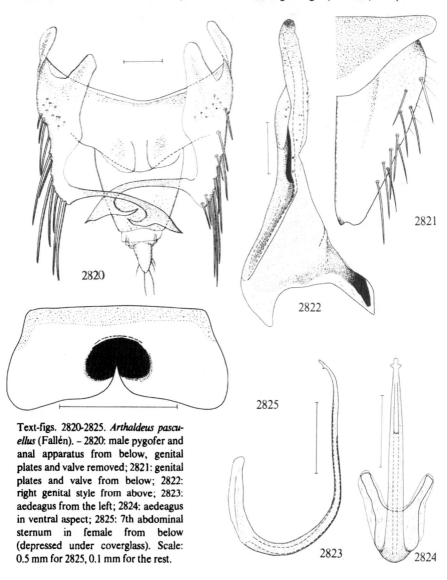

Text-figs. 2820-2825. *Arthaldeus pascuellus* (Fallén). – 2820: male pygofer and anal apparatus from below, genital plates and valve removed; 2821: genital plates and valve from below; 2822: right genital style from above; 2823: aedeagus from the left; 2824: aedeagus in ventral aspect; 2825: 7th abdominal sternum in female from below (depressed under coverglass). Scale: 0.5 mm for 2825, 0.1 mm for the rest.

860

Distribution. Very common in Denmark, even in small islands like Christiansø in the Baltic. – Very common in Sweden, Sk.-T.Lpm. – Widespread, apparently common in Norway, Ø, AK-Fi. – Very common in southern and central East Fennoscandia, Al, Ab, N-ObN, Ks; Vib, Kr. – Widespread in Europe, also in Algeria, Altai Mts., Kazakhstan, Tuva, Mongolia, m. and w. Siberia; Nearctic region.

Biology. In "Salzstellen, Flachmooren, Wäldern, Waldlichtungen, Wiesen" (Kuntze, 1937). In the *Juncus Gerardi-Festuca* zone and drier meadow area of seashores, also in moist sloping meadows, peaty meadows, tall-sedge bogs (Linnavuori, 1952a). In the "Molinio-Arrhenatheretea" (Marchand, 1953). "Eine ausserordentlich eurytope Art" (Strübing, 1955). Hibernation takes place in the egg stage (Müller, 1957). In Central Europe 2 generations (Remane, 1958; Schiemenz, 1969b). Mainly in mesophilous and hygrophilous biotopes (Schiemenz, 1969b). On grasses and *Juncus Gerardi* (Schaefer, 1973). "Nowadays most numerous on grass leys, pastures and moist meadows on wasteland" (Raatikainen & Vasarainen, 1976). Adults from end of June-October.

416. *Arthaldeus striifrons* (Kirschbaum, 1868)
   Plate-fig. 211, text-figs. 2826-2829.

Jassus (Deltocephalus) striifrons Kirschbaum, 1868: 139.

Resembling *pascuellus,* slightly more elongate, colour usually more purely light green or

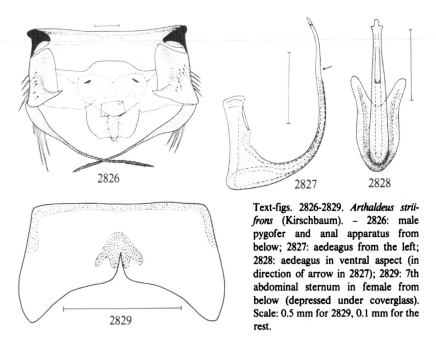

2826

2827   2828

2829

Text-figs. 2826-2829. *Arthaldeus strii-frons* (Kirschbaum). – 2826: male pygofer and anal apparatus from below; 2827: aedeagus from the left; 2828: aedeagus in ventral aspect (in direction of arrow in 2827); 2829: 7th abdominal sternum in female from below (depressed under coverglass). Scale: 0.5 mm for 2829, 0.1 mm for the rest.

yellow-green. Brownish or fuscous transverse streaks of frontoclypeus medially interrupted by an ivory-white longitudinal band (Plate-fig. 211). Male pygofer and anal apparatus as in Text-fig. 2826, genital plates and styles much as in *pascuellus*, aedeagus as in Text-figs. 2827, 2828. 7th abdominal sternum as in Text-fig. 2829. Overall length of males 3.0-3.6 mm, of females 3.5-4.4 mm. – Nymphs as in *pascuellus*, but wedge-shaped whitish spot on frontoclypeus extending over (after narrowing) to the longitudinal band of vertex; anterior margin of head forms a sharp angle (Vilbaste, 1982).

Distribution. Rare in Denmark, so far known from EJ: Tebbestrup Bakker; SZ: Vordingborg; LFM: Nagelsti; and B: Saltuna and Ypnasted. – Rare, locally common in southern Sweden, found in Sk.: Landskrona and Ven; Bl.: Jämjö, Torhamn; Öl.: several localities; Gtl.: Visby; Roma; Norrlanda, Hammars. – So far not found in Norway. – Rare and sporadic in East Fennoscandia, found in Ab: Pargas; Raisio; Korpo; N: Borgå. – Widespread in Europe, also recorded from Tunisia, Georgia, Moldavia.

Biology. On somewhat damp meadows, seashore meadows, etc. (Ossiannilsson, 1947b). Among *Trifolium* Lindberg, 1947). "Common in a tussocky meadow, where *Trifolium repens* grows" (Linnavuori, 1952a). In the *Arrhenatheretum* with *Juncus* (Schwoerbel, 1957). Probably univoltine (Remane, 1960). Adults in July and August.

Genus *Rosenus* Oman, 1949

Rosenus Oman, 1949: 170.
 Type-species: *Deltocephalus cruciatus* Osborn & Ball, 1898, by original designation.
Arctotettix Linnavuori, 1952c: 185.
 Type-species: *Deltocephalus abiskoensis* Lindberg, 1926, by original designation.

"Head wider than pronotum, anterior margin clearly defined but not angled; crown relatively flat or slightly convex, median length about one and one-half times length next eye. Pronotum short. Forewing with appendix small, inner anteapical cell closed basally. Male plates large, narrowed distally but truncate; spine-like setae uniseriate. Tenth segment of male broad basally, short, narrowed abruptly distally, sclerotized dorsally only as a narrow distal band. Pygofer of male setose, ventral margin thickened and terminating in a sharp, spine-like process which curves mesad. Aedeagus simple, dorsal apodeme flattened, short; shaft flattened distally and irregularly dentate laterally; gonopore terminal. Color sordid stramineous with fuscous and brown marks" (Oman, l.c.). In Fennoscandia one species.

417. *Rosenus abiskoensis* (Lindberg, 1926)
Text-figs. 2830-2836.

Deltocephalus abiskoensis Lindberg, 1926: 112.
Rosenus cruciatus auct., nec Osborn & Ball, 1898.

Above dirty yellow, shining, with brownish markings. Head not or little longer than pronotum, vertex with 5 pairs of brownish or fuscous dots: on each side a triangular or clavate spot near apex, a streak medially of ocellus, a transverse elongate spot behind these, and two relatively elongate spots at caudal border. Frontoclypeus with confluent brown transverse streaks, sutures of face and a longitudinal streak on anteclypeus dark. pronotum with six more or less distinct dark longitudinal streaks, sometimes also with small dark spots frontally; scutellum with dark lateral triangles and an interrupted dark longitudinal band. Fore wings in male distinctly longer, in female a little shorter than abdomen, semi-transparent with light veins here and there dark-bordered. Venter blackish, abdomen with narrow light segmental borders, apices of tibiae and tarsal seg-

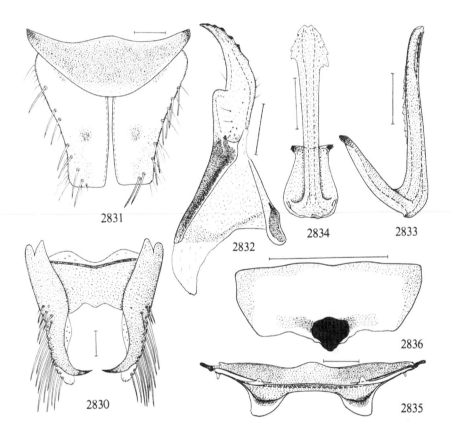

Text-figs. 2830-2836. *Rosenus abiskoensis* (Lindberg). – 2830: male pygofer from below (anal apparatus not considered); 2831: genital plates and valve from below; 2832: right genital style in male from above; 2833: aedeagus from the left; 2834: aedeagus in ventral aspect; 2835: 2nd abdominal sternum in male from above; 2836: 7th abdominal sternum in female from below (depressed under coverglass). Scale: 0.5 mm for 2836, 0.1 mm for the rest.

ments dark, hind tibiae dark spotted. Male pygofer lobes as in Text-fig. 2830, genital valve and plates as in Text-fig. 2831, style as in Text-fig. 2832, aedeagus as in Text-figs. 2833, 2834, 2nd abdominal sternum in male as in Text-fig. 2835, 7th abdominal sternum in female as in Text-fig. 2836. Overall length of males 2.3-2.6 mm, in females 2.6-3.3 mm. - Nymphs much as in *Sorhoanus*, pattern of dorsal surface of body distinct; frontoclypeus brown, with more or less distinct light transverse streaks and narrow median line, lower margin also light, anteclypeus light with brown median longitudinal band.

Distribution. Not found in Denmark. - Sweden: rare in northern Lapland, found in Lu.Lpm.: Sarek (Poppius); T.Lpm.: Abisko (F. Nordström, Lindberg, Ossiannilsson). - Norway: Fn: Björnesbukta 10 km N. Lakselv 12.VII.1947, 1 ♂ (Paul Ardö). - East Fennoscandia: very rare, found in Li: Utsjoki, Kevo 12.VII.1972, 19.VII.1972 and 1.VIII.1973 (Koponen and Ojala). - Mongolia, Tuva, m. and n. Siberia, n. Ural, Kamchatka, Maritime Territory. - Neartic region.

Biology. On *Dryas* spp. (Vilbaste, 1980). Adults in July and August.

Genus *Sorhoanus* Ribaut, 1947

Sorhoanus Ribaut, 1947: 85.
Type-species: *Cicada assimilis* Fallén, 1806, by original designation.

As *Arthaldeus*. Macrosetae of genital plates arranged in a lateral row. Male pygofer lobes each with a short and stout apical process. Connective short. Aedeagus with apical appendages. Phallotreme ventral, subapical. Fore wings with green or greenish yellow pigment. In Denmark and Fennoscandia two species.

Key to species of *Sorhoanus*

1   Larger, 3.5-4.75 mm. Apical and subapical cells of fore
     wings rarely dark-bordered. Veins of hind wings light.
     Shaft of aedeagus in lateral aspect S-curved (Text-fig. 2847).
     7th abdominal sternum in female as in Text-fig. 2851 .... 419. *assimilis* (Fallén)
-    Smaller, 3.1-3.75 mm. Apical and subapical cells of fore wings
     partly fuscous bordered. Veins of hind wings fuscous. Shaft of

Text-figs. 2837-2846. *Sorhoanus xanthoneurus* (Fieber). - 2837: right pygofer lobe in male from the right; 2838: genital plates and valve from below; 2839: right genital style from above; 2840: aedeagus from the left; 2841: aedeagus and connective in ventral aspect; 2842: 1st abdominal sternum in male from above; 2843: same in frontal aspect; 2844: 2nd and 3rd abdominal sterna in male from above; 2845: 7th abdominal sternum in female from below (depressed under coverglass); 2846: caudal margin of 7th abdominal sternum in female (another specimen). Scale: 0.5 mm for 2845 and 2846, 0.1 mm for the rest.

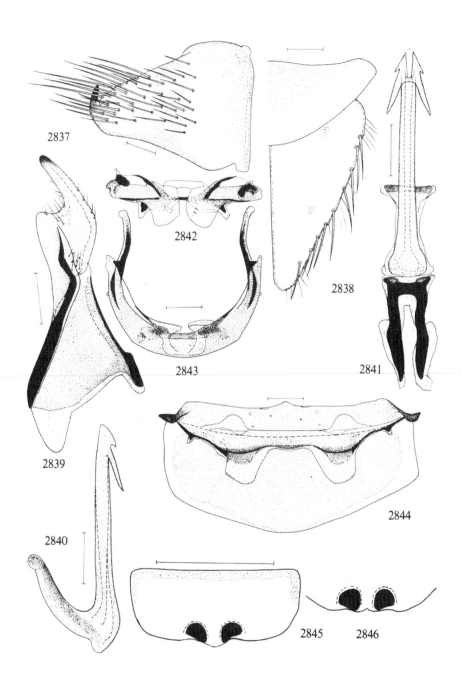

2837

2842

2838

2841

2843

2839

2844

2840

2845

2846

865

aedeagus straight (Text-fig. 2840). 7th abdominal sternum in
female as in Text-figs. 2845, 2846 .... 418. *xanthoneurus* (Fieber)

418. *Sorhoanus xanthoneurus* (Fieber, 1869)
Text-figs. 2837-2846.

Deltocephalus xanthoneurus Fieber, 1869: 219.

Resembling *S. assimilis,* smaller, colour more sordid greenish yellow, dark markings
often more distinct and more extended. Male pygofer lobes as in Text-fig. 2837, genital
plates and valve as in Text-fig. 2838, styles as in Text-fig. 2839, aedeagus and connective
as in Text-figs. 2840, 2841, 1st abdominal sternum in male as in Text-figs. 2842, 2843,
2nd and 3rd abdominal sterna in male as in Text-fig. 2844, 7th abdominal sternum in
females as in Text-figs. 2845, 2846. Overall length of males 3.1-3.4 mm, of females 3.3-
3.75 mm. – Abdomen of nymphs with two wide brownish longitudinal bands, each containing two rows of whitish spots; spots of median rows may coalesce and appear as 4
longitudinal bands. Frontoclypeus brown or dark brownish yellow with pale arched
lines only in upper region, a large unicoloured darker spot present on lower part. Pattern of dorsal surface of body distinct; frontoclypeus brown; intersegmental cuticle of
abdomen mostly red;double spots of abdomen equidistant from one another and from
lateral margin of the band (Vilbaste, 1982).

Distribution. Very rare in Denmark, known only from WJ: Hørbylunde 21.VII.1966
(Trolle). – Widespread, not uncommon in Sweden, Sk.-Nb. – Norway: found in Ø, HEs,
HEn, VAy, Ri, STi. – Comparatively common in southern and central Fennoscandia,
found in Ab, N, Ka, Sb, Kb, Om, Ks; Kr. – England, Scotland, France, Spain,
Netherlands, German D.R. and F.R., Austria, Bohemia, Poland, Estonia, Latvia,
Lithuania, n. Russia, Altai Mts., Mongolia, Tuva, Kamchatka; Nearctic region.

Biology. Tyrphobiont, present in all types of bogs and marshes (Linnavuori, 1952a).
Hibernates in egg-stage, 1 generation (Remane, 1958). "Hochmoorbewohner, ganz
überwiegend im Inneren der Moore, im Bereich der *Sphagnum-Eriophorum vaginatum-*
Bestände" (Wagner & Franz, 1961). In raised peat bogs, in fens (Vilbaste, 1974). Adults
in July-September.

419. *Sorhoanus assimilis* (Fallén, 1806)
Text-figs. 2847-2851.

Cicada assimilis Fallén, 1806: 22.
Deltocephalus fuscosignatus Dahlbom, 1851: 195 (sec. spec. typ.).

Light green or greenish yellow, shining. Head in dorsal aspect in males about as long as,
in females somewhat longer than pronotum medially, vertex with or without a pair of
short brownish oblique streaks near apex and a still shorter streak behind each ocellus.
Frontoclypeus often with brownish yellow transverse streaks. Pronotum unicolorous or

with indistinct dark patches. Fore wings semi-transparent with yellow veins, apical cells sometimes faintly dark-bordered. Veins of hind wings light. Thoracic venter largely light, legs light, femora often with dark patches, tibiae dark spotted. Abdomen in male dorsally black with yellow lateral and segmental borders, ventrally light with a broad black longitudinal band interrupted by yellow segmental margins, genital valve with a basal black patch, genital plates each with a small black spot near base and another just distally of middle. Female abdomen dorsally black with broad yellow lateral margins, ventrally light with a black patch basally and two black spots on distal border of 7th abdominal sternum (Text-fig. 2851). Aedeagus as in Text-figs. 2847, 2848, 1st abdominal sternum in male as in Text-fig. 2849, 2nd and 3rd abdominal sterna in male as in Text-fig. 2850. Overall length (♂♀) 3.5-4.75 mm. – Nymphs as in *xanthoneurus*, pattern on dorsal surface of body quite indistinct; nymph almost unicoloured;

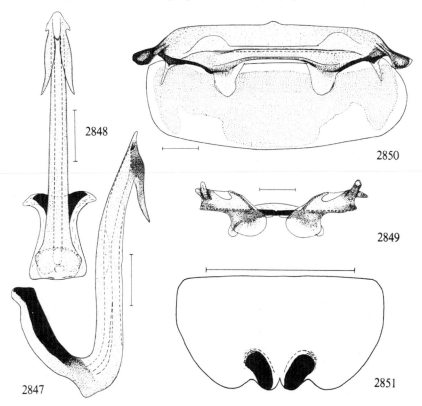

Text-figs. 2847-2851. *Sorhoanus assimilis* (Fallén). – 2847: aedeagus from the left; 2848: aedeagus in ventral aspect; 2849: 1st abdominal sternum in male from above; 2850: 2nd and 3rd abdominal sterna in male from above; 2851: 7th abdominal sternum in female from below (depressed under coverglass). Scale: 0.5 mm for 2851, 0.1 mm for the rest.

frontoclypeus brownish yellow; intersegmental cuticle of abdomen concolorous; of the double spots of abdomen, the lateral ones lie very close to lateral margin of the band (Vilbaste, 1982).

Distribution. Widespread but rather uncommon in Denmark. – Widespread, not very common in Sweden, Sk.-Nb. – Recorded from a few localities in southern Norway by Siebke (1874) but these records have not been revised. – Comparatively scarce and sporadic in southern and central East Fennoscandia, Al, Ab, N, St, Ta, Sb, Kb; Vib. – Spain, France, Belgium, Netherlands, German D.R. and F.R., Austria, Italy, Bohemia, Moravia, Slovakia, Bulgaria, Hungary, Romania, Yugoslavia, Poland, Estonia, Latvia, Lithuania, s. Russia, Ukraine, Moldavia, Georgia, Tunisia.

Biology. "Auf Mooren, in der Ebene vorwiegend auf Flachmooren" (Wagner & Franz, 1961). In moist meadows, in fens (Vilbaste, 1974). Adults in July-September.

Genus *Lebradea* Remane, 1959

Lebradea Remane, 1959: 386.
Type-species: *Lebradea calamagrostidis* Remane, 1959, by original designation.

Resembling *Sorhoanus,* somewhat more elongate, vertex a little shorter and more truncate, shorter than pronotum. Macropterous, fore wings longer than abdomen. Dorsal chaetation of fore tibiae varying. Male pygofer lobes (Text-fig. 2852) ventro-apically with a stout black process, apically with a membraneous appendage. Genital plates shorter than pygofer lobes, their macrosetae arranged in a lateral row; plates also with a conspicuous lateral fringe of long fine hairs. Styles with apical apophysis hook-like, distal margin serrate. Aedeagus apically with a pair of processes or appendages, phallotreme ventral, subterminal. 7th abdominal sternum in females short, not completely covering 8th sternum, caudal margin with a long, almost parallel-sided process (Text-fig. 2858). In Europe two, in Fennoscandia one species.

Key to species of *Lebradea*

1   Apical appendages of aedeagus long, longer than width of
     shaft (Text-fig. 2856) .... 420. *flavovirens* (Gillette & Baker)
–   Apical appendages of aedeagus short, about 1/2 width of
     shaft .... *calamagrostidis* Remane, 1959.

420. *Lebradea flavovirens* (Gillette & Baker, 1895)
     Text-figs.2852-2858.

Deltocephalus flavovirens Gillette & Baker, 1895: 87.
Thamnotettis (sic) *karafutonis* Matsumura, 1911: 29.
Lebradea icarus Ossiannilsson, 1976: 31.

868

Elongate, light green or yellowish green. In general aspect resembling *Elymana* spp. Anteclypeus narrowing towards apex. Frontoclypeus with brown transverse streaks or largely brown. Vertex anteriorly with arched "interocellar lines" extending from apex to each ocellus. Pronotum, scutellum and fore wings light. Thoracic venter light, abdominal dorsum largely black, venter largely light. Female pygofer light, saw-case (or its basal 2/3) black. Male pygofer lobes as in Text-fig. 2852, genital valve and plates

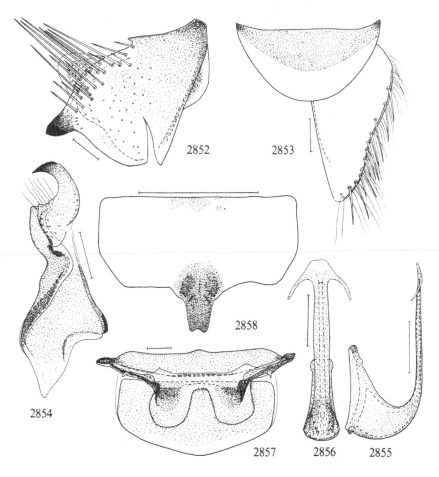

Text-figs. 2852-2858. *Lebradea flavovirens* (Gillette & Baker). – 2852: right pygofer lobe in male from the right; 2853: left genital plate and valve from below; 2854: right genital style from above; 2855: aedeagus from the left; 2856: aedeagus in ventral aspect; 2857: 2nd and 3rd abdominal sterna in male from above; 2858: 7th abdominal sternum in female from below (depressed under coverglass). Scale: 0.5 mm for 2858, 0.1 mm for the rest.

as in Text-fig. 2853, style as in Text-fig. 2854, aedeagus as in Text-figs. 2855, 2856. 2nd and 3rd abdominal sterna in male as in Text-fig. 2857. 7th abdominal sternum in female as in Text-fig. 2858. Overall length of one male 4.12 mm, of two females 4.62 and 4.85 mm.

Distribution. Not found in Denmark, Sweden and Norway. – East Fennoscandia: found in Ab: Lojo (Krogerus); N: Helsinge, Råby, Vanda 1970 (G. Söderman), Helsingfors, Drumsö (Söderman), Tvärminne (Albrecht), Strömfors (Albrecht). – Mongolia, Tuva, n. and m. Siberia, Kamchatka, Kurile Islands, Sakhalin, Korean Peninsula, Maritime Territory. – Nearctic region.

Biology. "Lives on *Calamagrostis* spp. and is therefore found in dry as well as in marshy habitats" (Vilbaste, 1980). Adults in June and July.

Genus *Cosmotettix* Ribaut, 1942

Cosmotettix Ribaut, 1942: 267.
  Type-species: *Jassus (Jassus) caudatus* Flor, 1861, by original designation.
Palus De Long & Sleesman, 1929: 85 (Preoccupied).
  Type-species: *Flexamia (Palus) delector* De Long & Sleesman, 1929, by original designation.
Airosus Ribaut, 1952: 126 (subgenus).
  Type-species: *Cicada costalis* Fallén, 1806, by original designation.

Comparatively small, fairly elongate species. Head wider than pronotum, frontally angular, transition between vertex and frontoclypeus rounded angular. Frontoclypeus elongate. Anterior tibiae with one seta in inner, 4 in outer dorsal row. Male anal tube dorsally well developed. Pygofer lobes reaching far beyond apices of genital plates. Lateral margins of genital plates concave. Macrosetae of genital plates arranged along lateral borders. Connective with branches approximately parallel. A transverse sclerotized and heavily pigmented strand connecting dorsal end of aedeagal socle with anal tube. Phallotreme terminal. In Denmark and Fennoscandia 5 species.

Key to species of *Cosmotettix*

1 Ventral margin of male pygofer without process. 7th abdominal sternum in female with a median incision (*Cosmotettix* s. str.) 2
– Ventral margin of male pygofer with a pointed process (Text-figs. 2891, 2892). 7th abdominal sternum in female without median incision (Subgenus *Airosus* Ribaut). Fore wings near apex each with a pair of well delimited black spots (Plate-fig. 208). Head without distinct dark markings .... 425. *costalis* (Fallén)
2 (1) Head without distinct dark markings frontally. Ground-colour brownish .... 424. *panzeri* (Flor)
– Head on transition between vertex and face with distinct black

421. *Cosmotettix (Cosmotettix) caudatus* (Flor, 1861)
Plate-fig. 205, text-figs. 2859-2867.

Jassus (Jassus) caudatus Flor, 1861a: 351.
Deltocephalus scriptifrons J. Sahlberg, 1871:345.

Whitish yellow with orange tinge to pale brownish yellow, shining. Head frontally on each side with a whitish or light yellow transverse clavate marking, its thicker end enclosing ocellus, "handle" directed towards apex of head. This light marking is enclosed in a more or less complete black frame as in Text-fig. 2859. Sutures of face usually dark. Anteclypeus somewhat varying in shape. Vertex and pronotum especially in females with indistinct dirty yellow or brownish yellow longitudinal bands, two on vertex, 4 or 6 on pronotum. Fore wings with light veins, cells sometimes with dark streaks, apical part more or less distinctly fumose. Males are smaller, more elongate and more bright orange-yellow than females, these being more brownish yellow. Fore wings in males considerably longer, in females slightly longer than abdomen. Thoracic venter largely light, legs light, hind tibiae black black spotted. Abdomen in males dorsally black with broadly light lateral margins, ventrally light with a black longitudinal band on basal segments, in females dorsally light with four black longitudinal bands or rows of spots, ventrally light with a basal black median patch. Male pygofer and anal apparatus as in Text-fig. 2860, genital valve and plates as in Text-fig. 2861, style as in Text-fig. 2862, aedeagus (with transverse strand) as in Text-figs. 2863-2865, 2nd and 3rd abdominal sterna in male as in Text-fig. 2866, 7th abdominal sternum in female as in Text-fig. 2867. Overall length of males 3.7-4.0 mm, of females 4.0-4.4 mm.

Distribution. Very rare in Denmark, found only in NEZ: Ruderhegn 29.VIII.1915 (Oluf Jacobsen). – Widespread but local in Sweden, found Sk.-Vb. – Norway: HEs: Kirkenes, 30.VIII.1961 (Holgersen). – East Fennoscandia: scarce and sporadic in southern and central parts, found in Al, Ab, N, Ta, Oa, Kb, Om; Kr. – England,

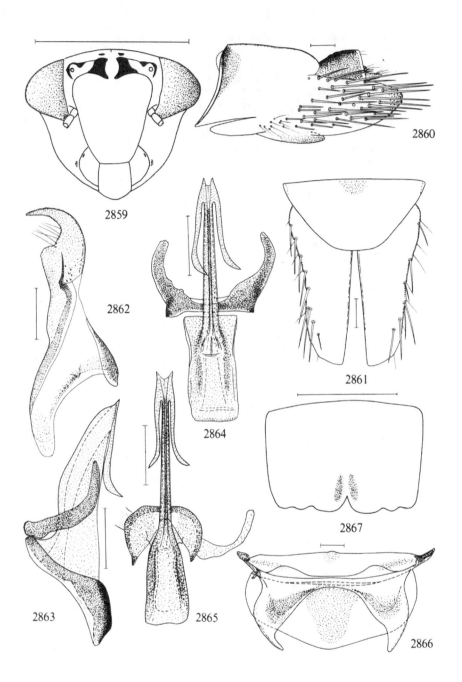

2859

2860

2861

2862

2863

2864

2865

2866

2867

France, German D.R. and F.R., Moravia, Latvia, Estonia, n. and m. Russia, Altai Mts., Kazakhstan.

Biology. On *Carex vesicaria* (Sahlberg, 1871). Tyrphophilous, in tall-sedge bogs (Linnavuori, 1952a). Adults in July-October.

422. *Cosmotettix (Cosmotettix) edwardsi* (Lindberg, 1924)
Plate-fig. 206, text-figs. 2868-2875.

Thamnotettix edwardsi Lindberg, 1924a: 28.

Resembling *caudatus*, differing mainly in markings of head (Plate-fig. 206, text-fig. 2868), by structure of male genitalia and shape of 7th abdominal sternum in female (Text-fig. 2875). Male pygofer lobes as in Text-fig. 2869, genital valve and plates as in Text-fig. 2870, styles as in Text-fig. 2871, aedeagus as in Text-figs. 2872, 2873, 2nd and 3rd abdominal sterna in male as in Text-fig. 2874. Overall length (♂♀) 3.4-4.2 mm.

Distribution. Very rare in Denmark, one female found in B: Bastemose 7.VIII.1966 (Trolle). – Scarce and sporadic in Sweden, found in Gtl.: Tingstäde 3.VIII.1935 (Ossiannilsson); Sdm.: Hyndevad 10.VII.1936 (Ehnbom); Upl.: Värmdö, vicinity of St. Björknäs 1946 (Ossiannilsson), Uppsala, Saltmossen 10.VIII.1952, and Vaksala, Jälla 12.VII.1953 (Ossiannilsson); Hls.: Edsbyn, Hobergstjärn 1.VIII.1980 (Henriksson); Äng.: Ådalsliden, Krångesjön 20.VII.1966 (Ossiannilsson); Jmt.: Rödön, Krokom 27.VII.1933 (Fahlander). – Not recorded from Norway. – Rare in East Fennoscandia, found in Ab: Raisio; N: Tvärminne; Ta: Lammi; St: Säkylä; Sb: Kiuruvesi; Kb: Joensuu; Kontiolahti; Hammaslahti; also in Oa (Raatikainen & Vasarainen, 1973). – Estonia, Kazakhstan, Kirghizia.

Biology. "Auf Sumpfwuesen mit *Carex lasiocarpa*" (Lindberg, 1947). Tyrphobiont, in tall-sedge bogs (Linnavuori, 1952a). Adults in July and August.

423. *Cosmotettix (Cosmotettix) evanescens* Ossiannilsson, 1976
Text-figs. 2876-2883.

Cosmotettix evanescens Ossiannilsson, 1976: 31.

Resembling *C. caudatus*. Vertex anterior with a more or less distinct pale transverse stripe enclosing ocelli and framed by a pair of medially interrupted transverse black stripes (Text-fig. 2876). A broad pale longitudinal band medially on vertex; pronotum

---

Text-figs. 2859-2867. *Cosmotettix caudatus* (Flor). – 2859: face of female; 2860: pygofer and anal apparatus in male from the left; 2861: genital plates and valve from below; 2862: right genital style from above; 2863: aedeagus with transverse strand from the left; 2864: same in ventral aspect; 2865: same of another specimen in ventral aspect; 2866: 2nd and 3rd abdominal sterna in male from above; 2867; 7th abdominal sternum in female from below (depressed under coverglass). Scale: 1 mm for 2859, 0.5 mm for 2867, 0.1 mm for the rest.

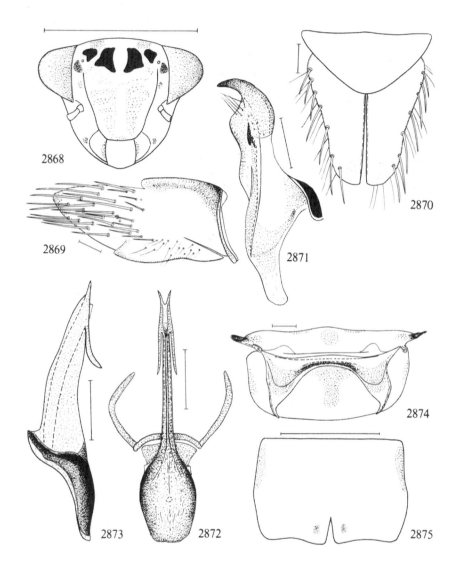

Text-figs. 2868-2875. *Cosmotettix edwardsi* (Lindberg). – 2868: face of female. – 2869: right pygofer lobe in male from the right; 2870: male genital plates and valve from below; 2871: right genital style from above; 2872: aedeagus and transverse strand in ventral aspect; 2873: aedeagus from the left (transverse strand not considered); 2874: 2nd and 3rd abdominal sterna in male from above; 2875: 7th abdominal sternum in female from below (depressed under coverglass). Scale: 1 mm for 2868, 0.5 mm for 2875, 0.1 mm for the rest.

874

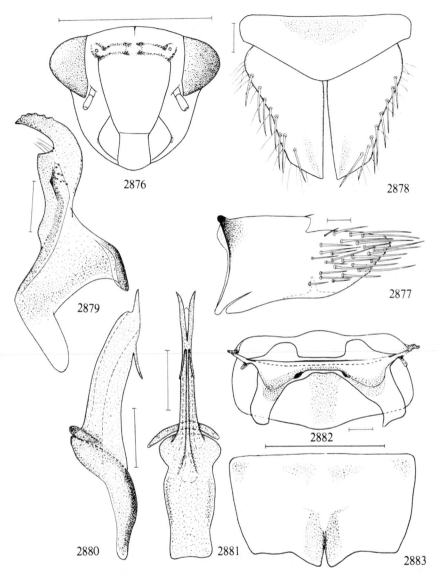

Text-figs. 2876-2883. *Cosmotettix evanescens* Ossiannilsson. – 2876: face of female; 2877: left pygofer lobe in male from the left; 2878: genital plates and valve from below; 2879: right genital style from above; 2880: aedeagus from the left; 2881: aedeagus in ventral aspect; 2882: 2nd and 3rd abdominal sterna in male from above; 2883: 7th abdominal sternum in female from below (depressed under coverglass). Scale: 1 mm for 2876, 0.5 mm for 2883, 0.1 mm for the rest.

with three pale longitudinal stripes; these markings on head and pronotum are sometimes almost entirely obliterated. Male pygofer lobes as in Text-fig. 2877, genital plates and valve as in Text-fig. 2878, style as in Text-fig. 2879, aedeagus as in Text-figs. 2880, 2881, 2nd and 3rd abdominal sterna in male as in Text-fig. 2882, 7th abdominal sternum in female as in Text-fig. 2883. Overall length of males 3.5-3.6 mm, of females 4.0-4.2 mm.

Distribution. So far only recorded from Sweden. Rare: found in Hall.: Tönnersjö, Hilleshult 8.VIII.1934; Sm.: Vrå 3.VIII.1934, Annerstad 29.VII., 30.VII.1934; Ög.: Vånga, Grensholmen 23.VIII.1932; Vg.: Ryd, Blängs mosse 10.VII.1953 (holotype); Upl.: Uppsala, vicinity of Norby 28.VIII.1949 (Ossiannilsson). Bo Henriksson found the present species in several localities in Hls.: Edsbyn in July and September 1972-1980, and in Hls.: Los, Kvarnsberg 28.VII.1980.

Biology not studied. Adult in July-September.

424. *Cosmotettix (Cosmotettix) panzeri* (Flor, 1861)
   Plate-fig. 209, text-figs. 2884-2890.

Jassus (Deltocephalus) panzeri Flor, 1861a: 265.
Deltocephalus concaviceps Lindberg, 1924a: 20.

Brownish yellow to almost chestnut brown, shining. Head frontally obtusely angular, as long as (♀) or almost as long (♂) as pronotum, vertex without markings. Frontoclypeus with only indistinctly darker transverse streaks, ocelli lighter than surrounding surface. Anteclypeus roughly parallel-sided. Pronotum with vestigial light longitudinal bands. Scutellum often a little lighter than head and pronotum. Fore wings somewhat longer than abdomen, veins light, dark-bordered, apical part of fore wing darker. Venter largely greyish yellow to brownish yellow to brownish, abdominal dorsum and median part of venter black. Male pygofer lobes strongly setose (Text-fig. 2884), genital plates and valve as in Text-fig. 2885, styles as in Text-fig. 2886, aedeagus and transverse strand as in Text-figs 2887, 2888, 2nd and 3rd abdominal sterna in male as in Text-fig. 2889, 7th abdominal sternum in female as in Text-fig. 2890. Overall length of males 3.1-3.6 mm, of females 3.5-4.0 mm.

Distribution. Scarce in Denmark, known from a few localities in EJ and NEZ. – Widespread, comparatively uncommon in Sweden, Sk.-P.Lpm. – Norway: found in HEs: Eidskog 7.IX.1974 (Hågvar & al.), Magnor 30.VIII.1961 (Holgersen); Ry: Kopervik 25.VII.1952 (Holgersen). – Comparatively scarce and sporadic in southern and central East Fennoscandia, Al, Ab-St, Sa, Kb; Vib, Kr.– England, Scotland, France, German D.R. and F.R., Bohemia, Poland, Estonia, Latvia, Lithuania, n. Russia.

Biology. In "Hochmooren; an *Eriophorum*" (Kuntze, 1937). Tyrphobiont; in tall-sedge bogs (Linnavuori, 1952a). "Tyrphobiont, in Hochmooren" (Strübing, 1955). On *Eriophorum* (Schiemenz, 1976). Hibernates in the egg-stage, 1 generation (Remane, 1958). Adults in end of June-September.

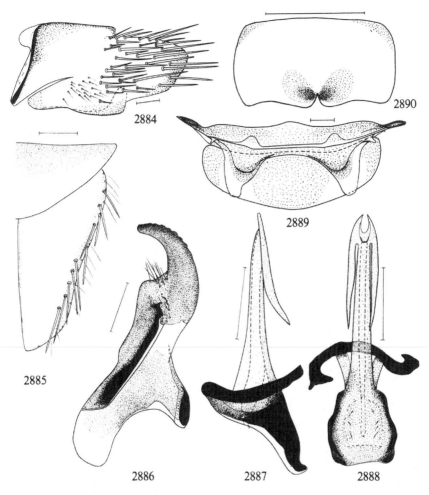

Text-figs. 2884-2890. *Cosmotettix panzeri* (Flor). – 2884: left pygofer lobe in male from the left; 2885: genital plates and valve from below; 2886: right genital style from above; 2887: aedeagus from the right (with transverse strand); 2888: aedeagus with transverse strand in ventral aspect; 2889: 2nd and 3rd abdominal sterna in male from above; 2890: 7th abdominal sternum in female from below (depressed under coverglass). Scale: 0.5 mm for 2890, 0.1 mm for the rest.

425. *Cosmotettix (Airosus) costalis* (Fallén, 1826)
    Plate-fig. 208, text-figs. 2891-2898.

Cicada costalis Fallén, 1826: 32.

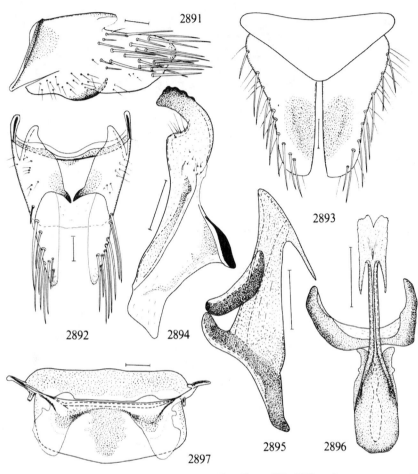

Text-figs. 2891-2898. *Cosmotettix costalis* (Fallén). – 2891: left pygofer lobe in male from the left; 2892: male pygofer in ventral aspect (genital plates, valve and anal apparatus removed); 2893: genital plates and valve from below; 2894: right genital style from above; 2895: aedeagus with transverse strand from the left; 2896: same in ventral aspect; 2897: 2nd and 3rd abdominal sterna in male from above; 2898: 7th abdominal sternum in female from below (depressed under coverglass). Scale: 0.5 mm for 2898, 0.1 mm for the rest.

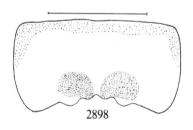

Deltocephalus bipunctipennis Boheman, 1845b: 156.

Whitish to yellow white to light orange-coloured, shining. Head frontally in male rounded obtuse-angular, somewhat shorter than pronotum, unicolorous light yellow or orange-coloured, in female more acute-angular, slightly longer than pronotum, unicolorous yellowish white. Pronotum and scutellum entirely light. Fore wings longer than abdomen, apically with markings: 1st and 4th apical cells each with a black spot, the cell proximally of 1st apical cell often with a less distinctly defined fuscous spot, and with a milky-white spot enclosing each transverse vein. Veins in apical part of fore wing partly dark-bordered, 2nd and 3rd apical cells distally dark, 2nd and 4th apical cells each with a hyaline spot; a hyaline spot present also in distal part of median subapical cell. Thoracic venter light, hind tibiae black spotted, apically black. 1st and 3rd tarsal segments apically black. Abdominal tergum with 3 black longitudinal bands; venter in male with a black median longitudinal band near base and segmental black transverse streaks, genital plates with a large joint black spot; abdominal venter in female largely light, 7th sternum (Text-fig. 2898) with a pair of black spots at caudal border. Male pygofer lobes as in Text-figs. 2891, 2892, genital valve and plates as in Text-fig. 2893, styles as in Text-fig. 2894, aedeagus and transverse strand as in text-figs. 2895, 2896, 2nd and 3rd abdominal sterna in male as in text-fig. 2897. Overall length of males 3.0-3.3 mm, of females 3.2-3.6 mm.

Distribution. Fairly common in Denmark. – Fairly common and widespread in Sweden, Sk.-Nb. – Norway: only found in VE: Hem 1.VIII.1976 (Holgersen). – Fairly common in southern and central East Fennoscandia, found in Al, Ab-Oa, Kb, Om; Vib. – England, German D.R. and F.R., Bohemia, Estonia, Latvia, Lithuania, n. and m. Russia, Kazakhstan, Mongolia.

Biology. In "Flachmooren" (Kuntze, 1937). In "Gross-seggenwiesen" (Remane, 1958). In damp meadows and fens (Vilbaste, 1974). "On alluvial fens and in wet meadows" (Raatikainen & Vasarainen, 1976). Adults in June-August.

Genus *Boreotettix* Lindberg, 1952

Boreotettix Lindberg, 1952: 145.
Type-species: *Cosmotettix serricauda* Kontkanen, 1949, by original designation.

Body elongate. Head frontally angular in female, rounded in male, transition between vertex and face rounded angular. Head wider than pronotum, in male somewhat shorter than, in female as long as the latter. Anteclypeus narrowing towards apex. Frontoclypeus comparatively short. Anterior tibiae with 3-4 setae in inner dorsal row. Macropterous in both sexes. Male pygofer longer than genital plates, without appendages, ventral part of apical margin serrate, strongly sclerotized. Genital plates with macrosetae and fine hairs arranged along lateral margins. Connective elongate, parallel-sided, stem short. Apical apophysis of style bird's-head-shaped, "crown" and "neck" curled. Shaft of aedeagus slender, with dorsal and lateral carinae, phallotreme terminal. Caudal margin of 7th abdominal sternum in female with a median incision. In Europe one species.

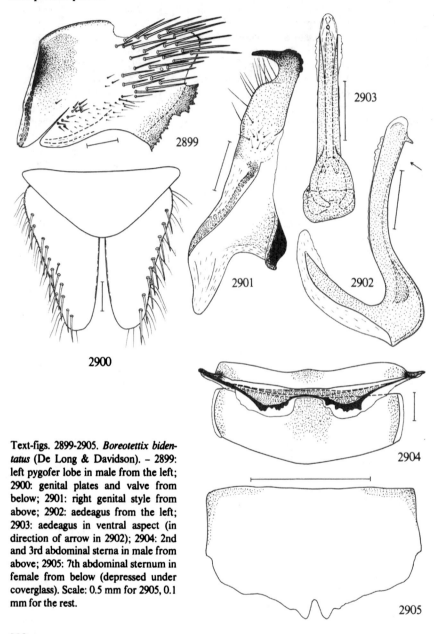

Text-figs. 2899-2905. *Boreotettix bidentatus* (De Long & Davidson). – 2899: left pygofer lobe in male from the left; 2900: genital plates and valve from below; 2901: right genital style from above; 2902: aedeagus from the left; 2903: aedeagus in ventral aspect (in direction of arrow in 2902); 2904: 2nd and 3rd abdominal sterna in male from above; 2905: 7th abdominal sternum in female from below (depressed under coverglass). Scale: 0.5 mm for 2905, 0.1 mm for the rest.

426. *Boreotettix bidentatus* (De Long & Davidson, 1935)
Text-figs. 2899-2905.

Laevicephalus bidentata De Long & Davidson, 1935: 169.
Cosmotettix serricauda Kontkanen, 1949b: 41.

Pale yellowish white; head, pronotum and scutellum more yellowish. Vertex frontally and frontoclypeus laterally of median line with fine arched brownish transverse streaks, more strongly marked in female, sometimes confluent on lower part of frontoclypeus. Vertex in females often with diffuse longitudinal streaks, median line light. Fore wings whitish, veins yellow. Abdominal dorsum in male largely black, with broad yellow lateral margin, in female largely light, venter in both sexes largely light. Male pygofer lobes as in Text-fig. 2899, genital valve and plates as in Text-fig. 2900 (but median margins of plates contiguous in repose), styles as in Text-fig. 2901, aedeagus as in Text-figs. 2902, 2903, 2nd and 3rd abdominal sterna in male as in Text-fig. 2904, 7th abdominal sternum in female as in Text-fig. 2905. Overall length of males 3.3-3.5 mm, of females 3.7-3.9 mm.

Distribution. Not found in Denmark and Norway. – Very rare in central Sweden, found in Dlr.: Ore, Östanvik 9.VIII.1976, in Rättvik, Råberget 4.VII.1977 and 17.VII 1978, and in Mora, Bonäs near Orsasjön 29.VI.1978 (Tord Tjeder); Hls.: Bjuråker, Råka 20.VII.1972 (H. Waldén), Edsbyn, Mucketjärn 4.VII.1978, and Edsbyn, Ämnatjärnsvägen 8.VIII.1978 (Bo Henriksson); Med.: Selånger 17.VII.1969 (Gyllensvärd). – Very rare in East Fennoscandia, found in Oa: Laihia, Kuppaarla 1959-62 (Raatikainen), also in Oa: Mustasaari and Ylistaro (Raatikainen & Vasarainen, 1973); Kb: Hammaslahti 22.VII., 27.VII.1948 (Kontkanen); ObN: Rovaniemi, Pisa 6-22.VII.1950 (Håkan Lindberg); LkW: Muonio 19.VII.1974 (Albrecht). – Kazakhstan, Tuva; Nearctic region.

Biology. ¹Auf einem seggenreichen Uferweissmoor" (Kontkanen, 1949b). In oatfields (Raatikainen, 1971). Also in barley, wheat and winter rye (Raatikainen & Vasarainen, 1971). "Occurred stactered in the fields and in the undergrowth of deciduous forests in their vicinity. . . The flying period was 23.VI.-17.VII." (Raatikainen & Vasarainen, 1973). "Its natural food-plants are grasses. . . . It was commonest in small cultivated clearings on humus and peat soils, which were usually moist or wet" (Raatikainen & Vasarainen, 1976).

Genus *Mocuellus* Ribaut, 1947

Mocuellus Ribaut, 1947: 83.
  Type-species: *Deltocephalus collinus* Boheman, 1850, by original designation.
Erzaleus Ribaut, 1952: 296 (subgenus).
  Type-species: *Jassus (Deltocephalus) metrius* Flor, 1861, by original designation.

Small leafhoppers. Body elongate. Head usually shorter than pronotum. Male pygofer lobes longer than genital plates, without processes. Connective short, branches converging towards apex. Shaft of aedeagus arising from ventral part of socle, with appendages. Style on outside near apex with beak-like prolongation directed laterad. 7th abdominal sternum in female with 2 shallow caudal incisions. In Denmark and Fennoscandia two subgenera, each with one species.

Key to species of *Mocuellus*

1   Anterior tibiae with 4 setae in inner, 4 in outer dorsal row. Phallotreme ventral, situated considerably proximally of apex of shaft (*Mocuellus* s.str.). Usually brachypterous. Antennae as long as vertex and pronotum together. Shaft of aedeagus with short, simple appendages (Text-fig. 2911) .... 427. *collinus* (Boheman)
\-   Anterior tibiae with one seta in inner, 4 in outer dorsal row. Phallotreme subapical (Subgenus *Erzaleus* Ribaut). Always macropterous. Antennae much longer than vertex, pronotum and scutellum together. Appendages of aedeagus branched (Text-fig. 2918) .... 428. *metrius* (Flor).

427. *Mocuellus (Mocuellus) collinus* (Boheman, 1850)
Plate-fig. 210, text-figs. 2906-2912.

Deltocephalus collinus Boheman, 1850: 261.
Deltocephalus aridellus Boheman, 1850: 263.
Athysanus lateralis J. Sahlberg, 1871: 281.

Upper side light greenish, shining. Head above brownish yellow, frontally on each side with an arched brownish transverse streak belonging to frontoclypeus; caudally of these streaks on each side a short brown oblique streak near median line at apex of head. Disk of vertex indistinctly brownish spotted. Face with brown transverse streaks and black sutures. Pronotum, scutellum and fore wings usually unicolorous green, fore wings apically fumose; pronotum sometimes with dark longitudinal stripes; colour in autumn often transforming into brownish, on pronotum arranged as 2-6 broad longitudinal bands, on fore wings bordering the pale veins or almost filling cells (f. *lateralis* Sahlberg). Wing dimorphous. In brachypters, fore wings apically rather acuminate, not reaching apex of abdomen (Plate-fig. 210); in macropters apically evenly rounded, longer than abdomen. Thoracic venter largely light, legs light, fore and middle femora often transversely dark banded, tibiae dark spotted. Anterior abdominal terga black, caudal terga light, dark spotted, in the autumn form with four brown longitudinal bands; abdominal venter light, each sternite with a median black patch. Male pygofer as in Text-figs. 2906, 2907, anal apparatus as in Text-fig. 2906, genital valve and plates as in Text-fig. 2908, styles as in Text-fig. 2909, aedeagus as in Text-figs. 2910, 2911, 7th abdominal sternum in female as in Text-fig. 2912. Overall length of males 3.0-4.1 mm, of females 3.6-4.7 mm. – Nymphs with meso- and especially metanotum uniformly dark, longitudinal bands indistinct, pattern of anterior body region brownish yellow (Vilbaste, 1982).

882

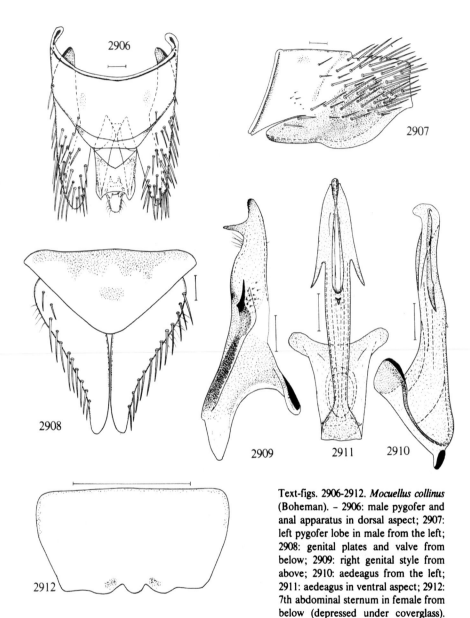

Text-figs. 2906-2912. *Mocuellus collinus* (Boheman). – 2906: male pygofer and anal apparatus in dorsal aspect; 2907: left pygofer lobe in male from the left; 2908: genital plates and valve from below; 2909: right genital style from above; 2910: aedeagus from the left; 2911: aedeagus in ventral aspect; 2912: 7th abdominal sternum in female from below (depressed under coverglass). Scale: 0.5 mm for 2912, 0.1 mm for the rest.

883

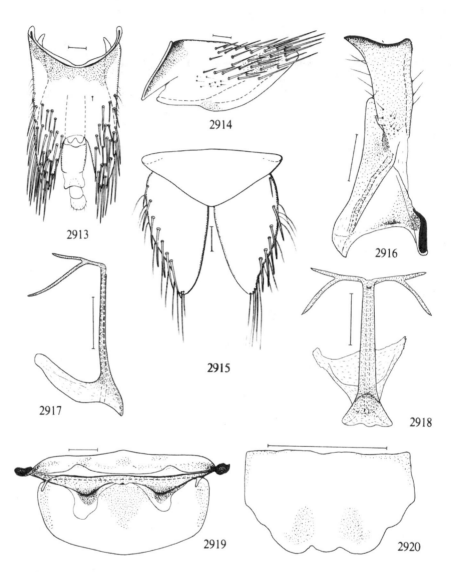

Text-figs. 2913-2920. *Mocuellus metrius* (Flor). – 2913: male pygofer and anal apparatus from above; 2914: left pygofer lobe in male from outside; 2915: male genital plates and valve from below; 2916: right genital style from above; 2917: aedeagus from the left; 2918: aedeagus in ventral aspect; 2919: 2nd and 3rd abdominal sterna in male from above; 2920: 7th abdominal sternum in female from below (depressed under coverglass). Scale: 0.5 mm for 2920, 0.1 mm for the rest.

Distribution. Fairly common in Denmark. – Fairly common in southern and central Sweden, Sk.-Upl., Vstm., Hls., Med. – Not found in Norway. – Rare in East Fennoscandia, only found in Kr: Simpele (O. Siitonen). – Widespread in Europe, also Madeira, Moldavia, Georgia, Kazakhstan, Altai Mts., Mongolia, Tuva, m. and w. Siberia, Kirghizia, Maritime Territory.

Biology. In "Binnendünen, Sandfeldern, besonnten Hängen, Heiden, Wiesen" (Kuntze, 1937). Hibernation in egg-stage, 1 generation (Remane, 1958), two generations (in central Europe, Schiemenz, 1969b). Often with *Elytrigia repens*, adults in June-September.

428. *Mocuellus (Erzaleus) metrius* (Flor, 1861)
    Text-figs. 2913-2920.

Jassus (Deltocephalus) metrius Flor, 1861a: 264.
Thamnotettix alismatis Haupt, 1933: 21.

Light yellow to greenish yellow, shining. Body more slender than in *collinus*. Head frontally rounded obtuse-angular, shorter than pronotum, longer and more acute in female than in male, colour often orange-yellow. Fore wings fairly transparent, about 4 times as long as broad, veins concolorous. Thoracic venter largely light, hind tibiae black dotted. Abdominal tergum black with light segmental borders and broadly light lateral margins, venter in male light with black transverse streaks, in female light with black lateral streaks and spots. Male pygofer and anal apparatus as in Text-figs. 2913, 2914, genital valve and plates as in Text-fig. 2915 (but plates medially contiguous in repose), styles as in Text-fig. 2916, aedeagus as in Text-figs. 2917, 2918, 2nd and 3rd abdominal sterna in male as in Text-fig. 2919, 7th abdominal sternum in female as in Text-fig. 2920. Overall length of males 3.7-4.2 mm, of females 3.9-4.5 mm.

Distribution. Rather scarce in Denmark, mainly found in the eastern parts. – Rather scarce and sporadic in southern Sweden, Sk.-Upl., Vstm., Hls. – Rare in Norway, only found in Ø: Kjölberg 24.VIII.1960 by Holgersen. – Rare in East Fennoscandia, only one male found in Oa: Ylistaro 6.VII.1959 (Raatikainen & Vasarainen, 1973). – England, Ireland, France, Belgium, German D.R. and F.R., Austria, Switzerland, Bohemia, Slovakia, Poland, Estonia, Latvia, n. and m. Russia, Ukraine, Kazakhstan, Altai Mts., m. Siberia, Kirghizia.

Biology. On *Alisma plantago* (Haupt, 1933). On *Baldingera arundinacea* (Ribaut, 1952). Hibernation in egg-stage, one generation (Remane, 1958). "Auf Sumpfrändern und auf sumpfigen Wiesen; wahrscheinlich an Gräsern" (Wagner & Franz, 1961). Adults in July-September.

Corrections and additions

P. 7, line 16 from below: "a and b" to be deleted.
    line 14 from below: for "1962" read 1963.
p. 17, Text-fig. 14.Usually there are 3 subapical cells in the fore wing of *Populicerus,* cf.
    Text-fig. 1035.
p. 18, line 2 from below: for "*Scleroracus russeolus*" read *Ophiola russeola.*

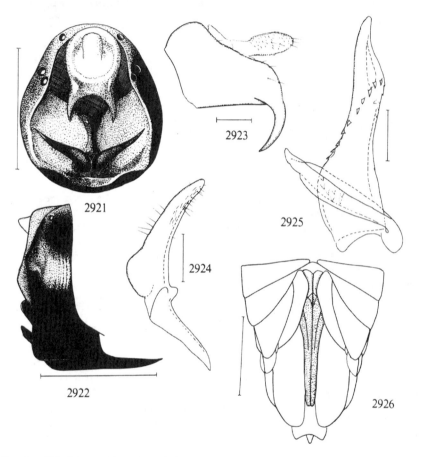

Text-figs. 2921-2926. – 2921: *Achorotile longicornis* (Sahlberg), male pygofer from behind; 2922: same from the right; 2923: male anal tube from the left; 2924: right genital style obliquely from outside; 2925: aedeagus from the left; 2926: *Paraliburnia adela* (Flor) (specimen from Moravia, Lauterer leg.), caudal part of female abdomen from below. Scale: 1 mm for 2926, 0.5 mm for 2921 and 2922, 0.1 mm for for the rest.

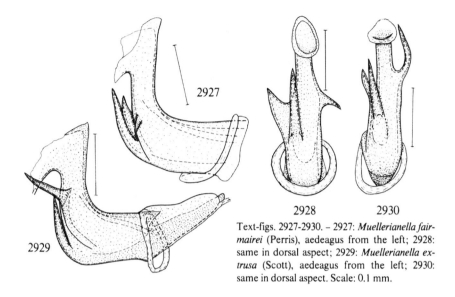

Text-figs. 2927-2930. – 2927: *Muellerianella fair-mairei* (Perris), aedeagus from the left; 2928: same in dorsal aspect; 2929: *Muellerianella ex-trusa* (Scott), aedeagus from the left; 2930: same in dorsal aspect. Scale: 0.1 mm.

p. 20, line 3 from below: for "*Macrostelis*" read *Macrosteles*.

p. 27, line 20: "a and b" to be deleted.

p. 33. Text-figs. 29 and 30 are placed upside down.

p. 57. Text-fig. 90 is placed upside down.

p. 68, line 1 and 3: for "1849" read 1850.

p. 80, line 8 from below: for "1850" read 1851.

p. 82, l. 6 and l. 8: for "1850" read 1851.

p. 83. Males of *Achorotile longicornis* were collected in Hls.: Voxna and Edsbyn by Bo Henriksson who kindly sent me some specimens. The male genitalia of this species are illustrated in Text-figs. 2921-2925.

p. 122, Text-fig. 342. Dr. Lauterer of Brno kindly sent me a female of *Paraliburnia adela* from Moravia. The caudal part of the abdominal venter of this specimen is illustrated in Text-fig. 2926.

p. 134, line 1: for "1860" read 1861.

p. 141, line 14-15 from below: for "*flexuosa*" read *caespitosa*. Booij (1981, 1982) demonstrated that *Muellerianella fairmairei* auct. consists of two species, *M. fairmairei* (Perris), and *M. extrusa* (Scott, 1871: 194). Both species belong to the fauna of Denmark and Fennoscandia. Our Text-figs. 414-423 refer to *fairmairei;* see also Text-figs. 2927 and 2928. The shape of the aedeagus of *extrusa* is illustrated (Text-figs. 2929, 2930).

p. 141, line 2 from below: delete "EJ, LFM, NEZ, B".
line 1 from below: delete "Bl., Sm., Gtl., Vstm.".

p. 143, line 1. The Norwegian records have not been revised.

    line 2: for "St, Ta, Tb, and Kr" read N: near Tvärminne..

    For distribution of *fairmairei* and *extrusa* outside Fennoscandia and Denmark, see Booij (l.c.). The distribution so far known of these species in Denmark and Fennoscandia is given in the Catalogue. The normal hostplant of *extrusa* is *Molinia coerulea* (Booij, 1981).

p. 187, line 16 from below: add Raatikainen, 1967.

p. 194, line 14 from below: for "(Raatikainen & Vasarainen, 1962)" read (Raatikainen & Vasarainen, 1976).

p. 211, line 20: for "(Schwoerbel, 1956)" read (Schwoerbel, 1957).

p. 217. Males of *Cixidia lapponica* were found in Hls.: Edsbyn, by Bo Henriksson, who kindly sent me some specimens. The male genitalia are illustrated in Text-figs. 2931-2934.

p. 219. Text-figs. 720-724 were drawn after a male from Dalmatia. Dr. Dlabola of Praha has kindly informed me that they represent *Issus novaki* Dlabola, 1959b:

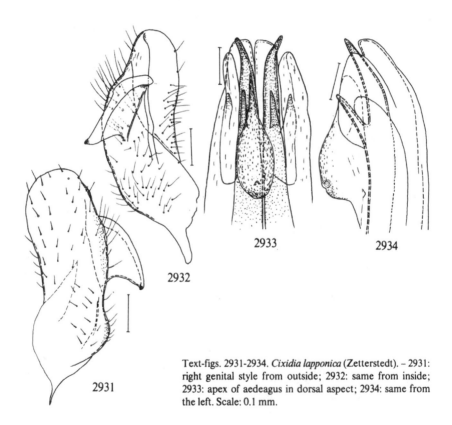

2933

2934

2932

2931

Text-figs. 2931-2934. *Cixidia lapponica* (Zetterstedt). – 2931: right genital style from outside; 2932: same from inside; 2933: apex of aedeagus in dorsal aspect; 2934: same from the left. Scale: 0.1 mm.

152, and has sent me males of the true *Issus muscaeformis* from Montenegro. The genitalia of one of these males are illustrated in Text-figs. 2935-2939.

p. 274, line 21: for "(Nuorteva, 1951)" read (Nuorteva, 1951a).

p. 318, line 6: for "rounded in" read rounded than in.

p. 363, line 6: for "Fabricius, 1758: 523" read Fabricius, 1798: 523.

p. 387. Text-fig. 1248 is placed upside down.

p. 528, line 8 from below: for "Doulglas" read Douglas.

p. 530, line 12: for "1894: 77" read 1894: 46.

p. 549, line 13: for "Dworakowska, 1970a" read Dworakowska, 1970b.

p. 554, line 12 from below: for "Dlabola, 1858" read Dlabola, 1958.

p. 556, line 2: for "considerately" read considerably.

For "Plate-fig. 184. *Scleroracus decumanus*" read Plate-fig. 184. *Ophiola decumana*.

For "Plate-fig. 185. *Scleroracus transversus*" read Plate-fig. 185. *Ophiola transversa*.

For "Plate-fig. 186. *Scleroracus paludosus*" read Plate-fig. 186. *Ophiola paludosa*.

For "Plate-fig. 204. *Diplocolenus abdominalis*" read Plate-fig. 204. *Verdanus abdominalis*.

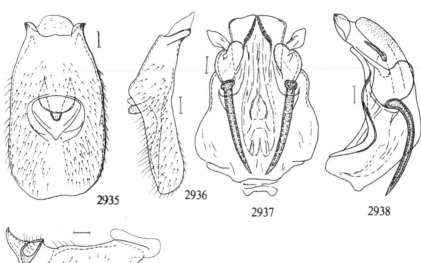

Text-figs. 2935-2939. *Issus muscaeformis* (Schrank) (specimen from Montenegro, Dlabola leg. et det.). – 2935: male anal apparatus from above; 2936: male anal tube from the right; 2937: right genital style from outside; 2938: aedeagus in ventral aspect; 2939: aedeagus from the left. Scale: 0.1 mm.

889

PLATES

Plate-fig. 157. *Grypotes puncticollis* (H.-S.), × 12.

Plate-fig. 158. *Neoaliturus fenestratus* (H.-S.) ♂, × 15.

Plate-fig. 159. *Coryphaelus gyllenhalii* (Fall.), × 10.

Plate-fig. 160. *Balclutha punctata* (F.), × 13.

Plate-fig. 161. *Macrosteles variatus* (Fall.) ♀, × 11.

Plate-fig. 162. *Macrosteles sexnotatus* (Fall.), head from above, × 22.

Plate-fig. 163. *Macrosteles horvathi* (Wagn.) ♂, × 15.

Plate-fig. 164. *Sonronius dahlbomi* (Zett.) ♀, × 10.

Plate-fig. 165. *Deltocephalus pulicaris* (Fall.) ♂, × 20.

Plate-fig. 166. *Deltocephalus maculiceps* (Boh.), head from above, × 22.

Plate-fig. 167. *Doratura stylata* (Boh.) ♂, × 18.

Plate-fig. 168. *Platymetopius guttatus* Fieb., × 10.

Plate-fig. 169. *Lamprotettix nitidulus* (F.), × 10.

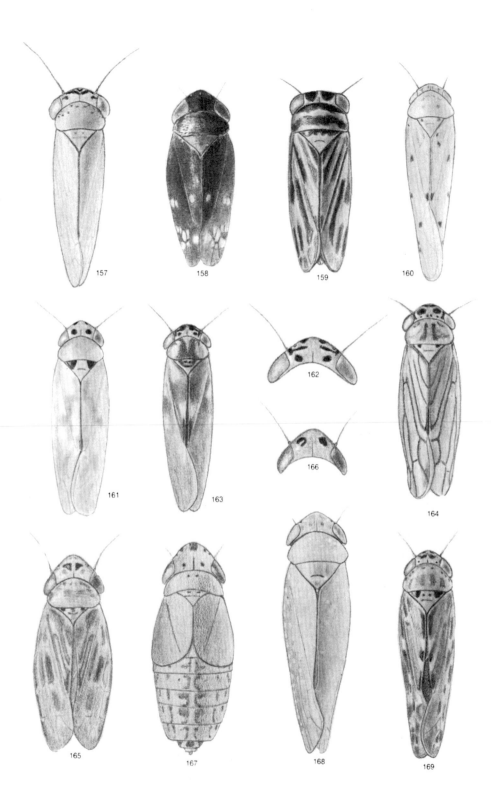

157   158   159   160

161   162   164

163   166

165   167   168   169

Plate-fig. 170. *Allygus mixtus* (F.), × 10.

Plate-fig. 171. *Graphocraerus ventralis* (Fall.), × 10.

Plate-fig. 172. *Rhytistylus proceps* (Kbm.), face, × 15.

Plate-fig. 173. *Rhopalopyx preyssleri* (H.-S.) ♂, × 13.

Plate-fig. 174. *Cicadula persimilis* (Edw.), head from above, × 22.

Plate-fig. 175. *Cicadula quinquenotata* (Boh.), head from above, × 19.

Plate-fig. 176. *Cicadula intermedia* (Boh.), face, × 22.

Plate-fig. 177. *Mocydiopsis attenuata* (Germ.), × 11.

Plate-fig. 178. *Hesium domino* (Reut.), × 9.

Plate-fig. 179. *Pithyotettix abietinus* (Fall.), × 9.

Plate-fig. 180. *Macustus grisescens* (Zett.), × 10.

Plate-fig. 181. *Athysanus argentarius* Metc., × 9.

Plate-fig. 182. *Athysanus quadrum* Boh. ♂, × 10.

Plate-fig. 183. *Stictocoris picturatus* (C. Sahlb.) ♂, × 11.

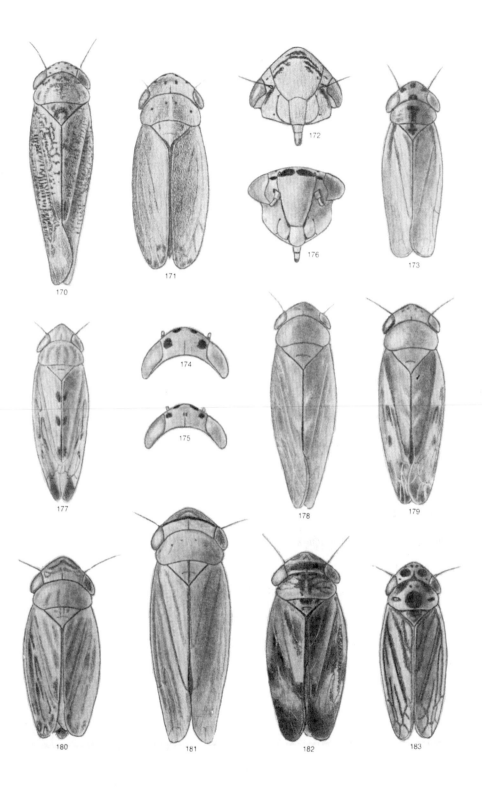

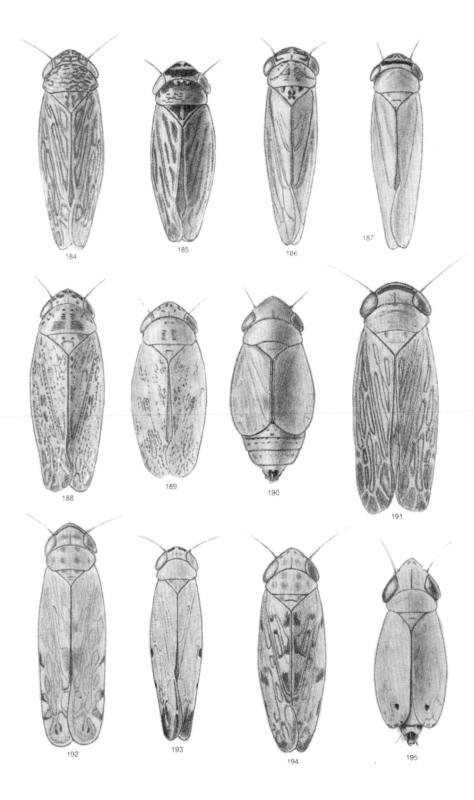

184   185   186   187

188   189   190   191

192   193   194   195

Plate-fig. 196. *Psammotettix alienus* (Dahlb.), × 13.

Plate-fig. 197. *Ebarrius cognatus* (Fieb.), × 17.

Plate-fig. 198. *Adarrus multinotatus* (Boh.), head from above, × 18.

Plate-fig. 199. *Errastunus ocellaris* (Fall.) ♀, × 12.

Plate-fig. 200. *Turrutus socialis* (Fl.), × 17.

Plate-fig. 201. *Jassargus flori* (Fieb.), × 18.

Plate-fig. 202. *Diplocolenus bohemani* (Zett.), v. *calceolatus* (Boh.) ♂, × 13.

Plate-fig. 203. *Diplocolenus bohemani* (Zett.) f. *typica,* left fore wing, × 11.

Plate-fig. 204. *Diplocolenus abdominalis* (F.) ♂, × 13.

Plate-fig. 205. *Cosmotettix caudatus* (Fl.), head from above, × 22.

Plate-fig. 206. *Cosmotettix edwardsi* (Lindb.), head from above, × 23.

Plate-fig. 207. *Hardya tenuis* (Germ.), head from above, × 22.

Plate-fig. 208. *Cosmotettix costalis* (Fall.), × 17.

Plate-fig. 209. *Cosmotettix panzeri* (Fl.), × 16.

Plate-fig. 210. *Mocuellus collinus* (Boh.) ♂, × 12.

Plate-fig. 211. *Arthaldeus striifrons* (Kbm.), face, × 19.

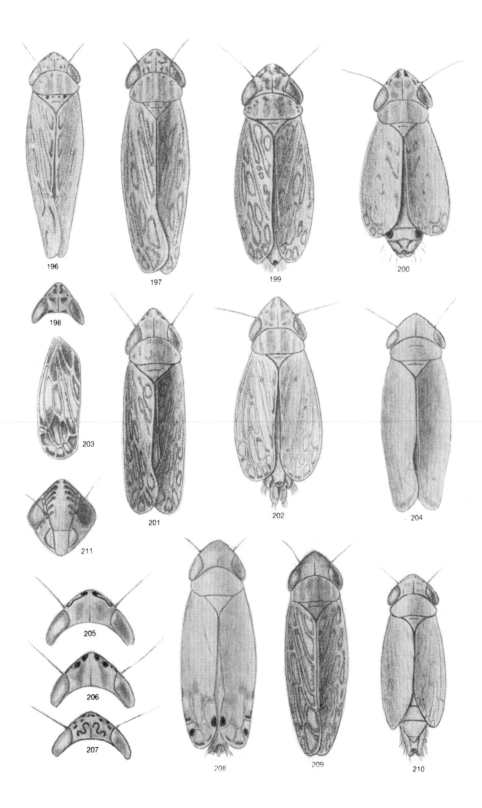

Plate-fig. 212. *Opsius stactogalus* Fieb. ♀, × 10.

Plate-fig. 213. *Macrosteles septemnotatus* (Fall.) ♀, × 12.

Plate-fig. 214. *Sagatus punctifrons* (Fall.) ♀, × 10.

Plate-fig. 215. *Metalimnus formosus* (Boh.) ♀, × 12.

Plate-fig. 216. *Metalimnus marmoratus* (Fl.) ♂, × 12.

Plate-fig. 217. *Cicadula frontalis* (H.-S.) ♀, × 7.5.

Plate-fig. 218. *Macrosteles cyane* (Boh.) ♀, × 12.

Plate-fig. 219. *Platymetopius undatus* (DeGeer) ♀, × 9.

Plate-fig. 220. *Idiodonus cruentatus* (Panz.) ♀, × 10.

Plate-fig. 221. *Colladonus torneellus* (Zett.) ♀, × 9.

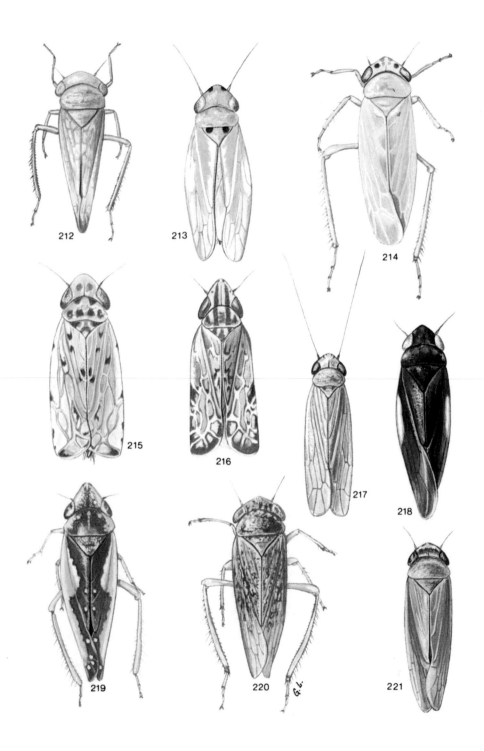

212   213   214

215   216   217   218

219   220   221

Catalogue

		N. Germany	G. Britain	SJ	EJ	WJ	NWJ	NEJ	F	LFM	SZ	NWZ	NEZ	B	Sk.	Bl.
Cixius cunicularius (L.)	1	●	●	●	●	●		●	●	●		●	●	●	●	●
C. nervosus (L.)	2	●	●	●	●	●	●	●	●	●	●					
C. distinguendus Kbm.	3	●	●													
C. similis Kbm.	4	●	●		●	●	●	●		●				●		
C. stigmaticus (Germ.)	5								●	●			●			
C. cambricus China	6		●													
Tachycixius pilosus (Ol.)	7	●	●	●	●	●			●	●	●		●		●	●
Pentastiridius leporinus (L.)	8	●	●	●					●							
Kelisia pallidula (Boh.)	9	●	●		●	●	●			●	●		●	●		
K. sabulicola Wagn.	10	●	●												●	●
K. ribauti Wagn.	11		●		●	●	●	●	●	●			●	●		
K. monoceros Rib.	12															●
K. guttula (Germ.)	13	●	●	●											●	●
K. vittipennis (J.Sahlb.)	14	●	●		●		●		●	●		●				
Anakelisia fasciata (Kbm.)	15	●	●					●	●	●		●				
A. perspicillata (Boh.)	16	●	●		●						●					
Stenocranus minutus (F.)	17	●	●					●	●	●	●	●	●	●	●	●
S. major (Kbm.)	18	●	●													●
Delphacinus mesomelas (Boh.)	19	●	●		●	●		●	●	●		●	●	●	●	●
Ditropis pteridis (Spin.)	20	●	●	●	●		●		●						●	●
Eurysa lineata (Perr.)	21	●	●											●		
Eurysula lurida (Fieb.)	22	●	●					●							●	●
Stiroma bicarinata (H.-S.)	23	●	●	●					●	●	●	●	●	●		●
S. affinis Fieb.	24	●	●	●	●			●	●	●	●	●	●	●	●	●
Stiromoides maculiceps (Horv.)	25															
Achorotile albosignata (Dahlb.)	26														●	●
A. longicornis (J.Sahlb.)	27															
Euconomelus lepidus (Boh.)	28	●	●	●	●	●	●	●	●	●			●	●		●
Conomelus anceps (Germ.)	29	●	●		●	●	●	●	●	●	●		●	●	●	●
Delphax crassicornis (Panz.)	30	●			●	●	●	●		●						
D. pulchellus (Curt.)	31	●	●	●		●	●	●	●	●	●				●	●
Euides speciosa (Boh.)	32	●	●	●	●											●
Chloriona smaragdula (Stål)	33	●	●				●	●	●	●	●	●	●	●		●
C. chinai Oss.	34														●	
C. dorsata Edw.	35		●			●		●	●	●	●					●
C. glaucescens Fieb.	36	●	●	●				●	●	●		●				●
C. vasconica Rib.	37	●	●				●									●
Megamelus notula (Germ.)	38	●	●	●	●	●	●		●		●		●	●	●	●

904

SWEDEN

	Hall.	Sm.	Öl.	Gtl.	G. Sand.	Ög.	Vg.	Boh.	Dlsl.	Nrk.	Sdm.	Upl.	Vstm.	Vrm.	Dlr.	Gstr.	Hls.	Med.	Hrj.	Jmt.	Ång.	Vb.	Nb.	Ås. Lpm.	Ly. Lpm.	P. Lpm.	Lu. Lpm.	T. Lpm.
1	●	●	●	●	●	●	●	●	●	●	●	●	●	●	●	●	●	●	●	●	●	●	●	●	●	●	●	●
2	●	●				●	●	●			●	●	●			●	●	●	●	●		●				●		
3	●	●		●		●	●				●	●					●											
4	●	●	●			●	●	●	●	●			●	●	●	●			●	●	●	●			●	●	●	●
5																												
6																												
7				●																								
8		●										●			●		●						●					
9		●	●	●			●	●			●	●																
10	●			¿																								
11	●	●	●	●			●	●			●	●			●		●			●								
12		●	●	●			●						●															
13		●	●	●			●				●			●	●	●					●							
14	●	●	●				●				●	●			●		●				●	●						
15		●										●																
16		●	●	●			●	●	●			●																
17		●	●	●			●			●	●	●	●	●														
18		●																										
19	●	●	●	●			●	●	●		●	●	●															
20		●																										
21			●																									
22			●																	●								
23	●	●				●	●	●			●	●	●	●	●	●			●	●	●	●	●	●	●			
24	●	●	●	●			●	●	●			●	●		●		●			●							●	
25																												
26		●		●			●	●				●	●		●		●				●						●	●
27												●																
28	●	●	●	●			●					●																
29	●	●	●			●	●	●	●	●	●	●	●	●	●	●												
30	●	●	●	●			●		●	●				●														
31				●																								
32							●										●											
33	●	●					●		●			●					●		●	●						●		
34							●	●			●			●			●						●					
35				●																								
36	●		●	●			●					●					●											
37							●					●					●											
38	●	●	●	●		●	●	●	●	●	●	●	●	●	●	●	●			●		●	●					

		Ø + AK	HE (s+n)	O (s+n)	B (ø+v)	VE	TE (y+i)	AA (y+i)	VA (y+i)	R (y+i)	HO (y+i)	SF (y+i)	MR (y+i)	ST (y+i)	NT (y+i)	Ns (y+i)
Cixius cunicularius (L.)	1	●	●	●			●	●	●	●	●	●	●			
C. nervosus (L.)	2		●						●	●	●	●	●			
C. distinguendus Kbm.	3								●	●	●	●				
C. similis Kbm.	4		●	●					●	●	●	●		●		
C. stigmaticus (Germ.)	5										●					
C. cambricus China	6															
Tachycixius pilosus (Ol.)	7															
Pentastiridius leporinus (L.)	8															
Kelisia pallidula (Boh.)	9	●						●	●							
K. sabulicola Wagn.	10															
K. ribauti Wagn.	11	●							●	●						●
K. monoceros Rib.	12				●											
K. guttula (Germ.)	13	●						●		●	●					
K. vittipennis (J.Sahlb.)	14	●							●	●						
Anakelisia fasciata (Kbm.)	15															
A. perspicillata (Boh.)	16															
Stenocranus minutus (F.)	17	●														
S. major (Kbm.)	18															
Delphacinus mesomelas (Boh.)	19															
Ditropis pteridis (Spin.)	20							●								
Eurysa lineata (Perr.)	21															
Eurysula lurida (Fieb.)	22							●								
Stiroma bicarinata (H.-S.)	23		●	●												●
S. affinis Fieb.	24		●	●		●				●						
Stiromoides maculiceps (Horv.)	25															
Achorotile albosignata (Dahlb.)	26		●			●										
A. longicornis (J.Sahlb.)	27															
Euconomelus lepidus (Boh.)	28			●					●	●						
Conomelus anceps (Germ.)	29	●	●				●	●	●	●	●					
Delphax crassicornis (Panz.)	30															
D. pulchellus (Curt.)	31							●								
Euides speciosa (Boh.)	32															
Chloriona smaragdula (Stål)	33															
C. chinai Oss.	34															
C. dorsata Edw.	35															
C. glaucescens Fieb.	36							●								
C. vasconica Rib.	37															
Megamelus notula (Germ.)	38	●			●	●	●		●	●						

	Nn(ø+v)	TR(y+i)	F(v+i)	F(n+ø)	Al	Ab	N	Ka	St	Ta	Sa	Öa	Tb	Sb	Kb	Om	Ok	Ob S	Ob N	Ks	LkW	LkE	Le	Li	Vib	Kr	Lr
1	●	●	●	●	●	●	●	●	●	●	●	●	●	●	●	●	●		●	●	●		●	●	●	●	●
2					●	●	●			●															●	●	
3					●	●	●			●	●														●	●	
4		●		●	●	●	●	●	●	●	●	●	●	●		●			●	●	●	●		●	●	●	●
5																											
6					●	●	●								●												
7																											
8					●	●	●	●							●											●	
9					●	●	●		●	●																	
10						●																					
11					●	●	●								●	●			●						●	●	
12					●	●	●					●															
13					●	●	●		●						●										●	●	
14						●	●		●	●	●				●	●											
15																											
16																											
17					●	●																					
18																											
19						●																					
20																											
21																											
22					●	●	●					●	●	●												●	
23					●	●	●		●	●	●	●	●	●	●		●	●	●						●	●	●
24			●		●	●	●			●	●	●					●	●					●			●	●
25												●															
26					●	●	●		●	●			●	●						●			●	●	●	●	●
27							●					●															
28					●	●	●		●			●					●	●									
29					●	●	●	●	●	●					●										●		
30					●	●	●								●												●
31					●	●	●		●																		
32					●	●				●					●										●		
33					●	●				●					●	●									●		
34					●	●	●			●	●				●	●									●		
35																											
36					●	●	●			●					●	●											
37																											
38					●	●	●		●	●	●	●		●	●	●			●	●					●	●	

				DENMARK												
		N. Germany	G. Britain	SJ	EJ	WJ	NWJ	NEJ	F	LFM	SZ	NWZ	NEZ	B	Sk.	Bl.
Unkanodes excisa (Mel.)	39	●		●	●				●				●		●	
Megadelphax sordidulus (Stål)	40	●														
M. haglundi (J.Sahlb.)	41															
Laodelphax striatellus (Fall.)	42	●	●												●	●
Paraliburnia adela (Fl.)	43	●	●			●					●	●	●			
P. clypealis (J.Sahlb.)	44	●	●													
Hyledelphax elegantulus (Boh.)	45	●	●	●	●	●			●	●	●	●	●	●	●	●
Megamelodes quadrimaculatus (Sign.)	46					●										
Calligypona reyi (Fieb.)	47	●	●	●												●
Delphacodes venosus (Germ.)	48	●	●	●				●				●	●	●		
D. capnodes (Scott)	49	●	●	●												
Gravesteiniella boldi (Scott)	50	●	●	●	●					●			●	●		
Muellerianella brevipennis (Boh.)	51	●	●	●					●	●			●	●		
M. fairmairei (Perr.)	52		●	●	●	●				●			●	●		
M. extrusa (Scott)	52a	●	●	●	●			●					●	●		●
Muirodelphax aubei (Perr.)	53	●	●	●	●	●	●	●				●	●			
Acanthodelphax denticauda (Boh.)	54	●	●		●					●		●	●			
Tyrphodelphax distinctus (Fl.)	55	●	●									●	●	●		
T. albocarinatus (Stål)	56													●		
Dicranotropis hamata (Boh.)	57	●	●	●	●				●	●	●	●	●	●	●	●
Florodelphax paryphasma (Fl.)	58	●	●													
F. leptosoma (Fl.)	59	●	●				●						●	●		
Kosswigianella exigua (Boh.)	60	●	●	●	●			●			●	●			●	●
Struebingianella lugubrina (Boh.)	61	●	●			●					●	●	●		●	●
S. litoralis (Reut.)	62		●													
Xanthodelphax flaveolus (Fl.)	63	●	●								●		●		●	●
X. stramineus (Stål)	64	●	●					●	●					●		
Paradelphacodes paludosa (Fl.)	65	●	●				●							●		
Oncodelphax pullulus (Boh.)	66	●	●				●			●		●				
Criomorphus albomarginatus Curt.	67	●	●	●	●	●	●	●	●			●	●	●		
C. moestus (Boh.)	68		●													
C. borealis (J.Sahlb.)	69															
Javesella pellucida (F.)	70	●	●	●	●	●	●	●			●	●	●	●	●	●
J. dubia (Kbm.)	71	●	●	●	●				●		●	●	●	●	●	
J. obscurella (Boh.)	72	●	●		●			●	●		●				●	
J. salina (Hpt.)	73	●														
J. discolor (Boh.)	74	●	●	●	●	●	●	●	●	●	●	●	●	●	●	●
J. simillima (Linnav.)	75															

	Hall.	Sm.	Öl.	Gtl.	G. Sand.	Ög.	Vg.	Boh.	Dlsl.	Nrk.	Sdm.	Upl.	Vstm.	Vrm.	Dlr.	Gstr.	Hls.	Med.	Hrj.	Jmt.	Ång.	Vb.	Nb.	Ås. Lpm.	Ly. Lpm.	P. Lpm.	Lu. Lpm.	T. Lpm.
39	●		●	●													●					●	●					
40				●		●					●	●	●		●		●	●		●		●						
41						●						●																
42		●	●	●		●	●					●			●		●											
43						●			●			●																
44												●																
45	●	●	●	●		●	●		●		●	●	●	●	●	●	●			●	●	●	●				●	
46																												
47		●	●			●																						
48	●	●	●	●		●	●			●	●	●	●		●	●	●											
49		●										●																
50	●			●	●																							
51		●	●	●		●	●				●	●	●		●	●	●											
52						●						●																
52a		●		●		●						●					●											
53	●	●	●			●																				2		
54		●	●			●						●		●						●						●		
55		●						●			●	●	●		●	●	●		●	●	●					●		
56						●											●			●	●							
57	●	●	●	●	●	●	●	●	●	●	●	●	●		●	●	●	●		●	●	●						
58		●	●	●		●						●	●				●											
59																												
60	●	●	●	●		●	●	●		●	●	●	●															
61		●	●			●	●					●	●		●													
62																												
63		●	●	●		●						●	●				●			●	●					●		
64		●	●			●	●	●	●											●								
65		●				●											●			●								
66		●	●			●	●						●	●														
67	●	●	●	●		●	●	●			●	●	●	●	●				●	●							●	
68						●						●			●					●							●	
69																				●	●			●		●	●	●
70	●	●	●	●	●	●	●	●			●	●	●	●	●	●	●			●	●	●	●	●	●	●	●	●
71	●	●	●	●					●	●	●	●	●		●		●				●						●	
72	●	●	●	●		●	●				●	●					●			●	●		●				●	●
73			●	●																								
74	●	●	●	●		●	●	●			●	●	●	●	●		●			●	●			●	●	●	●	●
75																												

		Ø+AK	HE (s+n)	O (s+n)	B (ø+v)	VE	TE (y+i)	AA (y+i)	VA (y+i)	R (y+i)	HO (y+i)	SF (y+i)	MR (y+i)	ST (y+i)	NT (y+i)	Ns (y+i)
Unkanodes excisa (Mel.)	39															
Megadelphax sordidulus (Stål)	40															
M. haglundi (J.Sahlb.)	41															
Laodelphax striatellus (Fall.)	42		●													
Paraliburnia adela (Fl.)	43															
P. clypealis (J.Sahlb.)	44															
Hyledelphax elegantulus (Boh.)	45	●	●	●	●	●				●	●		●	●		
Megamelodes quadrimaculatus (Sign.)	46															
Calligypona reyi (Fieb.)	47															
Delphacodes venosus (Germ.)	48	●														
D. capnodes (Scott)	49															
Gravesteiniella boldi (Scott)	50															
Muellerianella brevipennis (Boh.)	51	●	●													
M. fairmairei (Perr.)	52															
M. extrusa (Scott)	52a															
Muirodelphax aubei (Perr.)	53															
Acanthodelphax denticauda (Boh.)	54	●														
Tyrphodelphax distinctus (Fl.)	55		●							●	●			●		
T. albocarinatus (Stål)	56															
Dicranotropis hamata (Boh.)	57	●		●	●		●	●		●	●			●		
Florodelphax paryphasma (Fl.)	58															
F. leptosoma (Fl.)	59															
Kosswigianella exigua (Boh.)	60															
Struebingianella lugubrina (Boh.)	61															
S. litoralis (Reut.)	62															
Xanthodelphax flaveolus (Fl.)	63		●	●		●					●					
X. stramineus (Stål)	64															
Paradelphacodes paludosa (Fl.)	65															
Oncodelphax pullulus (Boh.)	66	●	●													
Criomorphus albomarginatus Curt.	67	●			●	●										
C. moestus (Boh.)	68															
C. borealis (J.Sahlb.)	69			●												●
Javesella pellucida (F.)	70	●	●	●	●	●	●				●	●	●	●	●	●
J. dubia (Kbm.)	71								●	●	●		●			
J. obscurella (Boh.)	72								●							●
J. salina (Hpt.)	73															
J. discolor (Boh.)	74	●	●	●			●	●	●	●	●		●	●		●
J. simillima (Linnav.)	75															

	Nn (ø+v)	TR (y+i)	F (v+i)	F (n+ø)	Al	Ab	N	Ka	St	Ta	Sa	Öa	Tb	Sb	Kb	Om	Ok	Ob S	Ob N	Ks	LkW	LkE	Le	Li	Vib	Kr	Lr
39					●		●	●	●								●		●						●		
40					●	●	●	●	●	●	●	●	●	●	●	●	●									●	
41																											
42					●	●											●						●		●	●	
43						●																					
44						●								●													
45		●			●	●	●	●	●	●	●	●	●		●	●	●	●					●		●	●	
46																											
47					●		●				●																
48						●	●								●	●									●	●	
49																											
50							●																				
51					●	●	●	●	●	●			●	●	●	●									●	●	
52							●																				
52a					●	●		●																			
53					●																						
54						●	●				●	●	●		●	●	●		●				●		●	●	
55															●		●	●	●				●	●			
56					●	●										●											
57					●	●	●	●			●	●	●	●	●										●	●	
58					●		●			●																	●
59																											
60					●	●																			●	●	
61					●	●	●	●		●															●	●	
62						●					●																
63						●	●				●		●	●					●								●
64					●	●	●	●				●	●	●	●	●									●	●	
65						●				●		●															
66					●	●	●			●		●		●	●										●	●	
67					●	●	●	●				●													●	●	
68							●					●				●					●	●			●	●	
69			●		●	●		●			●	●		●		●						●	●	●	●	●	●
70	●	●	●		●	●	●	●	●	●	●	●	●	●	●	●	●	●	●	●	●	●	●	●	●	●	●
71					●	●		●		●	●	●	●			●	●	●							●	●	
72					●	●	●	●		●		●	●			●	●	●					●		●	●	
73																											
74	●	●	●	●	●	●	●			●	●	●			●	●	●	●	●	●	●	●	●	●	●	●	●
75						●				●				●												●	

		N. Germany	G. Britain	SJ	EJ	WJ	NWJ	NEJ	F	LFM	SZ	NWZ	NEZ	B	Sk.	Bl.
J. bottnica Huld.	76															
J. forcipata (Boh.)	77	●	●	●	●			●	●			●	●	●	●	●
J. alpina (J.Sahlb.)	78															
J. stali (Metc.)	79															
Ribautodelphax collinus (Boh.)	80	●		●	●							●	●	●	●	●
R. angulosus (Rib.)	81		●					●								
R. pungens (Rib.)	82															
R. pallens (Stål)	83		●													
R. albostriatus (Fieb.)	84	●														●
Cixidia confinis (Zett.)	85															
C. lapponica (Zett.)	86															
Issus muscaeformis (Schrnk.)	87	●				●	●						●	●	●	
Ommatidiotus dissimilis (Fall.)	88	●			●	●	●	●	●					●	●	
Cicadetta montana (Scop.)	89		●											?		
Lepyronia coleoptrata (L.)	90	●		●							●		●	●	●	
Peuceptyelus coriaceus (Fall.)	91															
Neophilaenus exclamationis (Thunb.)	92	●	●	●	●	●	●	●		●				●	●	●
f. *diluta* (J.Sahlb.)	92a													?		
N. campestris (Fall.)	93	●	●		●	●		●						●		
N. lineatus (L.)	94	●	●		●	●	●	●	●	●	●		●	●	●	●
f. *aterrima* (J.Sahlb.)	94a	●		●												
f. *pulchella* (J.Sahlb.)	94b	●												●	●	●
f. *pallida* (Hpt.)	94c	●			●	●	●	●						●	●	
N. minor (Kbm.)	95	●														
Aphrophora corticea Germ.	96	●												●		
A. alni (Fall.)	97	●	●	●	●		●	●	●	●	●		●	●	●	●
A. salicina (Gze.)	98	●	●	●	●	●	●	●		●			●	●	●	●
A. costalis Mats.	99	●	●													
Philaenus spumarius (L.)	100	●	●	●	●	●	●	●	●	●	●	●	●	●	●	●
Centrotus cornutus (L.)	101	●	●		●	●	●		●	●		●	●	●	●	●
Gargara genistae (F.)	102	●	●													
Ulopa reticulata (F.)	103	●	●	●	●	●	●						●	●	●	●
Megophthalmus scanicus (Fall.)	104	●	●	●	●	●	●	●	●	●	●		●	●	●	●
Ledra aurita (L.)	105	●	●		●				●	●	●		●	●	●	●
Oncopsis flavicollis (L.)	106	●	●		●		●		●	●	●		●	●	●	●
O. carpini (J.Sahlb.)	107	●	●		●	●				●			●	●		
O. subangulata (J.Sahlb.)	108	●	●		●								●	●		
O. tristis (Zett.)	109	●	●		●	●			●	●			●	●	●	●

SWEDEN

	Hall.	Sm.	Öl.	Gtl.	G. Sand.	Ög.	Vg.	Boh.	Dlsl.	Nrk.	Sdm.	Upl.	Vstm.	Vrm.	Dlr.	Gstr.	Hls.	Med.	Hrj.	Jmt.	Ång.	Vb.	Nb.	Ås. Lpm.	Ly. Lpm.	P. Lpm.	Lu. Lpm.	T. Lpm.
76																	●											
77	●	●	●	●			●	●	●		●	●	●	●	●	●	●			●	●	●	●	●		●		●
78																	●				●				●		●	●
79															●		●				●							
80	●	●	●				●					●									●							
81							●	●				●																
82				●																								
83		●		●			●	●			●	●			●								●					
84		●		●							●	●		●	●	●			●	●	●	●	●		●		●	●
85				●	●							●		●	●								●					
86				●	●							●		●								●						
87	●	●	●						●		●																	
88	●	●					●	●	●		●	●	●	●														
89		●					●	●	●			●																
90	●		●				●	●		●	●	●			●													
91																												
92	●	●	●			●	●	●	●	●	●	●	●	●	●	●	●		●	●	●		●	●		●		
92a																							●	●				
93																												
94	●	●	●	●	●	●	●	●	●	●	●	●	●	●	●				●	●	●	●	●	●	●	●	●	●
94a		●																										
94b	●	●	●											●														
94c	●			●	●																							
95																												
96				●																								
97	●	●	●	●		●	●	●	●	●	●	●	●	●	●	●				●	●							
98	●	●	●	●									●															
99						●						●	●		●													
100	●	●	●	●		●	●	●	●	●	●	●	●	●	●	●		●	●	●	●	●	●	●	●	●	●	●
101	●	●	●	●		●	●	●	●	●	●	●	●	●	●	●				●	●							
102																												
103	●	●	●	●		●	●	●								●												
104	●	●	●	●		●	●	●							●						●							
105	●	●	●			●	●																					
106	●	●	●	●	●	●	●	●	●	●	●	●	●	●	●	●			●	●			●	●	●	●	●	●
107																												
108	●	●	●	●	●	●	●		●		●		●	●		●	●			●							●	
109	●	●	●	●	●	●	●	●	●	●	●	●	●						●	●	●	●	●			●	●	●

	No.	Ø+AK	HE (s+n)	O (s+n)	B (ø+v)	VE	TE (y+i)	AA (y+i)	VA (y+i)	R (y+i)	HO (y+i)	SF (y+i)	MR (y+i)	ST (y+i)	NT (y+i)	Ns (y+i)
J. bottnica Huld.	76															
J. forcipata (Boh.)	77							●	●	●						
J. alpina (J.Sahlb.)	78															●
J. stali (Metc.)	79															
Ribautodelphax collinus (Boh.)	80				●	●										
R. angulosus (Rib.)	81															
R. pungens (Rib.)	82															
R. pallens (Stål)	83		●	●												
R. albostriatus (Fieb.)	84	●	●	●												
Cixidia confinis (Zett.)	85															
C. lapponica (Zett.)	86															
Issus muscaeformis (Schrnk.)	87						●		●		●					
Ommatidiotus dissimilis (Fall.)	88															
Cicadetta montana (Scop.)	89	●			●	●										
Lepyronia coleoptrata (L.)	90	●			●				●							
Peuceptyelus coriaceus (Fall.)	91															
Neophilaenus exclamationis (Thunb.)	92	●	●	●	●		●	●	●		●	●	●	●		
f. *diluta* (J.Sahlb.)	92a															
N. campestris (Fall.)	93															
N. lineatus (L.)	94	●	●	●	●		●	●	●	●	●	●	●		●	●
f. *aterrima* (J.Sahlb.)	94a								●	●						
f. *pulchella* (J.Sahlb.)	94b										●					
f. *pallida* (Hpt.)	94c															
N. minor (Kbm.)	95															
Aphrophora corticea Germ.	96	●														
A. alni (Fall.)	97	●	●	●	●		●	●	●	●	●	●	●	●		
A. salicina (Gze.)	98															
A. costalis Mats.	99															
Philaenus spumarius (L.)	100	●	●	●	●	●	●	●	●	●	●	●	●	●	●	●
Centrotus cornutus (L.)	101	●	●	●	●	●	●	●	●	●						
Gargara genistae (F.)	102															
Ulopa reticulata (F.)	103	●	●		●		●	●	●	●						
Megophthalmus scanicus (Fall.)	104	●						●	●	●	●					
Ledra aurita (L.)	105	●														
Oncopsis flavicollis (L.)	106	●	●	●	●		●	●	●	●	●	●	●	●		
O. carpini (J.Sahlb.)	107															
O. subangulata (J.Sahlb.)	108			●			●		●	●	●					
O. tristis (Zett.)	109	●	●	●	●		●	●	●	●	●		●	●	●	●

	Nn(ø+v)	TR(y+i)	F(v+i)	F(n+ø)	Al	Ab	N	Ka	St	Ta	Sa	Oa	Tb	Sb	Kb	Om	Ok	ObS	ObN	Ks	LkW	LkE	Le	Li	Vib	Kr	Lr
76										•		•		•					•	•						•	
77					•	•	•	•	•	•	•		•	•	•	•			•	•	•	•	•			•	•
78		•																					•				
79						•	•			•		•	•				•		•	•	•				•	•	•
80						•	•			•	•		•		•											•	
81					•	•				•									•	•							
82																											
83			•		•	•	•			•	•				•		•			•	•					•	•
84			•			•	•			•	•		•	•	•		•			•	•	•			•	•	•
85						•						•	•														
86									•	•								•		•						•	
87																											
88						•	•		•	•				•											•	•	
89						•			•	•															•		
90						•	•	•	•	•															•		
91						•	•		•	•				•											•		
92		•	•		•	•	•	•	•	•	•		•	•	•		•	•	•	•					•	•	•
92a		•			•	•																					
93																									•		
94					•	•	•	•	•	•	•	•	•	•	•	•	•		•						•	•	
94a																											
94b																											
94c																											
95										•																	
96																											
97					•	•	•	•	•	•	•	•	•	•	•										•	•	
98																											
99					•	•	•	•	•		•	•	•												•		
100	•	•	•		•	•	•	•	•	•	•	•	•	•	•	•	•	•							•	•	•
101						•	•		•	•	•	•		•	•										•	•	
102																											
103					•	•	•	•	•	•			•		•											•	
104					•	•	•			•	•																•
105																											
106	•	•	•	•	•	•	•	•	•	•	•	•	•	•	•	•	•	•	•	•	•	•	•	•	•	•	•
107																											
108	•	•			•	•	•			•	•		•	•	•		•									•	
109	•	•			•	•	•			•	•	•	•	•	•	•	•								•	•	•

	№	N. Germany	G. Britain	SJ	EJ	WJ	NWJ	NEJ	F	LFM	SZ	NWZ	NEZ	B	Sk.	Bl.
O. appendiculata Wagn.	110	●													●	●
O. planiscuta (Thoms.)	111															
O. alni (Schrnk.)	112	●	●		●	●			●	●			●	●	●	●
Pediopsis tiliae (Germ.)	113	●	●	●					●	●			●	●		
Macropsis prasina (Boh.)	114	●	●	●	●		●	●	●	●				●	●	
M. infuscata (J.Sahlb.)	115	●	●					●		●					●	
M. cerea (Germ.)	116	●	●							●				●		●
M. fuscinervis (Boh.)	117	●	●						●				●	●	●	●
M. graminea (F.)	118															
M. impura (Boh.)	119	●	●	●			●	●	●			●	●	●	●	●
M. fuscula (Zett.)	120	●	●	●							●	●		●	●	●
M. scutellata (Boh.)	121	●	●							●		●		●		
M. megerlei (Fieb.)	122															
Hephathus nanus (H.-S.)	123	●														
Agallia consobrina Curt.	124	●	●	●	●											
A. venosa (Fourcr.)	125	●	●	●	●	●	●	●		●		●	●	●	●	●
A. ribauti Oss.	126	●	●													
A. brachyptera (Boh.)	127	●	●	●	●					●			●		●	
Rhytidodus decimusquartus (Schrnk.)	128	●														
Idiocerus stigmaticalis Lew.	129	●	●	●	●					●			●	●	●	●
I. lituratus (Fall.)	130	●	●	●	●	●	●	●	●				●	●	●	●
I. herrichii Kbm.	131	●	●											●		
Metidiocerus crassipes (J.Sahlb.)	132															
M. elegans (Fl.)	133	●	●											●	●	
Populicerus populi (L.)	134	●	●	●	●			●	●	●				●	●	●
P. laminatus (Fl.)	135	●	●	●	●			●	●	●					●	●
P. nitidissimus (H.-S.)	136	●	●											●	●	●
P. confusus (Fl.)	137	●	●	●	●	●	●							●	●	●
f. *nigricans* (Oss.)	137a															
P. albicans (Kbm.)	138	●	●	●	●				●	●	●			●	●	●
Tremulicerus tremulae (Estl.)	139	●	●							●				●	●	●
T. vitreus (F.)	140	●	●											●		
T. distinguendus (Kbm.)	141		●							●				●	●	
Stenidiocerus poecilus (H.-S.)	142	●	●													
Sahlbergotettix salicicola (Fl.)	143															
Iassus lanio (L.)	144	●	●	●	●			●	●	●				●	●	●
Batracomorphus allionii (Turt.)	145	●			●	●										
Eupelix cuspidata (F.)	146	●	●			●	●	●	●	●	●	●		●	●	●

916

	Hall.	Sm.	Öl.	Gtl.	G. Sand.	Ög.	Vg.	Boh.	Dlsl.	Nrk.	Sdm.	Upl.	Vstm.	Vrm.	Dlr.	Gstr.	Hls.	Med.	Hrj.	Jmt.	Ång.	Vb.	Nb.	Ås. Lpm.	Ly. Lpm.	P. Lpm.	Lu. Lpm.	T. Lpm.
110	●	●								●	●	●																
111															●		●		●	●	●	●	●		●	●	●	●
112	●	●	●			●	●	●	●	●	●	●	●			●	●			●								
113	●	●		●		●	●	●		●	●																	
114	●	●	●				●	●		●	●	●	●		●					●								
115	●		●			●	●	●		●	●	●	●			●	●											
116		●	●	●		●	●	●	●	●	●	●	●			●						●						
117	●	●		●		●				●	●	●	●			●	●			●								
118					*Hlm.*																							
119	●	●	●	●		●	●	●						●		●					●	●			●			
120	●	●	●	●		●	●	●		●	●	●	●	●		●	●			●								●
121	●		●																									
122																												
123																												
124		●																										
125	●	●	●	●		●	●			●	●	●	●		●		●	●		●	●	●	●					
126		●	●			●				●	●	●			●					●								
127	●	●	●	●		●	●	●	●	●	●	●	●	●		●	●											
128			●																									
129	●	●				●	●			●	●	●			●	●												
130	●	●	●	●		●	●	●	●	●				●		●	●			●	●					●		●
131		●										●																
132															●		●				●		●					
133	●	●				●	●				●	●			●	●	●	●		●	●	●	●			●		●
134	●	●	●	●	●	●	●	●	●		●	●	●	●	●	●	●			●	●	●						●
135		●					●			●	●	●	●	●	●	●		●		●	●	●						●
136												●																
137	●	●	●	●		●	●	●	●	●	●	●	●	●	●	●	●	●		●	●	●	●					
137a															●					●		●						●
138	●		●																									
139		●	●			●			●	●	●	●				●		●										
140																												
141																												
142	●					●			●		●				●		●					●						
143																												
144	●	●	●	●		●	●	●	●	●	●	●		●														
145																												
146	●	●	●	●		●	●	●		●	●	●			●													

		Ø+AK	HE (s+n)	O (s+n)	B (ø+v)	VE	TE (y+i)	AA (y+i)	VA (y+i)	R (y+i)	HO (y+i)	SF (y+i)	MR (y+i)	ST (y+i)	NT (y+i)	Ns (y+i)
O. appendiculata Wagn.	110															
O. planiscuta (Thoms.)	111				●		●	●		●	●		●			●
O. alni (Schrnk.)	112	●	●	●	●		●	●	●	●	●	●	●	●		●
Pediopsis tiliae (Germ.)	113				●											
Macropsis prasina (Boh.)	114	●														
M. infuscata (J.Sahlb.)	115															
M. cerea (Germ.)	116	●	●	●	●	●		●								
M. fuscinervis (Boh.)	117	●		●					●							
M. graminea (F.)	118	●														
M. impura (Boh.)	119							●	●	●						
M. fuscula (Zett.)	120	●	●		●				●	●	●					
M. scutellata (Boh.)	121															
M. megerlei (Fieb.)	122	●		●												
Hephathus nanus (H.-S.)	123															
Agallia consobrina Curt.	124															
A. venosa (Fourcr.)	125	●	●	●	●		●	●	●							
A. ribauti Oss.	126				●		●									
A. brachyptera (Boh.)	127	●	●		●		●									
Rhytidodus decimusquartus (Schrnk.)	128															
Idiocerus stigmaticalis Lew.	129	●		●	●				●							
I. lituratus (Fall.)	130	●						●	●	●						
I. herrichii Kbm.	131	●														
Metidiocerus crassipes (J.Sahlb.)	132															
M. elegans (Fl.)	133	●		●			●									
Populicerus populi (L.)	134	●	●	●	●	●	●	●	●	●	●	●	●	●	●	●
P. laminatus (Fl.)	135	●		●	●		●			●						
P. nitidissimus (H.-S.)	136	●														
P. confusus (Fl.)	137	●		●	●		●	●	●	●	●	●				
f. *nigricans* (Oss.)	137a															
P. albicans (Kbm.)	138															
Tremulicerus tremulae (Estl.)	139	●		●	●		●	●	●							
T. vitreus (F.)	140															
T. distinguendus (Kbm.)	141															
Stenidiocerus poecilus (H.-S.)	142	●					●									
Sahlbergotettix salicicola (Fl.)	143															
Iassus lanio (L.)	144	●						●	●	●						
Batracomorphus allionii (Turt.)	145															
Eupelix cuspidata (F.)	146	●		●	●		●		●	●						

	Nn(ø+v)	TR(y+i)	F(v+i)	F(n+ø)	Al	Ab	N	Ka	St	Ta	Sa	Öa	Tb	Sb	Kb	Om	Ok	ObS	ObN	Ks	LkW	LkE	Le	Li	Vib	Kr	Lr
110											●																
111	●	●							●	●				●					●	●							
112		●			●	●	●		●	●	●	●													●		
113						●																					
114																									●	●	
115					●	●	●		●	●	●		●	●	●	●				●		●			●	●	
116					●	●	●		●	●	●	●				●				●					●	●	
117					●	●				●																	●
118																											
119					●	●	●		●	●	●				●	●											
120						●	●		●	●	●	●	●	●			●								●	●	
121																											
122																											
123														●											●	●	
124																											
125					●	●	●	●	●	●	●	●			●	●			●						●	●	
126					●						●	●			●	●											
127					●	●	●			●	●		●	●	●										●	●	
128																											
129					●	●	●	●				●															
130					●	●	●	●	●	●	●	●		●	●				●	●					●	●	
131					●	●	●		●																	●	
132																											
133					●	●	●		●	●	●	●	●	●	●	●	●		●	●					●	●	
134					●	●	●		●	●	●	●	●			●				●					●	●	
135					●	●	●		●	●				●											●		
136						●	●																				
137					●	●	●		●	●	●	●			●					●					●	●	
137a										●																	
138																											
139	●				●	●	●	●		●															●	●	
140																											
141																											
142						●	●																				
143								●																			
144						●	●																				
145					●	●	●				●															●	
146					●	●	●	●	●	●	●				●	●									●	●	

		N. Germany	G. Britain	SJ	EJ	WJ	NWJ	NEJ	F	LFM	SZ	NWZ	NEZ	B	Sk.	Bl.
Aphrodes makarovi Zachv.	147	●	●	●	●	●	●	●	●	●		●	●	●	●	●
A. bicincta (Schrnk.)	148								●						●	●
Planaphrodes bifasciata (L.)	149	●	●	●	●	●	●	●	●	●	●			●	●	●
P. nigrita (Kbm.)	150	●														
P. trifasciata (Fourcr.)	151	●	●		●	●	●	●	●	●				●	●	●
Anoscopus albifrons (L.)	152	●	●												●	●
A. limicola (Edw.)	153	●	●											●		
A. serratulae (F.)	154	●	●		●			●								
A. albiger (Germ.)	155															
A. histrionicus (F.)	156	●	●	●	●		●			●		●				
A. flavostriatus (Don.)	157	●	●	●	●				●	●						
Stroggylocephalus agrestis (Fall.)	158	●	●	●				●						●	●	
S. livens (Zett.)	159	●	●		●									●	●	
Evacanthus interruptus (L.)	160	●	●	●	●		●	●	●	●	●	●	●	●	●	●
E. acuminatus (F.)	161	●	●	●	●				●			●	●	●	●	●
Bathysmatophorus reuteri J.Sahlb.	162															
Cicadella viridis (L.)	163	●	●	●	●	●	●	●	●	●	●	●	●	●	●	●
C. lasiocarpae Oss.	164													●		
Alebra albostriella (Fall.)	165	●	●	●	●		●	●	●	●	●		●		●	
A. wahlbergi (Boh.)	166	●	●		●		●	●	●					●		
Erythria aureola (Fall.)	167	●	●	●	●		●	●						●		
Emelyanoviana mollicula (Boh.)	168	●	●		●										●	●
Dikraneura aridella (J.Sahlb.)	169															●
D. variata Hardy	170	●	●	●	●		●	●								
Micantulina micantula (Zett.)	171															
M. pseudomicantula (Knight)	172															
Wagneriala minima (J.Sahlb.)	173			●												●
W. incisa (Then)	174															
Forcipata citrinella (Zett.)	175	●	●	●	●								●	●	●	●
F. forcipata (Flor)	176	●	●		●						●			●	●	●
Notus flavipennis (Zett.)	177	●	●		●		●	●	●	●	●			●	●	●
Empoasca vitis (Göthe)	178	●	●		●			●	●	●	●			●	●	●
E. solani (Curt.)	179	●	●											●	●	●
E. decipiens Paoli	180				●											
E. kontkaneni Oss.	181															
E. ossiannilssoni Nuort.	182															
E. apicalis (Flor)	183															
E. smaragdula (Fall.)	184	●	●	●	●				●					●	●	●

920

	Hall.	Sm.	Öl.	Gtl.	G. Sand.	Ög.	Vg.	Boh.	Dlsl.	Nrk.	Sdm.	Upl.	Vstm.	Vrm.	Dlr.	Gstr.	Hls.	Med.	Hrj.	Jmt.	Ång.	Vb.	Nb.	Ås. Lpm.	Ly. Lpm.	P. Lpm.	Lu. Lpm.	T. Lpm.
147	●	●	●	●	●	●	●	●	●	●	●	●	●	●		●	●			●	●		●					
148	●	●	●	●		●	●		●								●											
149	●	●	●	●		●	●	●	●	●	●	●	●	●	●	●				●	●	●	●		●		●	
150																					●					*LpUm*		●
151	●	●	●	●		●	●	●	●	●	●		●		●					●	●	●						
152	●	●	●	●	●	●	●			●	●	●		●														
153	●																											
154	●	●	●			●	●		●			●																
155																												
156		●	●	●	●																							
157	●	●	●	●		●	●	●	●	●	●	●	●	●	●	●												
158	●	●	●			●	●			●	●	●			●													
159	●	●				●								●		●				●		●			●	●	●	
160	●	●	●	●	●	●	●	●	●	●	●	●	●	●	●	●	●			●	●	●						
161		●	●	●		●	●	●		●	●	●	●	●	●	●				●	●							
162																						●					●	
163	●	●	●	●		●	●	●	●	●	●	●	●	●	●	●	●			●	●	●	●				●	
164		●		●		●	●			●	●				●							●						
165	●	●	●			●	●			●	●																	
166		●				●	●					●																
167		●				●	●	●		●	●											●			●			
168		●	●	●		●	●	●	●	●	●	●																
169															●		●	●				●						
170																												
171								●				●												*Lpl.*				
172																												
173		●						●				●																
174			●																									
175		●	●	●		●	●	●	●		●				●				●	●	●	●		●	●	●	●	●
176						●					●	●			●	●			●	●	●	●					●	
177	●	●	●	●		●	●	●	●	●	●	●			●		●	●		●	●	●			●	●	●	
178	●	●	●	●		●	●	●		●	●	●				●				●								
179	●	●	●	●		●				●		●			●													
180																												
181			●									●																
182												●																
183																												
184	●	●	●	●		●	●	●		●	●	●			●	●	●			●		●			●		●	

		Ø+AK	HE (s+n)	O (s+n)	B (ø+v)	VE	TE (y+i)	AA (y+i)	VA (y+i)	R (y+i)	HO (y+i)	SF (y+i)	MR (y+i)	ST (y+i)	NT (y+i)	Ns (y+i)
Aphrodes makarovi Zachv.	147	●	●	●	●	●	●	●	●	●	●	●				
A. bicincta (Schrnk.)	148	●	●							●						
Planaphrodes bifasciata (L.)	149	●		●	●		●	●	●	●	●	●	●	●	●	●
P. nigrita (Kbm.)	150															
P. trifasciata (Fourcr.)	151	●							●		●					
Anoscopus albifrons (L.)	152	●			●		●	●	●	●	●	●		●		
A. limicola (Edw.)	153															
A. serratulae (F.)	154															
A. albiger (Germ.)	155															
A. histrionicus (F.)	156	●														
A. flavostriatus (Don.)	157	●	●							●					●	
Stroggylocephalus agrestis (Fall.)	158	●						●								
S. livens (Zett.)	159	●	●													
Evacanthus interruptus (L.)	160	●	●	●	●	●	●	●		●	●	●	●	●	●	●
E. acuminatus (F.)	161	●		●	●	●	●	●	●	●	●	●	●			
Bathysmatophorus reuteri J.Sahlb.	162															
Cicadella viridis (L.)	163	●	●	●	●	●	●	●	●	●	●			●		
C. lasiocarpae Oss.	164															
Alebra albostriella (Fall.)	165	●		●				●	●							
A. wahlbergi (Boh.)	166		●		●											
Erythria aureola (Fall.)	167	●	●				●									
Emelyanoviana mollicula (Boh.)	168	●			●		●									
Dikraneura aridella (J.Sahlb.)	169						●									
D. variata Hardy	170															
Micantulina micantula (Zett.)	171		●													
M. pseudomicantula (Knight)	172															
Wagneriala minima (J.Sahlb.)	173															
W. incisa (Then)	174															
Forcipata citrinella (Zett.)	175				●			●		●	●					●
F. forcipata (Flor)	176		●	●	●		●							●		
Notus flavipennis (Zett.)	177	●	●						●	●				●		
Empoasca vitis (Göthe)	178	●		●	●		●	●	●	●	●	●		●		
E. solani (Curt.)	179	●														
E. decipiens Paoli	180															
E. kontkaneni Oss.	181															
E. ossiannilssoni Nuort.	182															
E. apicalis (Flor)	183															
E. smaragdula (Fall.)	184	●		●	●		●	●		●	●		●			

	Nn (ø+v)	TR (y+i)	F (v+i)	F (n+ø)	Al	Ab	N	Ka	St	Ta	Sa	Öa	Tb	Sb	Kb	Om	Ok	ObS	ObN	Ks	LkW	LkE	Le	Li	Vib	Kr	Lr
147					●	●	●	●	●	●	●	●	●	●	●	●									●	●	
148																											
149	●				●	●	●		●	●	●	●	●	●	●	●	●		●	●					●	●	
150							●			●	●	●		●													
151					●	●	●	●	●	●	●			●	●			●	●						●		●
152					●	●	●	●	●	●	●	●		●		●									●	●	
153																											
154																											
155																											
156					●	●	●			●	●				●										●		
157					●	●	●			●	●	2		●		●	●		●						●	●	
158					●	●	●			●	●	●	●												●	●	
159																		●		●						●	
160	●	●			●	●	●	●	●	●	●	●		●	●	●	●		●						●	●	
161					●	●	●		●	●	●			●	●	●									●	●	
162																			●		●						●
163					●	●	●	●	●	●	●	●	●	●	●	●	●	●							●	●	
164									●																		
165																										●	
166																											
167						●	●	●			●		●													●	
168																											
169						●	●	●		●		●		●	●					●			●		●	●	●
170						●	●			●	●																
171					●																						
172						●					●																
173											●																
174																											
175		●	●		●	●	●					●					●	●									
176						●					●	●													●		
177	●	●			●	●	●	●	●	●	●		●	●		●	●			●		●			●	●	●
178					●	●	●		●	●	●		●		●											●	
179						●	●																		●		
180																											
181						●	●			●	●				●												
182					●	●																					
183						●				●										●							
184	●				●	●	●	●	●	●	●		●	●			●		●							●	

		N. Germany	G. Britain	SJ	EJ	WJ	NWJ	NEJ	F	LFM	SZ	NWZ	NEZ	B	Sk.	Bl.
E. lindbergi Linnav.	185														●	
E. betulicola W.Wagn.	186	●	●												●	
E. strigilifera Oss.	187	●	●	●		●								●	●	●
E. virgator Rib.	188	●	●											●	●	●
E. volgensis Vilb.	189															
E. rufescens (Mel.)	190	●									●				●	●
E. butleri Edw.	191	●	●				●	●								
E. populi Edw.	192	●	●				●	●					●	●	●	●
E. sordidula Oss.	193															
E. abstrusa Linnav.	194	●														●
Kyboasca bipunctata (Osh.)	195															
Chlorita viridula (Fall.)	196		●	●					●					●	●	●
C. paolii (Oss.)	197	●														
C. dumosa (Rib.)	198															
Fagocyba cruenta (H.-S.)	199	●	●	●					●					●	●	●
F. douglasi (Edw.)	200	●	●	●				●	●					●	●	●
F. carri (Edw.)	201	●	●	●											●	●
Ossiannilssonola callosa (Then)	202	●	●	●					●	●	●			●		
Edwardsiana rosae (L.)	203	●	●	●				●	●		●			●	●	●
E. avellanae (Edw.)	204	●	●											●	●	●
E. staminata (Rib.)	205	●													●	●
E. stehliki Laut.	206															
E. crataegi (Dgl.)	207	●	●	●					●							
E. nigriloba (Edw.)	208	●	●													●
E. salicicola (Edw.)	209	●	●			●	●							●	●	●
E. alnicola (Edw.)	210	●	●					●							●	●
E. sociabilis (Oss.)	211	●		●										●	●	●
E. frustrator (Edw.)	212	●	●	●										●	●	●
E. ishidae (Mats.)	213			●					●							
E. prunicola (Edw.)	214	●	●	●											●	●
E. menzbieri Zachv.	215															
E. flavescens (F.)	216	●		●										●	●	●
E. plebeja (Edw.)	217	●	●	●	●											
E. kemneri (Oss.)	218															
E. candidula (Kbm.)	219	●	●													●
E. gratiosa (Boh.)	220	●		●					●		●			●	●	●
E. geometrica (Schrnk.)	221	●	●	●				●	●	●				●	●	●
E. tersa (Edw.)	222	●	●	●										●	●	●

924

	Hall.	Sm.	Öl.	Gtl.	G. Sand.	Ög.	Vg.	Boh.	Dlsl.	Nrk.	Sdm.	Upl.	Vstm.	Vrm.	Dlr.	Gstr.	Hls.	Med.	Hrj.	Jmt.	Ång.	Vb.	Nb.	Ås. Lpm.	Ly. Lpm.	P. Lpm.	Lu. Lpm.	T. Lpm.
185										●		●			●	●					●	●	●					
186								●				●			●								●					
187	●	●										●	●		●	●				●	●							
188		●		●								●			●	●				●								
189												●																
190												●		●														
191	●					●	●					●	●	●		●	●			●	●	●						
192	●		●			●						●	●		●													
193										●		●	●	●	●		●	●		●								
194												●																
195																												
196	●	●	●			●	●		●	●	●	●	●	●	●	●	●			●								
197																												
198		●																										
199							●																					
200	●		●	●		●	●			●		●	●			●				●		●						
201		●	●			●	●																					
202																												
203	●	●	●	●	●	●				●		●	●															
204						●																						
205						●																						
206																				●								
207		●																										
208								●																				
209		●		●		●					●	●											●					
210	●					●			●		●	●			●						●	●						
211		●	●			●				●	●	●																
212		●	●	●		●																						
213											●				●					●								
214				●		●																						
215																				●							●	
216		●																										
217	●			●																								
218																				●								
219																												
220			●																									
221	●	●	●			●	●	●		●	●	●			●		●	●		●	●	●	●				●	
222												●		●	●	●					●	●					●	

925

		Ø+AK	HE (s+n)	O (s+n)	B (ø+v)	VE	TE (y+i)	AA (y+i)	VA (y+i)	R (y+i)	HO (y+i)	SF (y+i)	MR (y+i)	ST (y+i)	NT (y+i)	Ns (y+i)
E. lindbergi Linnav.	185	●		●	●									●		
E. betulicola W.Wagn.	186									●	●	●				
E. strigilifera Oss.	187			●		●										
E. virgator Rib.	188															
E. volgensis Vilb.	189															
E. rufescens (Mel.)	190															
E. butleri Edw.	191															
E. populi Edw.	192			●												
E. sordidula Oss.	193			●												
E. abstrusa Linnav.	194															
Kyboasca bipunctata (Osh.)	195															
Chlorita viridula (Fall.)	196															
C. paolii (Oss.)	197															
C. dumosa (Rib.)	198															
Fagocyba cruenta (H.-S.)	199															
F. douglasi (Edw.)	200	●			●											
F. carri (Edw.)	201	●			●		●									
Ossiannilssonola callosa (Then)	202															
Edwardsiana rosae (L.)	203	●														
E. avellanae (Edw.)	204															
E. staminata (Rib.)	205															
E. stehliki Laut.	206															
E. crataegi (Dgl.)	207															
E. nigriloba (Edw.)	208															
E. salicicola (Edw.)	209									●						
E. alnicola (Edw.)	210											●				
E. sociabilis (Oss.)	211	●		●												
E. frustrator (Edw.)	212	●														
E. ishidae (Mats.)	213															
E. prunicola (Edw.)	214	●														
E. menzbieri Zachv.	215															●
E. flavescens (F.)	216															
E. plebeja (Edw.)	217															
E. kemneri (Oss.)	218															
E. candidula (Kbm.)	219															
E. gratiosa (Boh.)	220															
E. geometrica (Schrnk.)	221	●		●	●		●		●	●	●	●	●			●
E. tersa (Edw.)	222													●		●

926

	Nn(ø+v)	TR(y+i)	F(v+i)	F(n+ø)	Al	Ab	N	Ka	St	Ta	Sa	Öa	Tb	Sb	Kb	Om	Ok	Ob S	Ob N	Ks	LkW	LkE	Lc	Li	Vib	Kr	Lr
185						●				●			●				●		●								
186																											
187					●	●	●			●				●	●											●	
188					●	●	●								●												
189																											
190																											
191					●	●	●				●				●			●	●	●			●			●	
192					●	●	●				●			●					●								
193					●	●	●			●	●		●	●	●				●								
194						●																					
195						●																					
196												●															
197					●	●	●						●												●	●	
198										●																	
199																											
200					●	●	●	●		●																	
201																											
202																											
203					●		●																				
204								●																			
205																									●		
206																											
207					●	●	●																				
208																											
209						●				●																	
210						●	●			●				●													
211					●	●	●				●				●												
212						●	●																				
213						●				●				●													
214						●				●																	
215		●				●																					
216																											
217						●																					
218																											
219					●																						
220																											
221					●	●	●			●	●		●	●	●		●	●							●	●	
222	●	●																									

	N. Germany	G. Britain	SJ	EJ	WJ	NWJ	NEJ	F	LFM	SZ	NWZ	NEZ	B	Sk.	Bl.
E. soror (Linnav.)	223															
E. bergmani (Tullgr.)	224	●	●		●									●	●	●
E. hippocastani (Edw.)	225	●	●		●			●	●					●	●	●
E. lethierryi (Edw.)	226	●	●													
Eupterycyba jucunda (H.-S.)	227	●	●											●	●	●
Linnavuoriana sexmaculata (Hdy.)	228	●	●											●	●	●
L. decempunctata (Fall.)	229	●	●		●									●		●
ssp. intercedens (Linnav.)	229a															
Ribautiana ulmi (L.)	230	●	●		●		●	●		●				●	●	●
R. tenerrima (H.-S.)	231	●	●		●				●					●	●	●
R. scalaris (Rib.)	232		●		●									●		
Typhlocyba quercus (F.)	233	●	●		●			●	●	●				●	●	●
T. bifasciata (Boh.)	234	●	●	●	●						●	●		●	●	●
Eurhadina pulchella (Fall.)	235	●	●	●	●		●	●	●	●				●	●	●
E. concinna (Germ.)	236	●	●		●					●				●	●	●
E. ribauti W.Wagn.	237	●	●		●											
E. kirschbaumi W.Wagn.	238	●	●													
E. untica Dlab.	239	●	●		●				●	●				●		●
Eupteryx atropunctata (Gze.)	240	●	●		●				●	●				●	●	●
E. origani Zachv.	241	●	●													
E. aurata (L.)	242	●	●		●				●	●				●	●	●
E. signatipennis (Boh.)	243	●	●		●				●	●	●			●		●
E. artemisiae (Kbm.)	244	●	●		●					●	●			●		●
E. urticae (F.)	245	●	●		●								●	●	●	●
E. cyclops Mats.	246	●	●		●						●			●	●	●
E. calcarata Oss.	247	●												●	●	
E. stachydearum (Hdy.)	248	●	●		●					●				●	●	●
E. collina (Flor)	249															
E. thoulessi Edw.	250	●	●				●								●	●
E. vittata (L.)	251	●	●		●					●		●		●	●	●
E. notata Curt.	252	●	●		●			●	●	●	●			●	●	●
E. tenella (Fall.)	253	●	●												●	
Aguriahana germari (Zett.)	254	●	●		●						●			●	●	●
A. pictilis (Stål)	255				●											
A. stellulata (Burm.)	256	●	●								●			●	●	
Alnetoidia alneti (Dhlb.)	257	●	●		●			●	●	●	●			●	●	●
Hauptidia distinguenda (Kbm.)	258														●	
Zyginidia pullula (Boh.)	259				●									●	●	

	Hall.	Sm.	Öl.	Gtl.	G. Sand.	Ög.	Vg.	Boh.	Dlsl.	Nrk.	Sdm.	Upl.	Vstm.	Vrm.	Dlr.	Gstr.	Hls.	Med.	Hrj.	Jmt.	Ång.	Vb.	Nb.	Ås. Lpm.	Ly. Lpm.	P. Lpm.	Lu. Lpm.	T. Lpm.
223																				●								
224										●		●					●		●	●	●					●	●	●
225	●	●	●	●						●		●	●		●													
226				●																								
227		●				●	●				●	●																
228	●	●	●	●		●	●			●	●	●	●		●	●	●				●	●				●	●	
229	●	●				●	●	●		●	●	●				●	●	●	●	●	●						●	●
229a											●	●																
230	●	●	●	●	●	●	●			●	●	●	●	●	●	●	●				●							
231		●	●																									
232																												
233		●	●			●	●	●	●	●	●	●				●												
234		●									●	●	●															
235	●	●	●	●	●	●					●	●	●															
236			●	●								●																
237	●	●				●						●																
238																												
239																												
240	●	●	●	●		●	●				●	●	●															
241		●																										
242	●	●									●	●																
243	●	●	●				●	●	●	●																		
244				●								●																
245		●	●		●																							
246	●	●				●	●	●		●	●	●	●		●		●	●		●			●				●	
247	●	●						●			●	●	●															
248		●	●				●																					
249																												
250																												
251	●	●	●	●		●	●	●		●	●	●	●		●					●								
252	●	●	●	●		●	●	●		●	●	●			●				●	●						●		
253					●							●																
254		●	●	●	●	●	●			●	●	●	●	●	●	●	●		●	●	●	●	●	●	●	●	●	●
255								●			●	●	●		●			●			●	●	●					
256		●									●	●	●		●													
257	●	●	●	●		●	●		●	●	●	●	●				●	●			●	●	●					
258		●	●									●																
259			●	●		●	●			●	●	●																

929

		NORWAY														
		Ø+AK	HE (s+n)	O (s+n)	B (ø+v)	VE	TE (y+i)	AA (y+i)	VA (y+i)	R (y+i)	HO (y+i)	SF (y+i)	MR (y+i)	ST (y+i)	NT (y+i)	Ns (y+i)
E. soror (Linnav.)	223															●
E. bergmani (Tullgr.)	224	●		●								●	●			
E. hippocastani (Edw.)	225	●						●								
E. lethierryi (Edw.)	226	●														
Eupterycyba jucunda (H.-S.)	227															
Linnavuoriana sexmaculata (Hdy.)	228	●	●				●	●	●	●	●	●	●			
L. decempunctata (Fall.)	229		●								●	●	●			
ssp. *intercedens* (Linnav.)	229a		●					●								
Ribautiana ulmi (L.)	230	●	●								●					
R. tenerrima (H.-S.)	231															
R. scalaris (Rib.)	232							●		●						
Typhlocyba quercus (F.)	233	●				●		●	●	●						
T. bifasciata (Boh.)	234															
Eurhadina pulchella (Fall.)	235	●							●	●						
E. concinna (Germ.)	236	●						●	●	●						
E. ribauti W.Wagn.	237					●										
E. kirschbaumi W.Wagn.	238	●						●								
E. untica Dlab.	239															
Eupteryx atropunctata (Gze.)	240	●														
E. origani Zachv.	241															
E. aurata (L.)	242									●						
E. signatipennis (Boh.)	243	●				●		●		●						
E. artemisiae (Kbm.)	244															
E. urticae (F.)	245							●	●							
E. cyclops Mats.	246	●	●			●	●				●			●		●
E. calcarata Oss.	247															
E. stachydearum (Hdy.)	248															
E. collina (Flor)	249															
E. thoulessi Edw.	250															
E. vittata (L.)	251	●		●												
E. notata Curt.	252	●									●	●				
E. tenella (Fall.)	253															
Aguriahana germari (Zett.)	254	●	●	●	●	●					●					●
A. pictilis (Stål).	255	●		●												
A. stellulata (Burm.)	256		●													
Alnetoidia alneti (Dhlb.)	257	●	●	●	●	●	●		●	●	●	●	●			
Hauptidia distinguenda (Kbm.)	258															
Zyginidia pullula (Boh.)	259															

#	Nn(ø+v)	TR(y+i)	F(v+i)	F(n+ø)	Al	Ab	N	Ka	St	Ta	Sa	Öa	Tb	Sb	Kb	Om	Ok	Ob S	Ob N	Ks	LkW	LkE	Le	Li	Vib	Kr	Lr
223	●										●									●							
224	●	●			●	●	●			●	●	●		●	●	●					●						
225						●	●																				
226																											
227																											
228					●	●	●				●		●	●	●	●									●	●	●
229				●	●	●	●			●	●	●			●										●		
229a					●	●	●				●															●	
230					●	●	●	●		●	●	●													●		
231																											
232																											
233																											
234						●																					
235					●	●	●	●		●	●				●										●	●	
236						●	●		●	●																	
237						●																					
238						●																					
239																											
240						●	●			●	●					●									●	●	
241						●	●				●	●	●	●													
242																											
243						●	●			●	●					●										●	
244																											
245						●																					
246		●			●	●	●	●	●	●	●	●	●	●	●	●			●								
247					●	●																					
248																											
249																									●		
250																											
251					●	●	●	●	●	●		●														●	●
252					●	●	●	●		●				●											●	●	●
253						●	●	●		●					●										●	●	
254	●	●			●	●	●	●	●	●	●	●		●	●					●					●	●	
255						●	●			●	●	●			●										●		
256						●	●				●			●											●		
257					●	●	●				●	●	●	●	●	●									●		
258																											
259					●																						

		N. Germany	G. Britain	SJ	EJ	WJ	NWJ	NEJ	F	LFM	SZ	NWZ	NEZ	B	Sk	Bl
Z. mocsaryi (Horv.)	260															●
Zygina flammigera (Fourcr.)	261	●	●		●			●	●	●	●		●	●	●	●
Z. angusta Leth.	262	●	●												●	
Z. rosea (Flor)	263	●														
Z. tiliae (Fall.)	264	●	●		●									●	●	●
Z. ordinaria (Rib.)	265	●	●													
Z. rosincola (Cer.)	266				●											
Z. suavis Rey	267	●	●												●	
Z. schneideri (Günth.)	268															
Z. salicina Mitj.	269							●								
Z. rubrovittata (Leth.)	270	●	●		●	●		●						●	●	●
Z. hyperici (H.-S.)	271	●	●		●				●				●			
Arboridia parvula (Boh.)	272		●						●							
Grypotes puncticollis (H.-S.)	273	●	●	●	●	●					●	●		●	●	●
Opsius stactogalus Fieb.	274	●	●													
Neoaliturus fenestratus (H.-S.)	275	●														
Coryphaelus gyllenhalii (Fall.)	276	●												●		
Balclutha punctata (F.)	277	●	●	●	●	●			●	●	●			●	●	●
B. rhenana Wagn.	278	●												●		
B. calamagrostis Oss.	279															
B. lineolata (Horv.)	280															
Macrosteles septemnotatus (Fall.)	281	●	●	●	●			●	●	●			●	●	●	●
M. oshanini Razv.	282	●	●													
M. variatus (Fall.)	283	●	●		●			●		●	●		●	●	●	●
M. sexnotatus (Fall.)	284		●		●								●	●		
M. ossiannilssoni Lindb.	285		●		●			●					●			
M. alpinus (Zett.)	286		●													
M. cristatus (Rib.)	287	●	●										●	●	●	●
M. fascifrons (Stål)	288															
M. laevis (Rib.)	289	●	●										●	●	●	
M. fieberi (Edw.)	290	●	●													●
M. lividus (Edw.)	291	●	●										●	●		●
M. viridigriseus (Edw.)	292	●	●		●					●					●	●
M. quadripunctulatus (Kbm.)	293	●	●										●	●		
M. sordidipennis (Stål)	294	●	●		●									●		
M. empetri (Oss.)	295															
M. frontalis (Scott)	296	●	●													
M. horvathi (Wagn.)	297	●	●						●	●				●	●	

	Hall.	Sm.	Öl.	Gtl.	G. Sand.	Ög.	Vg.	Boh.	Dlsl.	Nrk.	Sdm.	Upl.	Vstm.	Vrm.	Dlr.	Gstr.	Hls.	Med.	Hrj.	Jmt.	Äng.	Vb.	Nb.	Ås. Lpm.	Ly. Lpm.	P. Lpm.	Lu. Lpm.	T. Lpm.
260				●																								
261	●	●	●			●	●	●		●	●	●	●		●													
262																												
263	●	●										●			●		●											
264	●	●	●			●		●		●		●			●	●					●							
265		●										●																
266		●	●			●						●																
267						●						●					●											
268						●						●																
269																												
270	●	●	●	●	●	●					●	●	●	●	●					●	●							
271		●	●	●		●	●			●	●	●																
272		●	●			●	●		●			●																
273	●	●	●	●		●	●	●		●	●	●			●													
274																												
275																												
276	●	●	●			●	●								●	●												
277	●	●	●	●	●	●	●	●	●	●	●	●	●	●	●	●	●	●		●	●		●					●
278					●		●					●																
279												●																
280												●																
281	●	●	●	●		●	●			●	●	●	●	●		●	●					●						
282		●	●	●								●																
283	●	●	●	●	●	●	●				●	●			●		●	●		●	●	●						
284		●	●	●		●	●		●	●		●			●													
285						●			●						●		●							●				
286																				●	●			●			●	●
287	●	●	●			●	●			●	●				●	●	●	●		●	●	●	●	●	●	●	●	●
288																●				●								
289	●	●	●	●		●	●	●		●	●	●	●	●	●	●	●	●		●	●	●	●			●	●	●
290						●				●					●		●			●	●	●		●	●			
291			●				●										●	●										
292		●	●			●		●							●		●					●						
293	●																											
294			●						●																			
295															●					●	●			●		●		
296												●			●		●			●	●	●	●				●	●
297		●	●			●		●	●			●			●		●			●	●							●

933

		Ø+AK	HE (s+n)	O (s+n)	B (ø+v)	VE	TE (y+i)	AA (y+i)	VA (y+i)	R (y+i)	HO (y+i)	SF (y+i)	MR (y+i)	ST (y+i)	NT (y+i)	Ns (y+i)
Z. mocsaryi (Horv.)	260															
Zygina flammigera (Fourcr.)	261	●							●							
Z. angusta Leth.	262							●								
Z. rosea (Flor)	263	●		●												
Z. tiliae (Fall.)	264	●	●							●						
Z. ordinaria (Rib.)	265															
Z. rosincola (Cer.)	266	●														
Z. suavis Rey	267															
Z. schneideri (Günth.)	268	●														
Z. salicina Mitj.	269															
Z. rubrovittata (Leth.)	270			●				●	●	●						
Z. hyperici (H.-S.)	271	●														
Arboridia parvula (Boh.)	272		●													
Grypotes puncticollis (H.-S.)	273	●		●				●								
Opsius stactogalus Fieb.	274	●														
Neoaliturus fenestratus (H.-S.)	275															
Coryphaelus gyllenhalii (Fall.)	276															
Balclutha punctata (F.)	277	●	●			●	●	●	●	●	●	●	●		●	●
B. rhenana Wagn.	278															
B. calamagrostis Oss.	279															
B. lineolata (Horv.)	280															
Macrosteles septemnotatus (Fall.)	281	●		●	●	●	●			●						
M. oshanini Razv.	282															
M. variatus (Fall.)	283	●		●	●											●
M. sexnotatus (Fall.)	284	●							●							
M. ossiannilssoni Lindb.	285		●					●					●			
M. alpinus (Zett.)	286	●	●				●				●	●				●
M. cristatus (Rib.)	287	●	●	●						●	●	●				●
M. fascifrons (Stål)	288															
M. laevis (Rib.)	289	●	●		●						●	●				●
M. fieberi (Edw.)	290	●								●	●					
M. lividus (Edw.)	291															
M. viridigriseus (Edw.)	292	●							●							
M. quadripunctulatus (Kbm.)	293															
M. sordidipennis (Stål)	294															
M. empetri (Oss.)	295											●				
M. frontalis (Scott)	296		●										●			
M. horvathi (Wagn.)	297	●					●			●						

934

	Nn(ø+v)	TR(y+i)	F(v+i)	F(n+ø)	Al	Ab	N	Ka	St	Ta	Sa	Öa	Tb	Sb	Kb	Om	Ok	ObS	ObN	Ks	LkW	LkE	Le	Li	Vib	Kr	Lr
260																											
261						●	●			●				●												●	
262																											
263						●	●			●															●	●	
264					●	●	●		●	●															●	●	
265																											
266																											
267						?																					
268																											
269																											
270					●	●	●	●	●	●	●	●															
271					●	●				●	●															●	
272					●	●					●															●	
273					●	●	●			●	●															●	
274																											
275																										●	
276					●	●	●		●	●			●	●		●	●									●	
277					●	●	●		●	●	●	●	●	●	●			●	●	●					●	●	●
278						●																					
279						●	●			●																	
280																											
281					●	●	●			●	●	●		●	●	●			●	●					●	●	
282																											
283		●			●	●	●	●	●	●				●	●		●		●						●	●	
284					●	●	●			●		●	●	●	●				●						●	●	●
285		●									●																
286	●	●	●	●	●									●				●	●		●	●	●		●		●
287		●	●		●					●				●	●	●			●								
288						●																					
289		●	●		●					●		●		●			●	●	●	●				●		●	●
290						●					●				●												
291	●				●	●	●	●																			
292					●	●	●	●	●			●													●		
293						●					●	●		●													
294					●	●	●					●															
295																●	●		●								
296						●					●		●	●	●		●								●	●	
297	●	●			●	●	●	●			●	●		●		●									●	●	

935

Species	No.	N. Germany	G. Britain	SJ	EJ	WJ	NWJ	NEJ	F	LFM	SZ	NWZ	NEZ	B	Sk.	Bl.
M. nubilus (Oss.)	298	●														
M. cyane (Boh.)	299	●	●						●						●	●
Sonronius dahlbomi (Zett.)	300		●												●	
S. binotatus (J.Sahlb.)	301														●	
S. anderi (Oss.)	302															
Sagatus punctifrons (Fall.)	303	●	●		●		●	●				●	●	●		
Deltocephalus pulicaris (Fall.)	304	●	●	●	●	●		●	●	●	●	●			●	●
D. maculiceps Boh.	305	●	●	●					●			●	●			
Endria nebulosa (Ball)	306													●		
Doratura stylata (Boh.)	307	●	●	●	●	●	●	●	●	●	●	●	●	●	●	●
D. exilis Horv.	308	●														
D. impudica Horv.	309	●	●				●	●				●	●	●		
D. homophyla (Fl.)	310	●		●	●		●	●		●			●		●	
Platymetopius undatus (De G.)	311		●		●	●	●						●			
P. major (Kbm.)	312	●			●											
P. guttatus Fieb.	313															●
Idiodonus cruentatus (Panz.)	314	●	●	●	●			●		●					●	●
Colladonus torneellus (Zett.)	315	●	●					●								
Lamprotettix nitidulus (F.)	316	●	●								●	●	●			
Allygus mixtus (F.)	317	●	●	●		●		●	●	●		●				
A. communis (Ferr.)	318	●		●							●		●			●
A. maculatus Rib.	319			●												●
A. modestus Scott	320			●												
Allygidius commutatus (Fieb.)	321	●	●	●			●	●	●			●		●		
Graphocraerus ventralis (Fall.)	322	●	●	●	●		●	●	●					●	●	●
Rhytistylus proceps (Kbm.)	323	●	●				●	●								
Hardya tenuis (Germ.)	324	●														
Rhopalopyx preyssleri (H.-S.)	325	●				●	●	●	●	●	●	●		●	●	●
Rh. adumbrata (C.Sahlb.)	326	●	●			●	●	●								
Rh. vitripennis (Fl.)	327	●	[●]							●				●		
Paluda flaveola (Boh.)	328							●						●	●	
Elymana sulphurella (Zett.)	329	●	●	●	●	●	●	●		●		●	●	●	●	●
E. kozhevnikovi (Zachv.)	330															
Cicadula quadrinotata (F.)	331	●	●		●				●			●	●	●		●
C. persimilis (Edw.)	332	●	●		●				●	●		●		●	●	●
C. albingensis Wagn.	333	●														
C. longiventris (J. Sahlb.)	334															
C. saturata (Edw.)	335	●	●		●										●	●

	Hall.	Sm.	Öl.	Gtl.	G. Sand.	Ög.	Vg.	Boh.	Dlsl.	Nrk.	Sdm.	Upl.	Vstm.	Vrm.	Dlr.	Gstr.	Hls.	Med.	Hrj.	Jmt.	Äng.	Vb.	Nb.	Ås. Lpm.	Ly. Lpm.	P. Lpm.	Lu. Lpm.	T. Lpm.
298																	●			●	●	●				●		●
299	●					●					●	●																
300						●					●	●		●	●	●	●	●	●			●	●	●	●			●
301		●													●		●	●										
302						●																						
303	●			●							●			●	●	●												
304	●	●	●	●		●	●	●			●	●	●	●	●	●		●	●	●	●	●	●			●	●	●
305				●																								
306				●																								
307	●	●	●	●	●	●	●	●	●	●	●	●	●	●	●			●	●		●						●	
308		●				●																						
309																												
310	●		●																									
311		●				●	●	●			●	●	●	●			●		●		●							
312		●																										
313						●					●	●																
314	●	●	●	●			●	●		●	●	●	●	●	●	●	●		●	●	●	●			●	●	●	●
315		●					●	●			●				●		●	●		●		●			●	●	●	●
316		●	●	●			●	●	●		●	●																
317	●	●	●	●			●	●	●		●	●				●												
318		●	●	●			●	●	●		●	●																
319																												
320																												
321		●	●	●			●	●	●	●	●	●	●	●	●	●												
322	●	●	●	●			●	●		●		●	●	●														
323	●																											
324		●	●	●			●	●			●	●																
325			●	●			●	●						●	●		●	●			●	●						
326	●						●	●									●	●										
327		●		●						●											●	●						
328	●	●	●	●			●	●		●	●	●	●	●	●	●	●	●			●	●	●					
329	●	●	●	●	●	●	●	●	●	●	●	●	●		●				●									
330			●	●										●	●													
331		●	●	●			●	●			●	●	●		●	●				●	●	●	●	●	●	●	●	●
332	●	●	●	●			●				●		●	●		●			●		●							
333																												
334																												
335		●	●	●			●				●	●	●	●		●	●	●				●						

Species	No.	Ø+AK	HE (s+n)	O (s+n)	B (ø+v)	VE	TE (y+i)	AA (y+i)	VA (y+i)	R (y+i)	HO (y+i)	SF (y+i)	MR (y+i)	ST (y+i)	NT (y+i)	Ns (y+i)
M. nubilus (Oss.)	298															
M. cyane (Boh.)	299	●														
Sonronius dahlbomi (Zett.)	300		●	●									●			
S. binotatus (J.Sahlb.)	301	●	●													
S. anderi (Oss.)	302															
Sagatus punctifrons (Fall.)	303	●			●	●										
Deltocephalus pulicaris (Fall.)	304	●	●	●	●	●	●	●	●	●	●			●	●	●
D. maculiceps Boh.	305															
Endria nebulosa (Ball)	306															
Doratura stylata (Boh.)	307	●		●	●		●	●	●							
D. exilis Horv.	308															
D. impudica Horv.	309															
D. homophyla (Fl.)	310															
Platymetopius undatus (De G.)	311	●			●			●								
P. major (Kbm.)	312															
P. guttatus Fieb.	313	●			●			●								
Idiodonus cruentatus (Panz.)	314	●	●	●	●		●	●	●	●	●			●		
Colladonus torneellus (Zett.)	315		●		●											
Lamprotettix nitidulus (F.)	316															
Allygus mixtus (F.)	317															
A. communis (Ferr.)	318															
A. maculatus Rib.	319															
A. modestus Scott	320															
Allygidius commutatus (Fieb.)	321	●			●	●		●	●		●	●				
Graphocraerus ventralis (Fall.)	322				●											
Rhytistylus proceps (Kbm.)	323															
Hardya tenuis (Germ.)	324															
Rhopalopyx preyssleri (H.-S.)	325	●														
Rh. adumbrata (C.Sahlb.)	326		●													
Rh. vitripennis (Fl.)	327															
Paluda flaveola (Boh.)	328	●	●													
Elymana sulphurella (Zett.)	329	●			●	●		●	●	●	●	●	●			
E. kozhevnikovi (Zachv.)	330															
Cicadula quadrinotata (F.)	331	●	●		●			●	●	●	●			●	●	●
C. persimilis (Edw.)	332	●		●	●	●		●						●		
C. albingensis Wagn.	333	●			●	●										
C. longiventris (J. Sahlb.)	334															
C. saturata (Edw.)	335	●	●		●		●									

	Nn (ø+v)	TR (y+i)	F (v+i)	F (n+ø)	Al	Ab	N	Ka	St	Ta	Sa	Oa	Tb	Sb	Kb	Om	Ok	ObS	ObN	Ks	LkW	LkE	Le	Li	Vib	Kr	Lr
298						●				●				●						●					●		
299						●	●				●																
300	●	●				●	●	●		●				●			●								●	●	●
301						●			●				●												●	●	
302						●																					
303						●	●	●	●	●				●	●										●	●	
304	●	●	●		●	●	●	●	●	●	●	●	●	●	●	●			●	●					●	●	●
305																											
306						●																					
307					●	●	●	●	●	●	●	●	●	●	●	●	●								●	●	●
308																											
309																											
310					●					●				●											●	●	
311					●	●	●			●	●	●		●											●	●	
312																											
313																											
314		●	●		●	●	●		●	●	●	●	●	●	●	●		●	●						●	●	
315		●	●	●		●	●		●	●			●	●	●	●		●	●	●	●	●		●	●	●	●
316																											
317					●	●	●			●			●		●										●	●	
318																											
319																											
320																											
321						●	●	●	●	●				●	●										●	●	
322						●	●		●	●			●	●											●	●	
323																											
324						●	●	●	●	●				●											●	●	
325						●	●	●	●	●			●	●											●	●	
326						●	●	●																			
327						●	●			●				●	●												●
328						●	●	●	●	●		●		●	●			●							●	●	
329						●	●	●	●	●		●		●	●										●	●	
330						●	●				●				●												
331	●	●			●	●	●	●	●	●		●		●	●	●		●	●	●			●		●	●	
332						●	●		●						●												●
333							●																				
334							●				●																
335						●	●			●	●			●	●											●	

		N. Germany	G. Britain	SJ	EJ	WJ	NWJ	NEJ	F	LFM	SZ	NWZ	NEZ	B	Sk.	Bl.
C. quinquenotata (Boh.)	336	●	●				●								●	●
C. nigricornis (J. Sahlb.)	337															
C. flori (J. Sahlb.)	338	●				●							●	●		●
C. intermedia (Boh.)	339	?	●													
C. ornata (Mel.)	340														●	●
C. frontalis (H.-S.)	341	●	●	●	●				●		●			●		
Mocydiopsis attenuata (Germ.)	342	●	●			●	●		●							●
M. parvicauda Rib.	343	●	●													
Speudotettix subfusculus (Fall.)	344	●	●	●	●	●			●	●	●	●				●
Hesium domino (Reut.)	345	●														●
Thamnotettix confinis (Zett.)	346	●	●	●	●	●	●		●	●	●	●	●	●	●	●
Th. dilutior (Kbm.)	347	●	●	●					●	●	●					
Pithyotettix abietinus (Fall.)	348	●			●	●			●	●						●
Perotettix orientalis (Anufr.)	349															
Colobotettix morbillosus (Mel.)	350	●														
Macustus grisescens (Zett.)	351	●	●		●		●	●		●		●	●	●		
Doliotettix lunulatus (Zett.)	352	●														
Athysanus argentarius Metc.	353	●	●								●	●	●	●		●
A. quadrum Boh.	354	●											●			
Stictocoris picturatus (C.Sahlb.)	355	●														
Ophiola decumana (Kontk.)	356	●	●	●	●			●					●		●	●
O. cornicula (Marsh.)	357	●	●													
O. russeola (Fall.)	358	●	●		●								●		●	●
O. transversa (Fall.)	359	●		●	●								●		●	●
O. paludosa (Boh.)	360													●		
Limotettix striola (Fall.)	361	●	●	●	●	●	●	●	●	●	●		●	●		●
L. atricapillus (Boh.)	362	●	●											●		
L. sphagneticus Emelj.	363															
L. ochrifrons Vilb.	364															
Laburrus impictifrons (Boh.)	365	●							●					●		
Euscelidius schenkii (Kbm.)	366	●	●		●				●	●	●			●		●	●
Conosanus obsoletus (Kbm.)	367	●	●			●	●		●	●			●		●	●
Euscelis incisus (Kbm.)	368	●	●	●	●	●	●	●	●	●	●			●	●	●
E. distinguendus (Kbm.)	369	●													●	●
E. ohausi Wagn.	370	●	●				●									
Ederranus sachalinensis (Mats.)	371															
E. discolor (J. Sahlb.)	372															
Streptanus aemulans (Kbm.)	373	●	●	●	●	●	●	●	●	●			●	●	●	●

	Hall.	Sm.	Öl.	Gtl.	G. Sand.	Ög.	Vg.	Boh.	Dlsl.	Nrk.	Sdm.	Upl.	Vstm.	Vrm.	Dlr.	Gstr.	Hls.	Med.	Hrj.	Jmt.	Ång.	Vb.	Nb.	Ås. Lpm.	Ly. Lpm.	P. Lpm.	Lu. Lpm.	T. Lpm.
336	●	●		●		●				●	●	●	●		●	●	●	●		●	●						●	
337		●				●						●					●											
338	●	●	●			●	●				●	●	●	●			●											
339			●									●			●		●	●	●	●	●	●	●	●	●	●		●
340		●		●		●	●			●	●	●	●			●	●											
341		●	●				●				●	●																
342		●																										
343												●																
344	●	●	●			●	●	●		●	●	●	●	●	●	●	●	●	●	●	●	●	●	●	●	●	●	●
345	●	●	●			●	●	●	●	●	●	●	●	●	●	●		●										
346	●	●	●			●	●	●	●	●	●	●	●	●	●	●	●	●		●	●	●	●	●	●	●	●	
347																												
348	●	●				●	●			●	●	●	●	●		●		●		●	●	●	●				●	
349													●		●		●											
350																												
351	●	●	●	●		●	●	●	●	●	●	●	●	●	●	●		●	●	●	●	●	●	●	●	●	●	●
352		●				●	●			●	●	●	●	●	●	●	●	●	●	●	●	●	●	●	●	●	●	●
353	●	●	●			●		●	●	●	●	●			●	●												
354		●	●			●					●	●																
355						●					●	●																
356	●	●		●	●	●					●	●			●			●			●	●	●		●			
357	●														●													
358	●	●	●	●	●			●	●		●	●			●		●	●			●	●		●	●	●		●
359	●	●	●	●	●	●				●					●		●	●	●		●	●						
360		●									●				●		●	●										
361	●	●	●	●	●	●	●				●				●		●	●		●		●	●					
362		●		●		●												●										
363																												
364																												
365																												
366	●	●	●	●	●	●	●			●	●	●				●												
367	●	●	●			●	●	●			●																	
368	●	●	●			●	●	●			●	●																
369	●			●		●				●																		
370																												
371																												
372																												
373	●	●	●	●	●	●	●			●	●	●	●	●	●		●			●							●	

Species	No.	Ø+AK	HE (s+n)	O (s+n)	B (ø+v)	VE	TE (y+i)	AA (y+i)	VA (y+i)	R (y+i)	HO (y+i)	SF (y+i)	MR (y+i)	ST (y+i)	NT (y+i)	Ns (y+i)
C. quinquenotata (Boh.)	336															
C. nigricornis (J. Sahlb.)	337															
C. flori (J. Sahlb.)	338															
C. intermedia (Boh.)	339	●	●	●	●	●	●	●	●				●			●
C. ornata (Mel.)	340															
C. frontalis (H.-S.)	341															
Mocydiopsis attenuata (Germ.)	342															
M. parvicauda Rib.	343															
Speudotettix subfusculus (Fall.)	344	●	●	●	●	●	●	●	●	●	●	●	●	●		●
Hesium domino (Reut.)	345	●		●	●	●				●						
Thamnotettix confinis (Zett.)	346	●	●	●	●	●	●	●	●	●	●	●	●	●	●	
Th. dilutior (Kbm.)	347															
Pithyotettix abietinus (Fall.)	348	●		●	●				●					●	●	
Perotettix orientalis (Anufr.)	349															
Colobotettix morbillosus (Mel.)	350															
Macustus grisescens (Zett.)	351	●	●	●	●	●		●	●	●	●	●		●	●	●
Doliotettix lunulatus (Zett.)	352	●	●	●	●	●	●	●		●				●	●	
Athysanus argentarius Metc.	353	●														
A. quadrum Boh.	354	●		●	●											
Stictocoris picturatus (C.Sahlb.)	355															
Ophiola decumana (Kontk.)	356															
O. cornicula (Marsh.)	357															
O. russeola (Fall.)	358		●	●							●					
O. transversa (Fall.)	359		·		●											
O. paludosa (Boh.)	360	●			●		●	●	●				●			
Limotettix striola (Fall.)	361	●			●				●	●	●					
L. atricapillus (Boh.)	362								●							
L. sphagneticus Emelj.	363															
L. ochrifrons Vilb.	364															
Laburrus impictifrons (Boh.)	365															
Euscelidius schenkii (Kbm.)	366	●			●											
Conosanus obsoletus (Kbm.)	367								●							
Euscelis incisus (Kbm.)	368															
E. distinguendus (Kbm.)	369	●	●		●											
E. ohausi Wagn.	370															
Ederranus sachalinensis (Mats.)	371															
E. discolor (J. Sahlb.)	372															
Streptanus aemulans (Kbm.)	373	●			●	●		●		●	●					

	Nn (ø+v)	TR (y+i)	F (v+i)	F (n+ø)	Al	Ab	N	Ka	St	Ta	Sa	Öa	Tb	Sb	Kb	Om	Ok	ObS	ObN	Ks	LkW	LkE	Le	Li	Vib	Kr	Lr
336					●	●	●			●	●	●		●											●	●	
337																									●		
338					●	●	●	●		●	●	●													●	●	
339	●	●	●				●		●	●	●	●		●			●	●		●				●	●	●	●
340						●				●				●													
341					●																						
342																											
343																											
344	●	●	●	●		●	●		●	●	●	●	●	●	●	●	●	●			●	●			●	●	
345						●	●			●	●			●											●	●	
346	●	●	●	●	●	●	●			●	●	●		●	●		●		●						●	●	●
347																											
348						●	●	●	●	●				●	●		●		●						●	●	
349																											
350						●							●	●													
351	●	●	●	●	●	●		●	●	●	●	●	●	●			●	●			●	●	●	●	●	●	
352		●	●	●		●	●	●	●	●	●	●	●	●		●		●	●	●						●	●
353						●	●	●						●													
354						●	●	●	●																●		
355						●	●	●	●	●				●											●	●	
356			●			●	●	●	●	●	●			●			●								●	●	
357						●	●	●	●	●	●	●		●		●	●								●	●	
358						●	●	●	●	●	●	●		●											●	●	
359						●	●	●	●	●	●	●	●	●											●	●	
360						●	●	●	●	●	●	●	●	●		●									●	●	
361						●	●	●	●	●	●	●	●	●		●	●								●	●	●
362						●	●	●	●	●	●			●												●	
363							●							●													
364														●													
365																									●		
366	●				●	●	●	●	●	●				●											●	●	
367																											
368																											
369						●	●		●																	●	
370																											
371									●																		
372								●	●																●	●	
373					●	●	●	●	●	●			●	●	●				●								

		N. Germany	G. Britain	SJ	EJ	WJ	NWJ	NEJ	F	LFM	SZ	NWZ	NEZ	B	Sk.	Bl.
S. sordidus (Zett.)	374	●	●	●	●	●	●	●	●	●	●	●		●	●	●
S. okaensis Zachv.	375	●												●		
S. confinis (Reut.)	376					●		●						●		
S. marginatus (Kbm.)	377	●	●	●	●	●	●	●						●	●	●
Paramesus obtusifrons (Stål)	378	●	●	●	●	●				●	●	●		●	●	●
Parapotes reticulata (Horv.)	379	●														●
Paralimnus phragmitis (Boh.)	380	●	●			●	●		●	●				●	●	●
P. rotundiceps (Leth.)	381															
Metalimnus formosus (Boh.)	382	●	●												●	
M. marmoratus (Fl.)	383															
Arocephalus punctum (Fl.)	384	●	●	●	●	●	●	●	●	●	●			●	●	
A. languidus (Fl.)	385	●														
A. longiceps (Kbm.)	386	●												●		
Psammotettix alienus (J. Sahlb.)	387	●				●		●	●		●	●				
P. cephalotes (H.-S.)	388	●	●			●	●	●				●	●			
P. confinis (Dahlb.)	389	●	●				●	●			●			●	●	●
P. sabulicola (Curt.)	390	●	●			●	●	●		●				●	●	●
P. nodosus (Rib.)	391	●	●			●	●	●						●	●	
P. putoni (Then)	392	●										●			●	
P. albomarginatus Wagn.	393	●	●			●										
P. excisus (Mats.)	394	●						●						●		
P. dubius Oss.	395														●	
P. frigidus (Boh.)	396		●													
P. lapponicus (Oss.)	397															
P. pallidinervis (Dahlb.)	398	●														
P. poecilus (Fl.)	399	●														
Ebarrius cognatus (Fieb.)	400		●													●
Adarrus multinotatus (Boh.)	401	●	●													
Errastunus ocellaris (Fall.)	402	●	●	●	●					●	●	●		●	●	●
Turrutus socialis (Fl.)	403	●	●		●			●								
Mongolojassus bicuspidatus (J. Sahlb.)	404															
Jassargus distinguendus (Fl.)	405	●	●	●	●	●	●	●	●	●	●			●	●	●
J. flori (Fieb.)	406	●	●	●	●				●	●				●	●	●
J. alpinus neglectus (Then)	407															
J. sursumflexus (Then)	408	●	●				●	●							●	●
J. allobrogicus (Rib.)	409	●					●	●	●							
Mendrausus pauxillus (Fieb.)	410															
Pinumius areatus (Stål)	411	●														

No.	Hall.	Sm.	Öl.	Gtl.	G. Sand.	Ög.	Vg.	Boh.	Dlsl.	Nrk.	Sdm.	Upl.	Vstm.	Vrm.	Dlr.	Gstr.	Hls.	Med.	Hrj.	Jmt.	Ång.	Vb.	Nb.	Ås. Lpm.	Ly. Lpm.	P. Lpm.	Lu. Lpm.	T. Lpm.
374	●	●	●	●		●	●	●	●	●	●	●	●	●	●	●	●	●	●	●	●	●	●	●		●	●	
375							●					●									●							
376		●	●	●		●						●																
377	●	●	●	●		●	●	●	●	●	●	●	●	●	●	●	●		●	●	●	●	●	●	●	●	●	●
378		●	●	●		●		●				●																
379																												
380		●		●	●		●				●			●		●												
381									●																			
382		●	●	●		●					●	●	●		●		●											
383																												
384	●	●	●	●		●	●			●	●	●	●	●	●		●			●	●	●						
385			●	●		●	●			●	●	●																
386																												
387	●	●	●	●	●	●	●			●	●	●	●	●										●				●
388		●	●	●	●	●	●				●	●		●						●								
389	●	●	●	●	●	●	●			●	●	●	●	●		●	●			●	●	●	●			●	●	
390	●		●	●																							●	
391	●	●	●	●	●	●	●	●				●			●		●			●		●				●	●	
392																												
393		●																										
394				●																								
395	●	●	●	●													●									●		
396																				●		●				●	●	
397																										●		
398			●	●																								
399																				●								
400				●		●	●	●			●	●	●		●	●	●	●	●	●	●					●		●
401			●	●																								
402		●		●							●					●				●	●		●				●	
403	●	●	●			●	●		*Hlm.*	●																		
404																												
405	●	●	●			●	●	●	●	●	●	●	●	●	●					●	●							
406	●	●	●	●		●	●			●	●	●	●	●	●	●	●	●	●	●						●	●	
407																						●					●	●
408	●	●	●						●			●	●		●											●		
409																						●						
410																				●								
411																												

		Ø+AK	HE (s+n)	O (s+n)	B (ø+v)	VE	TE (y+i)	AA (y+i)	VA (y+i)	R (y+i)	HO (y+i)	SF (y+i)	MR (y+i)	ST (y+i)	NT (y+i)	Ns (y+i)
S. sordidus (Zett.)	374	●	●	●	●	●	●	●		●	●	●		●	●	●
S. okaensis Zachv.	375															
S. confinis (Reut.)	376	●														
S. marginatus (Kbm.)	377	●	●	●	●	●		●	●	●	●	●	●	●		●
Paramesus obtusifrons (Stål)	378				●			●								
Parapotes reticulata (Horv.)	379															
Paralimnus phragmitis (Boh.)	380															
P. rotundiceps (Leth.)	381	●						●								
Metalimnus formosus (Boh.)	382															
M. marmoratus (Fl.)	383															
Arocephalus punctum (Fl.)	384	●	●					●								
A. languidus (Fl.)	385															
A. longiceps (Kbm.)	386															
Psammotettix alienus (J. Sahlb.)	387	●		●												
P. cephalotes (H.-S.)	388					●										
P. confinis (Dahlb.)	389	●	●	●	●	●		●	●	●	●	●	●	●		●
P. sabulicola (Curt.)	390															
P. nodosus (Rib.)	391								●	●						
P. putoni (Then)	392								●							
P. albomarginatus Wagn.	393															
P. excisus (Mats.)	394															
P. dubius Oss.	395		●		●		●			●				●		●
P. frigidus (Boh.)	396			●		●				●	●					
P. lapponicus (Oss.)	397			●						●						
P. pallidinervis (Dahlb.)	398															
P. poecilus (Fl.)	399															
Ebarrius cognatus (Fieb.)	400	●		●	●		●				●		●	●		
Adarrus multinotatus (Boh.)	401															
Errastunus ocellaris (Fall.)	402		●	●												●
Turrutus socialis (Fl.)	403															
Mongolojassus bicuspidatus (J. Sahlb.)	404															
Jassargus distinguendus (Fl.)	405	●			●			●								
J. flori (Fieb.)	406	●	●	●	●	●		●	●	●	●	●	●	●	●	●
J. alpinus neglectus (Then)	407															●
J. sursumflexus (Then)	408	●							●							
J. allobrogicus (Rib.)	409															
Mendrausus pauxillus (Fieb.)	410															
Pinumius areatus (Stål)	411															

	Nn (ø+v)	TR (y+i)	F (v+i)	F (n+ø)	Al	Ab	N	Ka	St	Ta	Sa	Öa	Tb	Sb	Kb	Om	Ok	ObS	ObN	Ks	LkW	LkE	Le	Li	Vib	Kr	Lr
374	●	●			●	●	●		●	●		●		●	●	●			●						●	●	
375					●	●	●												●					●			●
376					●	●	●					●															
377		●	●	●	●	●	●		●	●	●	●	●	●	●	●			●	●	●	●	●	●	●	●	●
378					●	●	●	●																			
379																											
380					●	●	●		●	●	●	●			●	●										●	
381																●											
382						●	●			●	●	●		●	●	●			●						●	●	
383					●	●	●			●	●	●			●										●	●	
384	●				●	●	●																			●	
385						●									●										●	●	
386																											
387						●	●			●	●	●	●	●	●	●										●	
388					●	●									●											●	
389	●	●			●	●	●			●		●		●	●	●				●					●		
390				●				●	●							●		●							●		●
391		●	●					●	●											●					●		
392																											
393						●										●											
394																											
395																											
396																								●		●	●
397																											
398		●			●				●																	●	●
399					●			●	●	●	●				●										●	●	
400		●			●				●							●											●
401																											
402		●			●				●											●					●	●	●
403				●																							
404																									●	●	
405																											
406		●			●	●	●		●	●	●	●	●	●	●					●					●	●	●
407	●	●	●																						●		
408														●					●								
409						●	●				●	●	●	●	●					●							
410																											
411																									●		

		N. Germany	G. Britain	SJ	EJ	WJ	NWJ	NEJ	F	LFM	SZ	NWZ	NEZ	B	Sk.	Bl.
Diplocolenus bohemani (Zett.)	412			●	●											
Verdanus abdominalis (F.)	413	●	[•]	●	●		●	●	●	●		●	●	●	●	●
V. limbatellus (Zett.)	414															
Arthaldeus pascuellus (Fall.)	415	●	●	●	●	●	●	●	●	●	●		●	●	●	●
A. striifrons (Kbm.)	416	●	●	●				●	●				●	●	●	
Rosenus abiskoensis (Lindb.)	417															
Sorhoanus xanthoneurus (Fieb.)	418	●	●		●										●	●
S. assimilis (Fall.)	419	●		●		●			●			●	●			
Lebradea flavovirens (Gill.Bak.)	420															
Cosmotettix caudatus (Fl.)	421	●	●									●		●	●	
C. edwardsi (Lindb.)	422												●			
C. evanescens Oss.	423															
C. panzeri (Fl.)	424	●	●	●								●		●		
C. costalis (Fall.)	425	●	●			●	●	●	●			●	●	●	●	
Boreotettix bidentatus (De L.Davids.)	426															
Mocuellus collinus (Boh.)	427	●	●	●	●		●	●	●		●		●	●	●	●
M. metrius (Fl.)	428	●	●	●					●	●		●	●	●	●	

948

SWEDEN

	Hall.	Sm.	Öl.	Gtl.	G. Sand.	Ög.	Vg.	Boh.	Dlsl.	Nrk.	Sdm.	Upl.	Vstm.	Vrm.	Dlr.	Gstr.	Hls.	Med.	Hrj.	Jmt.	Ång.	Vb.	Nb.	Ås. Lpm.	Ly. Lpm.	P. Lpm.	Lu. Lpm.	T. Lpm.
412		●				●	●				●	●	●		●	●	●											
413	●	●	●	●	●	●	●	●	●	●	●	●	●	●	●	●	●	●	●	●	●	●	●	●	●	●	●	●
414															●					●	●			●	●	●	●	●
415	●	●	●	●		●	●		●	●	●	●	●	●	●	●	●	●			●	●	●			●	●	●
416			●	●																								
417																											●	●
418	●	●								●	●		●		●	●				●	●	●						
419		●	●			●					●					●					●	●						
420																												
421		●		●		●	●				●	●			●	●	●					●						
422			●								●	●								●	●							
423	●	●				●	●									●	●											
424	●	●	●			●			●						●	●				●	●	●		●		●		
425		●	●	●		●	●			●	●	●			●	●	●				●	●						
426															●	●	●											
427	●	●	●	●		●		●		●	●	●	●			●	●											
428	●	●	●			●				●	●	●	●				●											

		Ø+AK	HE (s+n)	O (s+n)	B (ø+v)	VE	TE (y+i)	AA (y+i)	VA (y+i)	R (y+i)	HO (y+i)	SF (y+i)	MR (y+i)	ST (y+i)	NT (y+i)	Ns (y+i)
Diplocolenus bohemani (Zett.)	412			●												
Verdanus abdominalis (F.)	413	●	●	●	●	●	●	●	●	●	●	●	●		●	
V. limbatellus (Zett.)	414	●	●	●	●	●					●	●	●	●		●
Arthaldeus pascuellus (Fall.)	415	●	●	●	●	●		●	●	●	●		●	●		●
A. striifrons (Kbm.)	416															
Rosenus abiskoensis (Lindb.)	417															
Sorhoanus xanthoneurus (Fieb.)	418	●	●						●	●	●			●		
S. assimilis (Fall.)	419															
Lebradea flavovirens (Gill.Bak.)	420															
Cosmotettix caudatus (Fl.)	421		●													
C. edwardsi (Lindb.)	422															
C. evanescens Oss.	423															
C. panzeri (Fl.)	424		●													
C. costalis (Fall.)	425					●										
Boreotettix bidentatus (De L.Davids.)	426															
Mocuellus collinus (Boh.)	427															
M. metrius (Fl.)	428	●														

	Nn (ø+v)	TR (y+i)	F (v+i)	F (n+ø)	Al	Ab	N	Ka	St	Ta	Sa	Oa	Tb	Sb	Kb	Om	Ok	ObS	ObN	Ks	LkW	LkE	Le	Li	Vib	Kr	Lr
412						●	●	●	●	●	●			●	●										●	●	
413	●	●	●	●	●	●	●	●	●	●	●	●	●	●		●	●	●	●	●			●		●	●	●
414	●	●	●	●			●							●						●	●	●	●	●			●
415		●	●			●	●			●	●	●		●	●	●		●	●						●	●	
416						●	●																				
417																								●			
418						●	●	●						●	●	●			●							●	
419					●	●	●	●	●					●	●											●	
420						●	●																				
421					●	●	●		●		●			●	●											●	
422						●	●		●	●		●		●	●												
423																											
424						●	●	●	●	●		●		●											●	●	
425						●	●	●	●	●	●	●		●	●										●		
426												●		●				●	▲								
427																									●		
428											●																

Literature

Albrecht, A., 1977: Intressanta fynd av skinnbaggar och stritar i Finland (Heteroptera & Homoptera, Auchenorrhyncha). – Notul. ent., 57: 51-52.

Allen, A. A., 1966: *Eupteryx tenella* Fall. (Hem., Cicadellidae) in South-east London – a probable addition to the Kent fauna. – Entomologist's mon. Mag., 101: 104.

Amyot, C. J. B. & Serville, J. G. A., 1843: Histoire Naturelle des Insectes, Hémiptères. 676 pp. Paris.

Anufriev, G. A., 1970a: New genera of Palaearctic Dicraneurini (Homoptera, Cicadellidae, Typhlocybinae). – Bull. Acad. pol. Sci., Sér. biol., 18: 261-263.

– 1970b: Notes on *Empoasca kontkaneni* Oss. (Auchenorrhyncha, Cicadellidae) with description of a new species from the Far East. – Ibid., 18: 633-635.

– 1971: New and little-known leaf hoppers (Homoptera, Auchenorrhyncha) from the Far East of the U.S.S.R. and neighbouring countries. – Ént. Obozr., 50: 95-116. (In Russian).

– 1978: Les cicadellides de la Territoire Maritime. – Horae Soc. Ent. Unionis Soveticae 60, 216 pp. Leningrad. (In Russian).

Azrang, M., 1978: Studier av *Laodelphax striatellus* (Fallén, 1826) (Homoptera, Delphacidae). 162 pp. Uppsala.

Baker, C. F., 1925: Nomenclatorial notes on the Jassoidea, IV. – Philipp. J. Sci., 27: 537.

Ball, E. D., 1900: Additions to the western Jassid fauna. – Can. Ent., 32: 337-347.

– 1936: Some new genera of leafhoppers related to *Thamnotettix*. – Bull. Brooklyn ent. Soc., 31: 57-60.

Beirne, B. P., 1952: The Nearctic species of *Macrosteles* (Homoptera: Cicadellidae). – Can. Ent., 84: 207-232.

– 1956: Leafhoppers (Homoptera: Cicadellidae) of Canada and Alaska. – Can. Ent., 88, Suppl. 2: 5-177.

Boheman, C. H., 1845a: Nya svenska Homoptera. – K. svenska VetenskAkad. Handl., 1845: 21-63.

– 1845b: Nya svenska Homoptera. – Öfvers. K. VetenskAkad. Förh., 1845: 154-164.

– 1847a: Nya svenska Homoptera. – K. svenska VetenskAkad. Handl., 1847: 23-67.

– 1847b: Nya svenska Homoptera. – Öfvers. K. VetenskAkad. Förh., 1847: 263-266.

– 1850: Bidrag till Gottlands insekt-fauna. – K. svenska VetenskAkad. Handl., 1849: 195-267.

– 1851: Iakttagelser rörande några insekt-arters metamorfos. – Öfvers. K. VetenskAkad. Förh., 1850: 211-215.

– 1852: Entomologiska anteckningar under en resa i södra Sverige 1851. – K. svenska VetenskAkad. Handl., 1851: 53-211.

– 1864: Entomologiska anteckningar under en resa i norra Skåne och södra Halland år 1862. – Öfvers. K. VetenskAkad. Förh., 21: 57-85.

Booij, C. J. H., 1981: Biosystematics of the *Muellerianella* complex (Homoptera, Delphacidae), taxonomy, morphology and distribution. – Neth. J. Zool., 31: 572-595.

– 1982: Biosystematics of the *Muellerianella* complex (Homoptera, Delphacidae): host-plants, habitats and phenology. – Ecol. Ent., 7: 9-18.

Brander, T. & Huldén, L., 1971: Havaintoja Lounais-Hämeen Luonto, Forssa, 43: 16-18.

Brullé, G. A., 1832: Expédition scientifique de Morée Section des sciences Physiques III, 1:e Partie Zoologie, Deuxième Section des Animaux Articulés. 400 pp. Paris.

Burmeister, H. C. C., 1838a: Rhynchota No. 1. Genera Insectorum. Iconibus illustravit et descripsit, 1: pls. 10, 11, 17, 20. Berlin.
- 1838b: Rhynchota No. 2. Genera Insectorum. Iconibus illustravit et descripsit, 1: pls. 6, 12, 14, 15. Berlin.
- 1841: Rhynchota No. 6. Genera Insectorum. Iconibus illustravit et descripsit, 1: pl. 13. Berlin.
Cerutti, N., 1939: Les Typhlocybidae du Valais. - Bull. Murithienne, 56: 81-95.
China, W. E., 1935: A new species of *Cixius* (Homoptera, Cixiidae) from Snowden. - Entomologist's mon. mag., 71: 38-40.
- 1943: New and little-known species of British Typhlocybidae (Homoptera) with keys to the genera *Typhlocyba, Erythroneura, Dikraneura, Notus, Empoasca,* and *Alebra.* - Trans. Soc. Br. Ent., 8: 111-153.
Christian, P. J., 1953: A revision of the North American species of *Typhocyba* and its allies (Homoptera, Cicadellidae). - Kans. Univ. Sci. Bull., 35: 1103-1277.
Claridge, M. F. & Howse, P. E., 1968: Songs of some British *Oncopsis* species (Hemiptera: Cicadellidae). - Proc. R. ent. Soc. Lond. (A), 43: 57-61.
Claridge, M. F. & Reynolds, W. J., 1972: Host plant specificity, oviposition behaviour and egg parasitism in some woodland leafhoppers of the genus *Oncopsis* (Hemiptera Homoptera: Cicadellidae). - Trans. R. ent. Soc. Lond., 124: 149-166.
- 1973: Male courtship songs and sibling species in the *Oncopsis flavicollis* species group (Hemiptera: Cicadellidae). - J. Ent. (B), 42: 29-39.
Claridge, M. F., Reynolds, W. J., & Wilson, M. R., 1977: Oviposition behaviour and food plant discrimination in leafhoppers of the genus *Oncopsis.* - Ecol.Ent., 2: 19-25.
Claridge, M. F., & Wilson, M. R., 1976: Diversity and distribution patterns of some mesophyll-feeding leafhoppers of temperate woodland canopy. - Ibid., 1: 231-250.
- 1978: Seasonal changes and alternation of food plant preference in some mesophyll-feeding leafhoppers. - Oecologia, 37: 247-255.
Curtis, J., 1829: A guide to an arrangement of British insects; being a catalogue of all the named species hitherto discovered in Great Britain and Ireland. 256 pp. London.
- 1833: Characters of some undescribed genera and species indicated in the "Guide to an arrangement of British Insects". - Ent. Mag., 1: 186-199.
- 1836: *Acucephalus.* British Entomology, 13: pl. 620. London.
- 1837a: *Aphrodes.* British Entomology, 14: pl. 633. London.
- 1837b: *Eupteryx.* British Entomology, 14: pl. 640. London.
- 1846: *Eupteryx solani* (The potato frog-fly). - Gdnrs' Chron., 1846: 388.
Dahlbom, A. F., 1851: Anteckningar öfver Insekter, som blifvit observerade på Gottland och i en del af Calmare Län, under sommaren 1850. - K. svenska VetenskAkad. Handl., 1850: 155-229.
De Geer, C., 1741: Beskrifning uppå et Insect, som lefver uppå mäst alla Örter och Trän uti et hvitt Skum, ock kallas: *Cicada fusca, alis superioribus maculis albis, in spuma quadam vivens.* - Ibid., 2: 221-236.
- 1773: Des Cigales. Mémoires pour servir à l'Histoire des Insectes.3, 696 pp. Stockholm.
De Long, D. M., 1936: Some new genera of leafhoppers related to *Thamnotettix.* - Ohio J. Sci., 36: 217-219.
- 1937: The genera *Cyperana* and *Paluda* (Homoptera-Cicadellidae). - Am. Midl. Nat., 18: 225-236.
- & Caldwell, J. S., 1936: A new genus - *Forcipata* - and nine new species of Typhlocybine leafhoppers closely allied to *Dikraneura* (Cicadellidae: Homoptera). - Ann. ent. Soc. Am., 29: 70-77.
- & Davidsen, R. H., 1935: Some new North American species of deltocephaloid leafhoppers. - Can. Ent., 67: 164-172.

Distant, W. L., 1918: Rhynchota. Homoptera: Appendix. Heteroptera: Addenda. – Fauna Br. India, 7: 1-210.

Dlabola, J., 1944: IV. Attributio ad cognitionem homopterorum faunae. – Acta ent. bohemoslov., 41: 94-100.

– 1954: Krísi – Homoptera. – Fauna ČSR, 1, 340 pp. Praha.

– 1958: A reclassification of Palaearctic Typhlocybinae (Homopt., Auchenorrh.). – Acta Soc. ent. Čechoslov., 55: 43-57.

– 1959a: Neue paläarktische Zikaden der Fam. Meenoplidae und der Gattung *Handianus* Rib. (Homopt. Auchenorrhyncha). – Acta ent. Mus. Nat. Pragae, 33: 445-452.

– 1959b: Fünf neue Zikaden-Arten aus dem Gebiet des Mittelmeers. – Boll. Soc. ent. ital., 89: 150-155.

– 1963: Typen und wenig bekannte Arten aus der Sammlung H. Haupt mit Beschreibungen einiger Zikadenarten aus Sibirien (Homoptera). – Acta ent. Mus. Nat. Pragae, 35: 313-331.

– 1967: Ergebnis der 1. mongolisch-tschechoslovakischen entomologisch-botanischen Expedition in der Mongolei. Nr 1: Reisebericht, Lokalitätenübersicht und Beschreibungen neuer Zikadenarten (Homopt. Auchenorrhyncha). – Acta faun. ent. Mus. Nat. Pragae, 12: 1-34.

– 1974: Generische Gliederung der Unterfamilie Idiocerinae in der Paläarktis (Homoptera Auchenorrhyncha). – Ibid., 15: 59-68.

Donovan, E., 1799: The natural history of British Insects, 8: 1-88. London.

Dorst, H. E., 1937: A revision of the leafhoppers of the *Macrosteles* group (*Cicadula* of authors) in America north of Mexico. – Misc. Publs U. S. Dep. Agric., 271: 1-24.

Douglas, J. W., 1874: Captures of Hemiptera on the west coast of Scotland, with description of a new species. – Entomologist's mon. Mag., 11: 118.

– 1876: British Hemiptera-Homoptera. Additional species. – Ibid., 12: 203-204.

– & Scott, J., 1873: British Hemiptera: New species – Homoptera. – Ibid., 9: 210-212.

Drosopoulos, S., 1975: Some biological differences between *Muellerianella fairmairei* (Perris) and *M. brevipennis* (Boheman), a pair of sibling species of Delphacidae (Homoptera Auchenorrhyncha). – Ent. Ber., Amst., 35: 154-157.

– 1977: Biosystematic studies on the *Muellerianella* complex (Delphacidae, Homoptera Auchenorrhyncha). – Meded. LandbHoogesch. Wageningen, 77-14, 133 pp.

Dworakowska, I., 1970a: On the genus *Zygina* Fieb. and *Hypericiella* sgen. n. (Auchenorrhyncha, Cicadellidae, Typhlocybinae). – Bull. Acad. Pol. Sci. Sér. biol., 18: 559-567.

– 1970b: Three new genera of Erythroneurini (Auchenorrhyncha, Cicadellidae, Typhlocybinae). – Ibid., 18: 617-624.

– 1970c: On the genera *Zyginidia* Hpt. and *Lublinia* gen.n. (Auchenorrhyncha, Cicadellidae, Typhlocybinae). – Ibid., 18: 625-632.

– 1972: Revision of the genus *Aguriahana* Dist. (Auchenorrhyncha, Cicadellidae, Typhlocybinae). – Polskie Pismo ent., 273-312.

– 1973: *Baguoidea rufa* (Mel.) and some other Empoascinini (Auchenorrhyncha, Cicadellidae). – Bull. Acad. pol. Sci. Sér. biol., 21: 49-58.

– 1976: *Kybos* Fieb., subgenus of *Empoasca* Walsh (Auchenorrhyncha, Cicadellidae, Typhlocybinae) in Palaearctic. – Acta zool. cracov., 21: 387-463.

Edwards, J., 1878: Description of a new British *Typhlocyba*. – Entomologist's mon. Mag., 14: 248.

– 1881: An additional species of British Homoptera. – Ibid., 17: 224.

– 1885: Notes on British Typhlocybidae with diagnoses of two new species. – Ibid., 21: 228-231.

– 1888: British Hemiptera: Additional species. – Ibid., 24: 196-198.

– 1888b: Descriptions of four new species of *Typhlocyba*. – Ibid., 25: 157-158.

– 1888c: A synopsis of British Homoptera-Cicadina. Part II. – Trans. ent. Soc. London, 1888: 13--108.

- 1889: Fauna and flora of Norfolk. XX. Hemiptera. – Trans. Norfolk Norwich Nat. Soc., 4: 702-711.
- 1891: On the British species of the genus *Cicadula* (Zett.) Fieber. – Entomologist's mon. Mag., 27: 27-34.
- 1894: British Hemiptera: additions and corrections. – Ibid., 30: 101-106.
- 1898: Notes on the genus *Chloriona*, Fieber; with description of a new species. – Ibid., 34: 58-62.
- 1908a: On some British Homoptera hitherto undescribed or unrecorded. – Ibid., 44: 56-59.
- 1908b: On some British Homoptera hitherto undescribed or unrecorded. – Ibid., 44: 80-87.
- 1914: Additional species of British Typhlocybidae. – Ibid., 50: 168-172.
- 1915: On certain British Homoptera. – Ibid., 51: 206-211.
- 1919: A note on the British representatives of the genus *Macropsis* Lewis; with descriptions of two new species. – Ibid., 55: 55-58.
- 1920: New or little known species of British Cicadina. – Ibid., 56: 53-58.
- 1922: A generic arrangement of British Jassina. – Ibid., 58: 204-207.
- 1924: On some new or little-known British Cicadina. – Ibid., 60: 52-58.
- 1925: On an unrecognized species of *Typhlocyba*. – Ibid., 61: 64.
- 1926: On some new and little-known British Cicadina, with a table of the genus *Eupteryx*. – Ibid., 62: 52-56.
Emeljanov, A. F., 1962: New tribes of leaf-hoppers of the subfamily Euscelinae (Auchenorrhyncha, Cicadellidae). – Ént. Obozr., 41: 388-397. (In Russian).
- 1964: 1. Podotrjad Cicadinea (Auchenorrhyncha) – tsikadovye. Pp. 337-437 in: Bei-Bienko: Opredelitel nasekomych evropeiskoj chasti SSSR v pjati tomach. Moskva-Leningrad.
- 1966: New palearctic and certain nearctic cicads (Homoptera, Auchenorrhyncha). – Ént. Obozr., 45: 95-133. (In Russian).
- 1972: New palaearctic leafhoppers of the subfamily Deltocephalinae (Homoptera, Cicadellidae). – Ibid., 51: 102-111. (In Russian).
- 1975: Materials for the revision of the tribe Adelungiini (Homoptera, Cicadellidae). – Ibid., 54: 383-390. (In Russian).
- 1977: Leaf-hoppers (Homoptera, Auchenorrhyncha) from the Mongolian People's Republic based mainly on materials of the Soviet-Mongolian zoological expeditions (1967-1969). – Insects Mongolia, 5: 96-195. (In Russian).
Estlund, O., 1796: Entomologiska anmärkningar hörande til Fauna Svecica. – K. svenska VetenskAkad. Nya Handl., 17: 126-130.
Fabricius, J. C., 1775: Systema Entomologiae. – xxviii + 832 pp. Flensburgi et Lipsiae.
- 1777: Genera Insectorum. – Praefatio 1776. Editio 1777. 310 pp. Kilonii.
- 1787: Mantissa Insectorum. 2: 1-382. Hafniae.
- 1794: Entomologia systematica, emendata et aucta. 4: 1-472. Hafniae.
- 1798: Supplementum Entomologiae Systematicae. 572 pp. Hafniae.
- 1803: Systema Rhynchotorum. X + 314 pp. Brunsvigiae.
Fallén, C. F., 1805: Försök till de Svenska Cicad-Arternas uppställning och beskrifning. – K. svenska VetenskAkad. Nya Handl., 26: 229-253.
- 1806: Försök till de Svenska Cicad-Arternas uppställning och beskrifning. – Ibid., 27: 6-53.
- 1814: Specimen novam Hemiptera disponendi methodum exhibens. Lund, 26 pp.
- 1826: Cicadariae, earumque affines. Hemiptera Sveciae, 2: 80 pp. Londini/Gothorum.
Fennah, R. G., 1956: Fulgoroidea from southern China. – Proc. Calif. Acad. Sci., 4th Ser., 28: 441-527.
- 1963: New genera of Delphacidae (Homoptera: Fulgoroidea). – Proc. R. ent. Soc. Lond. (B), 32: 15-16.
Ferrari, P. M., 1882: Cicadaria Agri Ligustici. – Annali Mus. Civ. Stor. nat. Giacomo Doria, 18: 75-165.

Fieber, F. X., 1866a: Neue Gattungen und Arten in Homoptern (*Cicadina* Burm.). – Verh. zool.-bot.Ges. Wien, 16: 497-516.
- 1866b: Grundzüge zur generischen Theilung der Delphacini. – Ibid., 16: 517-534.
- 1868: Europäische neue oder wenig bekannte Bythoscopida. – Ibid., 18: 449-464.
- 1869: Synopse der europäischen Deltocephali. – Ibid., 19: 201-222.
- 1872: Katalog der europäischen Cicadinen. i-iv, 1-19. Wien.
- 1875: Les Cicadines d'Europe. Première partie. – Revue mag. Zool., (3) 3: 288-416.
- 1876: Les Cicadines d'Europe. Deuxième partie. – Ibid., (3) 4: 11-268.
- 1885: Description des Cicadines d'Europe des genres *Cicadula* et *Thamnotettix*. – Revue Ent., 4: 40-110.

Flor, G., 1861a: Die Rhynchoten Livlands in systematischer Ordnung beschrieben. Zweiter Theil. – Arch. Naturk. Liv. – Est. u. Kurlands 4: 1-567.
- 1861b: Neue *Typhlocyba*-Arten für Livland. – Ibid., 4: 632-637.

Fourcroy A. F., 1785: Entomologia parisiensis. 231 pp. Paris.

Froggatt, W. W., 1918: The apple-leaf Jassid (*Empoasca australis*). – Agric. Gaz. N. S. W., 29: 568-571.

Germar, E. F., 1818: Bemerkungen über einige Gattungen der Cicadarien. – Magazin Ent. (Germar), 3: 177-227.
- 1821: Bemerkungen über einige Gattungen der Cicadarien. – Ibid., 4: 1-106.
- 1822: Fauna Insectorum Europae, 7: pls. 19-20. Halae.
- 1830: Species Cicadarum enumeratae et sub genera distributae. *In:* Thon's entomologisches Archiv, 2: 37-57.
- 1831: Fauna Insectorum Europae, 14: pls. 11-15. Halae.
- 1833: Conspectus generum Cicadariarum. – Rev. Ent. (Silbermann), 1: 174-184.
- 1837: Fauna Insectorum Europae, 17: pls. 10-20. Halae.
- 1838: Ibid. 20: pls. 24-25. Halae.

Gillette, C. P., & Baker, C. F., 1895: A preliminary list of the Hemiptera of Colorado. – Bull. Colo. St. Univ. agric. Exp. Stn., 31: 1-137.

Gmelin, J. F., 1789: Insecta Hemiptera. Caroli a Linné Systema Naturae, 1 (4): 1517-2224.

Goeze, J. A. E., 1778: Entomologische Beyträge zu des Ritter Linné Zwölften Ausgabe des Natursystems, 2: 1-352.

Gravestein, W. H., 1953: Faunistische mededelingen over Cicaden 1. – Ent. Ber., Amst., 14: 280-281.
- 1965: New faunistic records on Homoptera-Auchenorrhyncha from the Netherlands North Sea Islands Terschelling. – Zool. Beitr. (N. F.), 11: 103-111.

Gyllensvärd, N., 1961: Några för Sverige nya eller sällsynta hemiptera. – Opusc. ent., 26: 255-256.
- 1963: Några för Sverige nya eller sällsynta hemiptera II. – Ibid., 28: 198-200.
- 1964: *Typhlocyba sundholmi* n.sp. (Hem. Hom.). – Ibid., 29: 170-173.
- 1965: Några för Sverige nya eller sällsynta Hemiptera III. – Ibid., 30: 227-230.
- 1969a: Några för Sverige nya eller sällsynta Hemiptera. IV. – Ibid., 34: 162-166.
- 1969b: Några för Sverige nya eller sällsynta Hemiptera V. – Ibid., 270-274.
- 1971: Några för Sverige nya eller sällsynta Hemiptera VI. – Ent. Tidskr., 92: 78-81.
- 1972: Några för Sverige nya eller sällsynta Hemiptera. VII. – Ibid., 93: 224-226.

Günthart, H., 1971: Kleinzikaden (Typhlocybinae) an Obstbäumen in der Schweiz. – Schweiz. Z. Obst- u. Weinbau, 107 (80): 285-306.
- 1974: Beitrag zur Kenntnis der Kleinzikaden (Typhlocybinae, Hom., Auch.) der Schweiz, 1. Ergänzung. – Mitt. schweiz. ent. Ges., 47: 15-27.

Göthe, H., 1875: Die Ursachen des Schwarzen Brenners an den Reben. – Wien. Landw. Ztg., 1875: 397-398.

Hackwitz, G. v., 1910: Entomologiska anteckningar. – Ent. Tidskr., 31: 243.

Halkka, O., Raatikainen, M., Vasarainen, A., & Heinonen, L., 1967: Ecology and ecological genetics of Philaenus spumarius (L.) (Homoptera). – Annls zool. fenn., 4: 1-18.

Hamilton, K. G. A., 1975: A review of the Northern Hemisphere Aphrodina (Rhynchota: Homoptera: Cicadellidae), with special reference to the Nearctic fauna. – Can. Ent., 107: 1009-1027.

Hardy, J., 1850: Descriptions of some new British Homopterous insects. – Trans. Tyneside Nat. Fld Cl., 1: 416-431.

Hassan, A. J., 1939: The biology of some British Delphacidae (Homopt.) and their parasites with special reference to the Strepsiptera. – Trans. R. ent. Soc. Lond., 89: 345-384.

Haupt, H., 1912: Neues und Kritisches über Arten und Varietäten einheimischer Homoptera. – Berl. ent. Z., 56: 177-196.

– 1917: Neue paläarktische Homoptera nebst Bemerkungen über einige schon bekannte. – Wien. ent. Ztg, 36: 229-262.

– 1919: Die europäischen Cercopidae Leach. (Blutströpfchen und Schaumzikaden). – Ent. Jb., 28: 152-172.

– 1924: Alte und neue Homoptera Mitteleuropas. – Konowia, 3: 285-300.

– 1925: Ueber eine Homopteren-Ausbeute von Mitttelwald und "Revision der Gattung Cicadula Zett.". – Mitt. münch. ent. Ges., 15: 9-40.

– 1929: Neueinteilung der Homoptera-Cicadina nach phylogenetisch zu wertenden Merkmalen. – Zool. Jb. (Syst.), 58: 173-286.

– 1933: Zwei neue Arten der Homoptera-Cicadina und synonymische Erörterungen über zwei schon bekannte Arten. – Mitt. dt. ent. Ges., 4: 18-26.

– 1935: Unterordnung Gleichflügler, Homoptera. In: Brohmer-Ehrmann-Ulmer: Tierwelt Mitteleur. 4(3): 115-262.

Heikinheimo, O., 1958: Surveys to the results of the investigations regarding the damage to oats in the year of 1957. – Maataloust. Aikakausk., 30: 199-200.

Heinze, K., 1959: Phytopathogene Viren und ihre Überträger. Berlin. 291 pp.

Hellén, W., 1936-1966: Verzeichnis der in Jahren 1931-1965 für die Fauna Finnlands neu hinzugekommenen Insektenarten.-Notul. ent., 16 (1936: in der Jahren 1931-1935), 21 (1942: in den Jahren 1936-1940), 26 (1946: in den Jahren 1941-1945), 32 (1952: in den Jahren 1946-1950), 36 (1956: in den Jahren 1951-1955), 41 (1961: in den Jahren 1956-1960), 1966 (in den Jahren 1961-1965).

Herrich-Schäffer, G. A. W., 1834a: Deutschlands Insecten, 122: 1-6. Regensburg.

– 1834b: Deutschlands Insecten, 124: 1-15.

– 1834c: Deutschlands Insecten, 125: 1-8.

– 1834d: Deutschlands Insecten, 126: 1-8.

– 1835: Nomenclator entomologicus, 1: i-iv, 1-116. Regensburg.

– 1836: Deutschlands Insecten, 143: 1-22.

– 1837: Deutschlands Insecten, 144: 1-16.

– 1838: Deutschlands Insecten, 164: 7-21.

Holgersen, H., 1944a: Norske sikader I. – Nytt Mag. Naturvid., 84: 205-218.

– 1944b: Norske sikader (Homoptera cicadina) II. – Bergens Mus. Årb., 9: 1-37.

– 1945: Norske sikader (Homoptera cicadina) III. De norske arter av slekten Thamnotettix Zett. 1839. – Norsk ent. Tidsskr., 7: 107-114.

– 1946a: Konservator Helliesens sikadesamling. – Stavanger Mus. Årb, 1946: 135-140.

– 1946b: Om en del norske fulgorider (Norske sikader, Homoptera Cicadina, IV). – Norsk ent. Tidsskr., 7: 149-155.

– 1949: The Norwegian species of Euscelis and related genera (Homoptera Cicadina). – Meddr

957

zool. Mus., Oslo, 56: 77-94.

- 1954: Norwegian leaf-hoppers of the genera *Macrosteles, Erotettix,* and *Balclutha* (Hom. Cicadina). - Norsk ent. Tidsskr., 9: 18-25.

Horváth, G., 1897: Homoptera nova ex Hungaria. - Természetr. Füz., 20: 620-643.

- 1903: Homoptera quinque nova ex Hungaria. - Annls hist.-nat. Mus. natn. hung., 1: 472-476.

- 1903a: Synopsis generis *Doratura* Sahlb. - Ibid., 1: 451-459.

- 1903b: Adnotationes synonymicae de Hemipteris palaearcticis. - Ibid., 1: 555-558.

- 1904: Insecta Heptapotamica a DD. Almásy et Stummer-Traunfels collecta. I. Hemiptera. - Ibid., 2: 574-590.

- 1910: Magyarországi új Homoptera. - Rovart. Lap., 17: 176-177.

Huldén, K., 1974: The *Javesella discolor* group (Homoptera, Delphacidae) of North Europe, with description of a new species. - Notul. ent., 54: 114-116.

- 1975: Faunistic notes from Finland I. - Ibid., 55: 86-88.

- 1982: Records of Heteroptera and Auchenorrhyncha (Hemiptera) from northern Norway. - Ibid., 62: 66-68.

-, Meinander, M., Nybom, O., & Silfverberg, H., 1977: Deletions from the Finnish fauna I. - Ibid., 57: 11-12.

Ikäheimo, K., & Raatikainen, M., 1961: *Calligypona obscurella* (Boh.), a new vector of the wheat striate mosaic and oat sterile-dwarf viruses. - J. Sci. Agric. Soc. Finl., 33: 146-152.

- 1963: *Dicranotropis hamata* (Boh.) (Hom., Araeopidae) as a vector of cereal viruses in Finland. - Annls agric. fenn., 2: 153-158.

International Commission on Zoological Nomenclature, 1961: Opinion 590. *Aphrophora* Germar, 1821 (Insecta, Hemiptera). - Bull. zool. Nom., 18: 109-111.

Jacobsen, O., 1915: Fortegnelse over danske Cicader. - Ent. Meddr, 10: 317-328.

- 1917: Nye danske Cicader. - Ibid., 11: 363-364.

Janković, L., 1966: Fauna Homoptera: Auchenorrhyncha Srbije II. - Bull. Mus. Hist. nat. Belgrade (B), 21: 137-166.

Jansson, A., 1925: Die Insekten-, Myriopoden- und Isopodenfauna der Gotska Sandön. 182 pp. Örebro.

- 1935: Supplement till Die Insekten-, Myriopoden- und Isopodenfauna der Gotska Sandön. - Ent. Tidskr., 56: 52-87.

Jensen-Haarup, A. C., 1912: Nye eller sjældne danske Tæger og Cikader. - Flora og Fauna, 14: 29-30.

- 1915-1918: Danmarks Cikader. - Ibid., 1915: 33-40, 81-88, 97-104, 137-144; 1916: 33-40, 105-112; 1917: 41-48, 65-72, 97-104; 1918: 33-40, 97-104, 137-144.

- 1917: Some new Delphacinae from Denmark. - Ent. Meddr, 11: 1-5.

- 1920: Cikader. - Danmarks Fauna, 20: 190 pp. København.

- 1920b: Hemipterological notes and descriptions. - Ent. Meddr, 13: 209-224.

-, & Lindberg, H., 1931: Halvvingar - Hemiptera. *In:* Insektfaunan inom Abisko nationalpark III. - K. svenska VetenskAkad. Skr. Naturskydd., 17: 37-42.

John, B. & Claridge, M. F., 1974: Chromosome variation in British populations of *Oncopsis* (Hemiptera: Cicadellidae). - Chromosoma, 46: 77-89.

Jürisoo, V., 1964: Agro-ecological studies on leafhoppers (Auchenorrhyncha, Homoptera) and bugs (Heteroptera) at Ekensgård Farm in the province of Hälsingland, Sweden. - Natn. Inst. Plant Prot. Contr., 13: 101, 147 pp. Stockholm.

Kanervo, V., Heikinheimo, O., Raatikainen, M., & Tinnilä, A., 1957: The leafhopper *Delphacodes pellucida* (F.) (Hom., Auchenorrhyncha) as the cause and distributor of the damage to oats in Finland. - Publ. Finn. State Agric. Res. Board, 160: 5-56.

Kirschbaum, C. L., 1858a: Die *Athysanus*-Arten der Gegend von Wiesbaden. 14 pp. Wiesbaden.

- 1858b: Ueber die Zertheilung der Gattung *Jassus* in mehrere Gattungen. - Jb. nassau. Ver. Naturk., 13: 355-358.
- 1868a: Die Gattung *Idiocerus* Lew. und ihre Europäischen Arten. 19 pp. Wiesbaden.
- 1868b: Die Cicadinen der Gegend von Wiesbaden und Frankfurt a.M. nebst einer Anzahl neuer oder schwer zu unterscheidender Arten aus anderen Gegenden Europas. - Jb. nassau. Ver. Naturk., 21-22, 204 pp.

Kisimoto, R., 1961: Rothamsted report, 1961: 107.

Klefbeck, E. & Tjeder, B., 1946: Insekter från södra Bohuslän. - Ent. Tidskr., 67: 198-209.

Knight, W. J., 1966: A re-description of *Dikraneura micantula* (Zett.) (Homoptera: Cicadellidae) and a closely related new species from southern Finland. - Ann. Mag. nat. Hist., (13), 8: 345-350.

Koblet-Günthart, M., 1975: Die Kleinzikaden *Empoasca decipiens* Paoli und *Eupteryx atropunctata* Goetze (Homoptera, Auchenorrhyncha) auf Ackerbohnen (*Vicia faba*). 125 pp. Zürich.

Kolenati, F., 1857: Meletemata Entomologica. - Bull. Soc. nat. Moscou, 30: 399-429.

Kontkanen, P., 1937: Einige Cicadinenfunde (Hem., Hom.) aus Finnland. - Annls Ent. Fenn., 3: 146-149.
- 1938: Zur Kenntnis der Cicadinenfauna von Nord-Karelien (Hem., Hom.). - Annls zool. Soc. Zool.-Bot. Fenn. "Vanamo", 5: No. 7: 1-37.
- 1947a: Beiträge zur Kenntnis der Zikadenfauna Finnlands. I. - Annls Ent. Fenn., 13: 113-124.
- 1947b: Beiträge zur Kenntnis der Zikadenfauna Finnlands. II. - Ibid., 13: 170-175.
- 1948: Über eine Zikadenausbeute aus der Umgebung von Prääshä (AK). - Ibid., 14: 115-120.
- 1949a: Beiträge zur Kenntnis der Zikadenfauna Finnlands. III. - Ibid., 14: 85-97.
- 1949b: Beiträge zur Kenntnis der Zikadenfauna Finnlands. IV. - Ibid., 15: 32-42.
- 1949c: Beiträge zur Kenntnis der Zikadenfauna Finnlands. V. - Ibid., 15: 93-95.
- 1950: Quantitative and seasonal studies on the leafhopper fauna of the field stratum on open areas in North Karelia. - Annls zool. Soc. Zool.-Bot. Fenn. "Vanamo", 13: No. 8: 1-91.
- 1952: Über das holarktische, boreale und östliche Faunenelement in der Zikadenfauna Finlands. - Trans. 9th Int. Congr. Ent., 1: 561-563.
- 1953a: On the sibling species in the leafhopper fauna of Finland (Homoptera, Auchenorrhyncha). - Soc. Vanamo Arch., 7: 100-106.
- 1953b: Beiträge zur Kenntnis der Zikadenfauna Finnlands. VII. - Annls Ent. Fenn., 19: 190-198.
- 1954: Studies on insect populations I. The number of generations of some leafhopper species in Finland and Germany. - Soc. Vanamo Arch., 8: 150-156.

Kristensen, N. P., 1965a: Cikaden *Eupteroidea stellulata* (Burmeister 1841) i Danmark (Hemiptera, Cicadellidae). - Flora og Fauna, 71: 81-82.
- 1965b: 11. Cikader (Homoptera auchenorrhyncha) fra Hansted-reservatet. - Ent. Meddr, 30: 269-287.

Kuntze, H. A., 1937: Die Zikaden Mecklenburgs, eine faunistisch-ökologische Untersuchung. - Arch. Naturgesch. (N. F.), 6: 299-388.

Latreille, P. A., 1804: Cicadaires: Cicadariae. *In:* Histoire naturelle générale et particulière des Crustacés et des Insectes, 12: 5-424. Paris.
- 1817: Homoptera Lat. *In:* Cuvier's le Règne Animal, 3: 400-408. Paris.

Lauterer, P., 1958: A contribution to the knowledge of the leaf-hoppers of Czechoslovakia (Hom. Auchenorrhyncha) II. - Acta Mus. Morav., 43: 125-136.
- 1980: New and interesting records of leafhoppers from Czechoslovakia (Homoptera, Auchenorrhyncha). - Ibid., 65: 117-140.

Le Peletier de Saint-Fargeau, A. L. M., & Audinet-Serville, J. G., 1825: Tettigomètre, *Tettigometra,* and Tettigone, *Tettigonia. In:* Olivier's Encyclopédie Méthodique, 10: 600-613. Paris.

Le Quesne, W. J., 1960a: Some modifications in the British list of *Delphacidae* (Hem.), including a new genus and a new species. – Entomologist, 93: 13-19, 29-35, 54-60.

- 1960b: Hemiptera (Fulgoromorpha). – Handbk Ident. Br. Insects, II, (3): 68 pp. London.
- 1961: An examination of the British species of *Empoasca* Walsh sensu lato (Hem., Cicadellidae) including some additions to the British list. – Entomologist's mon. Mag., 96: 233-239.
- 1961b: Taxonomic studies in the British and some European species of *Scleroracus* van Duzee (Hem., Cicadellidae). – Ibid., 97: 260-264.
- 1965: The establishment of the relative status of sympatric forms, with special reference to cases among the Hemiptera. – Zool. Beitr. (N. F.), 11: 117-128.
- 1968: *Macrosteles ossiannilssoni* (Hem., Cicadellidae), a new species previously confused with *M. sexnotatus* (Fallén). – Entomologist's mon. Mag., 103: 190-192.
- 1969: Hemiptera (Cicadomorpha – Deltocephalinae). – Handbk Ident. Br. Insects, II, 2 (b): 65-148. London.
- 1974: *Eupteryx origani* Zakhvatkin (Hem., Cicadellidae) new to Britain, and related species. – Entomologist's mon. Mag., 109: 203-206.
Lethierry, L. F., 1869: Catalogue des Hémiptères du Département du Nord. – Mém. Soc. Sci. Agric. Lille, 1868: 305-374.
- 1874: Catalogue des Hémiptères du Département du Nord. 2e édition. – Ibid., 1874: 205-312.
- 1885: Description de deux Cicadines nouvelles. – Revue Ent., 4: 111-112.
Lewis, R. H., 1834: Descriptions of some new genera of British Homoptera. – Trans. R. ent. Soc. Lond., 1: 47-52.
Lindberg, H., 1923a: Zur Kenntnis der paläarktischen Cicadina I. Cercopidae. – Notul. ent., 3: 34-43.
- 1923b: Zur Kenntnis der Cicadinengattung *Batracomorphus* Lew. – Ibid., 3: 68-71.
- 1924a: Anteckningar om Östfennoskandiens Cicadina. – Acta Soc. Fauna Flora fenn., 56: 1-49.
- 1924b: Cicadinenfunde aus Schweden. – Notul. ent., 4: 40-44.
- 1926: Hemipterfynd från nordligaste Norge och Sverige. – Ibid., 6: 109-113.
- 1932a: Die Hemipterenfauna Petsamos. – Meddr Soc. Fauna Flora fenn., 7: 193-235.
- 1932b: *Delphax crassicornis* Panz. (Hem. Hom.) in Finnland gefunden. – Notul. ent., 12: 38-40.
- 1935: Månadsmöte den 17. sept. 1935. – Ibid., 15: 115.
- 1935a: *Paralimnus rotundiceps* Leth., eine für Nordeuropa neue Cicade. – Ibid., 15: 31-35.
- 1935b: Über einige arktische und subarktische Hemipteren aus Fennoskandien. – Norsk ent. Tidsskr., 3: 382-394.
- 1937: Über einige nordische Delphaciden. – Notul. ent., 17: 59-62.
- 1937b: Märkligare insekt-fynd i Skåne. – Opusc. ent., 2: 136-137.
- 1937c: Die Ostfennoskandischen *Cicadula*-Arten. – Notul. ent., 17: 141-146.
- 1938: Die finnländischen Arten der *Thamnotettix quadrinotatus*-Gruppe. – Ibid., 18: 1-4.
- 1939: Der Parasitismus der auf *Chloriona*-Arten (Homoptera Cicadina) lebenden Strepsiptere *Elenchinus chlorionae* n.sp. sowie die Einwirkung derselben auf ihren Wirt. – Acta zool. fenn., 22: 1-179.
- 1947: Verzeichnis der ostfennoskandischen Homoptera Cicadina. – Fauna Fennica, 1: 1-81.
- 1949: On stylopisation of Araeopids. – Acta zool. fenn., 57: 1-40.
- 1952: *Empoasca borealis* n.sp. und *Boreotettix* (n.gen.) *serricauda* (Kontk.) (Hom. Cicad.) aus Nordfinnland. – Notul. ent., 32: 144-147.
- 1953: Hemiptera Insularum Canariensium (Systematik, Ökologie und Verbreitung der Kanarischen Heteropteren und Cicadinen. – Commentat. biol., 14 (1): 1-304.
- & Saris, N.-E., 1952: Insektfaunan i Pisavaara naturpark (Finland, Prov. Ob.). – Acta Soc. Fauna Flora fenn., 69 (2): 82 pp.
Lindroth, C. H., 1942: *Oodes gracilis* Villa. Eine thermophile Carabide Schwedens. – Notul. ent., 22: 109-157.

Lindroth, C. H., Andersson, H., Bödvarsson, H., & Richter, S. H., 1973: Surtsey, Iceland. The development of a new fauna, 1963-1970. Terrestrial invertebrates. – Ent. scand., Suppl. 5: 280 pp.

Lindsten, K., 1961: Studies on virus diseases of cereals in Sweden. – K. LantbrHögsk. Annlr, 27: 137-271. Uppsala.

– 1979: Planthopper vectors and plant disease agents in Fennoscandia. – Leafhopper vectors and plant disease agents, Chapter 4: 155-178. Academic Press.

– & Gerhardsen, B., 1971: Stråsädens bestockningssjuka – en ny och svårartad viros som under 1971 påträffats i Östergötland. – Växtskyddsnotiser, 35: 66-75.

– – 1973: Virusangrepp på stråsäd under senare år och en "prognos" för 1973. – Ibid., 37: 19-26.

Lindsten, K., Vacke, J., & Gerhardson, B., 1970: A preliminary report on three cereal virus diseases new to Sweden spread by Macrosteles- and Psammotettix-leafhoppers. – Not. Swedish Inst. Plant prot. Contr., 14: 128: 283-297.

Linnavuori, R., 1948: Neue oder bemerkenswerte Zikadenfunde aus Finnland, nebst Beschreibung einer neuen Art. – Annls Ent. fenn., 14: 45-48.

– 1949a: Havaintoja Etelä-Hämeen lude- ja kaskaseläimistöstä. – Ibid., 15: 63-71.

– 1949b: Hemipterologisches aus Finnland 3-5. – Ibid., 15: 146-156.

– 1950a: Hemipterologisches aus Finnland. II. – Ibid., 16: 182-188.

– 1950b: Lisähavaintoja Etelä-Hämeen nivelkärsäiseläimistöstä. – Ibid., 16: 122-125.

– 1951a: Hemipterological observations. – Ibid., 17: 51-65.

– 1951b: Calligypona leptosoma Fl. and C. albofimbriata (Sign.) Fieb. (Hom., Delphacidae). – Ibid., 17: 109-110.

– 1952a: Studies on the ecology and phenology of the leafhoppers (Homoptera) of Raisio (S. W. Finland). – Annls zool. Soc. Zool.bot. fenn., "Vanamo", 14, No. 6, 32 pp.

– 1952b: Records of Hemiptera from the province of Savo, E. Finland. – Annls Ent. Fenn., 18: 64-75.

– 1952c: Studies on some Palearctic Hemiptera. – Ibid., 18: 181-187.

– 1953a: Hemipterological studies. – Ibid., 19: 107-118.

– 1953b: Investigations of the Hemipterous Fauna of Finland. – Ibid., 19: 133-134.

– 1959a: Hemiptera III. – Animalia Fennica, 12: 1-244. Helsinki.

– 1969b: Hemiptera IV. – Ibid., 13: 1-312.

Linné, C. von, 1758: Systema Naturae. Editio decima, reformata, 1: 1-824. Stockholm.

– 1761: Fauna Suecica sistens animalia Sueciae regni. Editio altera, auctior. 579 pp. Stockholm.

– 1767: Systema naturae. Editio duodecima, reformata. 1 (2): 533-1327. Stockholm.

Lundblad, O., 1950: Studier över insektfaunan i Fiby urskog. – K. svenska VetenskAkad. Avh. Naturskydd., 6: 235 pp.

– 1954: Studier över insektfaunan i Uppsala universitets naturpark vid Vårdsätra. – Ibid., 8: 68 pp.

– & Olsson, A., 1954: Insektfaunan på Hallands Väderö. – Ibid., 9: 76 pp.

Marchand, H., 1953: Die Bedeutung der Heuschrecken und Schnabelkerfe als Indikatoren verschiedener Graslandtypen (Ein Beitrag zur Agrarökologie). – Beitr. Ent., 3: 116-162.

Marshall, Th. A., 1866: Homoptera at Rannock. – Entomologist's mon.mag., 3: 118-119.

Matsumura, S., 1902: Monographie der Jassinen Japans. – Természetr. Füz., 25: 353-404.

– 1903: Monographie der Cercopiden Japans. – J. Coll. Agric. Tohuku Imp. Univ., 2: 15-52.

– 1906: Die Cicadinen der Provinz Westpreussen und des östlichen Nachbargebiets. Mit Beschreibungen und Abbildungen neuer Arten. – Schr. naturf. Ges. Danzig (N. F.), 11: 64-82.

– 1911: Erster Beitrag zur Insekten-Fauna von Sachalin. – J. Coll. Agric. Tohuko Imp. Univ., 4: 1-145.

– 1932: A revision of the Palaearctic and Oriental Typhlocybid genera, with descriptions of new species and new genera. – Insecta matsum., 6: 93-120.

961

- 1935: Supplementary note to the revision of *Stenocranus* and allied species of Japan-Empire. – Ibid., 10: 71-78.
Mc Atee, W. L., 1926: Revision of the American leafhoppers of the Jassid genus *Typhlocyba*. – Proc. U.S. natn. Mus., 68: 1-47.
Melichar, L., 1896: Cicadinen (Hemiptera-Homoptera) von Mittel-Europa. 364 pp., 12 pl. Berlin.
- 1896b: Einige neue Homoptera-Arten und Varietäten. – Verh. zool.-bot. Ges. Wien, 46: 176-180.
- 1898: Eine neue Homopteren-Art aus Schleswig-Holstein. – Wien.ent. Ztg, 17: 67-69.
- 1900: Beitrag zur Kenntnis der Homopteren-Fauna von Sibirien und Transbaikal. – Ibid., 19: 33-45.
Metcalf, Z. P., 1922: On the genus *Elidiptera* (Homop.). – Can. Ent., 54: 263-264.
- 1943: Fulgoroidea Araeopidae (Delphacidae). *In:* General Catalogue of the Hemiptera, 3, 552 pp. Menasha, Wisc.
- 1952: New names in the Homoptera. – J. Wash. Acad. Sci., 42: 226-231.
- 1955: New names in the Homoptera. – Ibid., 45: 252-267.
- 1962: Cercopoidea Part 3 Aphrophoridae. *In:* General Catalogue of the Homoptera, 7, 600 pp. Baltimore.
Mitjaev, I. D., 1975: New species of Cicadinea (Homoptera) from Kazakhstan. – Ent. Obozr., 54: 577-586. (In Russian).
Morcos, G., 1953: The biology of some Hemiptera-Homoptera (Auchenorrhyncha). – Bull. Soc. Fouad I Ent., 37: 406-439.
Mulsant, M. E., & Rey, C., 1855: Description de quelques Hémiptères-Homoptères nouveaux ou peu connus. – Annls Soc. linn. Lyon, 2: 197-249, 426.
Murtomaa, A., 1967: Aster yellow-type virus infecting grasses in Finland. – Annls Agric. Fenn., 5: 324-333.
- 1969: Aster yellows på graminéer. – Nordisk jordbruksforskning, 51: 290-292.
Müller, H. J., 1942: Über Bau und Funktion des Legeapparates der Zikaden (Homoptera Cicadina). – Z. Morph. Ökol. Tiere, 38: 534-629.
- 1947: Saisondimorphismus bei Arten der Gattung *Euscelis* Brullé. – Bombus, 40: 173-174.
- 1951: Über das Schlüpfen der Zikaden (Homoptera Auchenorrhyncha) aus dem Ei. – Zoologica, 37 (103): 1-40.
- 1954: Der Saisondimorphismus bei Zikaden der Gattung *Euscelis* Brullé (Homoptera Auchenorrhyncha). – Beitr. Ent., 4: 1-56.
- 1955: Die Bedeutung der Tageslänge für die Saisonformenbildung der Insekten, insbesondere bei den Zikaden. – Ber. 7. Wanderversammlung deutscher Entomologen 8. bis 10. September 1954 in Berlin, 102-120.
- 1956: Homoptera. *In:* Sorauer: Handbuch der Pflanzenkrankheiten, V, 5. Aufl., 3 Lfg.: 150-359. Berlin & Hamburg.
- 1957: Über die Diapause von *Stenocranus minutus* Fabr. (Homoptera-Auchenorrhyncha). – beitr. Ent., 7: 203-226.
- 1957b: Die Wirkung exogener Faktoren auf die zyklische Formenbildung der Insekten, insbesondere der Gattung *Euscelis* (Hom. Auchenorrhyncha). – Zool. Jb., (Syst.), 85: 317-430.
- 1978: Strukturanalyse der Zikadenfauna (Homoptera Auchenorrhyncha) eines Rasenkatena Thüringens (Leutratal bei Jena). – Ibíd., 105: 258-334.
- 1981: Die Bedeutung der Dormanzform für die Populationsdynamik der Zwergzikade *Euscelis incisus* (Kbm.) (Homoptera Cicadellidae). – Ibid., 108: 314-334.
Nast, J., 1938: Homopterologische Notizen. III. Zur Morphologie von *Empoasca apicalis* (Flor). – Annls Mus. zool. pol., 13: 161-163.
- 1972: Palaearctic Auchenorrhyncha (Homoptera), an annotated check list. 550 pp. Warszawa.

– 1976: *Auchenorrhyncha (Homoptera)* of the Pieniny Mts. – Fragm. faun., 21: 145-183.

Nuorteva, P., 1948: Über *Empoasca apicalis* sensu Nast und *E. ossiannilssoni* sp.n. – Annls Ent. Fenn. 14: 99-100.

– 1951: Experimentelle Untersuchungen über die Nährpflanzenwahl einer oligophagen Zikade, *Aphrophora alni* (L.) (Hom., Cercopidae). – Ibid., 17: 10-17.

– 1951a: Ein Massenauftreten von *Oncopsis tristis* Zett. (Hom., Macropsidae) auf Birken nebst Beobachtungen über die Biologie der Art. – Ibid., 17: 162-166.

– 1952: Die Nährungspflanzenwahl der Insekten im Lichte von Untersuchungen an Zikaden. – Annls Acad. Sci. Fenn. (A) IV, Biologica, 19: 90 pp.

– 1952a: Über die Phänologie der baumbewohnenden Zikaden in Vuohiniemi (Kirschsp. Hattula) in Südfinnland. – Annls Ent. Fenn., 18: 198-203.

– 1955: *Typhlocyba rosae* (L.) (Hom., Typhlocybidae) found in Finland. – Ibid., 21: 195-198.

Obrtel, R., 1969: The insect fauna of the herbage stratum of lucerne fields in southern Moravia (Czechoslovakia). – Přírodov. Pr. Cesk.Akad. Věd. 3 (10): 1-49.

Okáli, I., 1960: Homoptera Auchenorrhyncha einiger Biotope in der Umgebung von Bratislava. – Acta Fac. Rerum nat. Univ. comen., Bratisl., Zoologia 4-6-8: 353-363.

Olivier, G. A., 1791: Encyclopédie méthodique, histoire naturelle. Insectes 6: 561-577. Paris.

Oman, P. W., 1949: The Nearctic leafhoppers (Homoptera: Cicadellidae). A generic classification and check list. – Mem. ent. Soc. Wash., 3: 1-253.

Osborn, H., & Ball, E. D., 1902: A review of the North American species of *Athysanus* (Jassidae). – Ohio Nat., 2: 231-256.

Oshanin, V. T., 1871: Homoptera. – Mém. Soc. Amis Sci. Nat. Moscou, 8: 194-213.

Ossiannilsson, F., 1934: Bidrag till kännedomen om Sveriges Homoptera Cicadina I. – Ent. Tidskr., 55: 129-139.

– 1935a: Eine neue Art der Gattung *Cicadula* (Hemiptera, Homoptera) aus Nordschweden. – Ibid., 56: 127-128.

– 1935b: Bidrag till kännedomen om Sveriges Homoptera Cicadina. II. – Ibid., 56: 129-137.

– 1936a: Zur Kenntnis einiger schwedischer Arten der Gattungen *Eupteryx* und *Typhlocyba* (Homoptera). – Ibid., 57: 254-261.

– 1936b: Über einige schwedischen Arten der Gattung *Cicadula* (Homoptera Cicadina). – Opusc. ent., 1: 6-11.

– 1936c: Einige Bemerkungen zur schwedischen Cicadinenfauna. – Ibid., 1: 47-51.

– 1937a: Zur Kenntnis der schwedischen Homopterenfauna mit Beschreibung der neuen Art *Erythroneura silvicola* Oss. – Ibid., 2: 19-27.

– 1937b: Über die Typen einiger von A. G. Dahlbom beschriebener Cicadinen. – Ibid., 2: 132-134.

– 1938a: Über Zetterstedts *Cicada lividella* und verwandte Arten der Gattung *Deltocephalus* (Homoptera, Cicadina) mit Beschreibungen von zwei neuen Arten. – Ibid., 3: 1-6.

– 1938b: Revision von Zetterstedts lappländischen Homopteren. I. Cicadina. – Ibid., 3: 65-79.

– 1939: Bidrag till kännedomen om Sveriges Hemiptera (Cicadina, Psyllina, Heteroptera). – Ibid., 4: 23-29.

– 1941a: Släktet *Cixius* Latr. i Sverige. – Ibid., 6: 1-5.

– 1941b: A new Swedish species of *Empoasca* (Hom.), *Empoasca strigilifera* n.sp. – Ent. Tidskr., 62: 198-199.

– 1941c: Några för Sverige nya eller hos oss föga beaktade Hemiptera (Het., Cic., Psyll., Aph.). – Opusc. ent., 6: 50-56.

– 1941d: Nomenclatorial remarks on some Swedish Cicadina, with description of a new species of the genus *Empoasca* Walsh. – Ibid., 6: 67-70.

– 1942a: Hemipterfynd i Stockholmstrakten. – Ibid., 7: 28-37.

– 1942b: Contributions to the knowledge of Swedish Cicadina. With description of a new species.

– Ibid., 7: 113-114.
– 1943a: Studier över de svenska potatisfältens insektfauna och dess betydelse för spridning av virussjukdomar I. Hemiptera, förekomst och utbredning. – Meddn St. VäxtskAnst., 39, 72 pp.
– 1943b: Hemipterologiska notiser. Opusc. ent., 8: 12-19.
– 1943c: The Hemiptera (Heteroptera, Cicadina, Psyllina) of the Tromsø Museum. A contribution to the knowledge of the Hemiptera of Norway. – Tromsø Mus. Årsh., 65: 1-38.
– 1944: Contributions to the knowledge of Swedish Cicadina II, with description of a new species. – Opusc. ent., 9: 14-16.
– 1945: Fem för Sveriges fauna nya Hemiptera. (Hemipterologiska notiser IV). – Ibid., 10: 36-38.
– 1946a: On the sound-production and the sound-producing organ in Swedish Homoptera auchenorrhyncha. A preliminary note. – Ibid., 11: 82-84.
– 1946b: Chloriona chinai n.sp. A new Swedish of Chloriona (Hom. Araeopidae). With remarks on the synonymy of Chloriona smaragdula (Stål). – Ibid., 11: 84-87.
– 1946c: Halvvingar. Hemiptera. Stritar Homoptera auchenorrhyncha. – Svensk insektfauna, 7: 1-150.
– 1946d: Två för Sveriges fauna nya stritarter (Hemipterologiska notiser VI). – Opusc.ent., 11: 156.
– 1947a: Om C. H. Lindroths isländiska stritar (Hemiptera Homoptera). – Ent. Tidskr., 68: 127-128.
– 1947b: Halvvingar. Hemiptera. Stritar Homoptera auchenorrhyncha. – Svensk insektfauna, 7: 151-270. Stockholm.
– 1948a: Hemiptera Homoptera Auchenorrhyncha (Cicadina). Catalogus insectorum Sueciae VIII. – Opusc.ent., 13: 1-25.
– 1948b: A new Swedish leafhopper, Macrosteles anderi n.sp. (Hem. Hom.). – Ibid., 13: 26.
– 1949a: Insect drummers. – Ibid., Suppl. 10, 146 pp. Lund.
– 1949b: A new leafhopper, Empoasca kontkaneni n.sp. from Finland. With a supplementary description of another Empoasca species. – Ibid., 14: 71-72.
– 1950a: On the wing-coupling apparatus of the Auchenorrhyncha (Hem. Hom.). – Ibid., 15: 127-130.
– 1950b: On the identity of Cicada spumaria Linnaeus (1758) (Hem. Hom.). With notes on the breeding-plants of three Swedish Cercopids. – Ibid., 15: 145-156.
– 1951a: Hemiptera, in Brinck & Wingstrand: The mountain fauna of the Virihaure area in Swedish Lapland. II. Special account. – K. fysiogr. Sällsk. Handl. (N. F.), 61: 51-59.
– 1951b: Homoptera aus einigen nordestländischen Inseln. – Opusc. ent., 16: 10-14.
– 1951c: On the shape of the apodemes of the second abdominal sternum of the males as a specific character in the genus Macrosteles Fieb. (Hom. Auchenorrhyncha). – Ibid., 16: 109-111.
– 1953a: On the music of some European leafhoppers (Homoptera auchenorrhyncha) and its relation to courtship. – Trans. 9th Int. Congr. Ent., 2: 139-142.
– 1953b: VIII. Hemiptera Homoptera Auchenorrhyncha, in: Catalogus insectorum Sueciae, Additamenta ad partes I-X. – Opusc. ent., 18: 106-108.
– 1954: Nomenclatorial notes on some Homoptera Auchenorrhyncha (Hem.). – Ent. Tidskr., 75: 117-127.
– 1955: Några för Sverige nya stritar (Hom. Auchenorrhyncha). – Ibid., 76: 131-133.
– 1961a: Balclutha calamagrostis, n.sp. A new Swedish leafhopper (Hem., Hom., Auchenorrh., – Opusc. ent., 26: 59-60.
– 1961b: Anmärkningar och tillägg till Sveriges hemipterfauna (Hemipterologiska notiser VIII). – Ibid., 26: 228-234.
– 1962: Hemipterfynd i Norge 1960. – Norsk ent. Tidsskr., 12: 56-62.
– 1971: Till kännedomen om Kullabergs halvvingar (Hemiptera). – Kullabergs Natur, 14, 55 pp.

- 1972: Till kännedomen om Abiskotraktens Hemiptera. – Ent. Tidskr., 93: 88-99.
- 1974: Hemiptera (Heteroptera, Auchenorrhyncha and Psylloidea). – Fauna of the Hardanger-vidda, No. 5: 13-35.
- 1976: Two new species of leafhoppers from Fennoscandia (Homoptera: Cicadelloidea). – Ent. scand., 7: 31-34.
- 1977: Mire invertebrate fauna at Eidskog, Norway. V. Auchenorrhyncha, Psylloidea, and Coccoidea (Hem.). – Norw. J. Ent., 24: 11-14.
Panzer, G. W. F., 1796: Fauna Insectorum Germanicae Initia. Deutschlands Insecten, 32: 8-10. Nürnberg.
- 1799: Faunae Insectorum Germanicae Initia. Deutschlands Insecten, 61: 12-19. Nürnberg.
Paoli, G., 1930: Caratteri diagnostici delle *Empoasca* e descrizione di nuove specie. – Atti Soc. tosc. Sci. nat., 39: 64-75.
- 1036: Alcune specie di *Empoasca* viventi in Egitto (Hemiptera-Homoptera). – Bull. Soc. R. Ent. Égypte, 1936: 144-151.
Pekkarinen, A. & Raatikainen, M., 1973: The Strepsiptera of Eastern Fennoscandia. – Notul. ent., 53: 1-10.
Perris, E., 1857: Nouvelles excursions dans les Grandes Landes. – Annls Soc. linn. Lyon, 4: 83-180.
Port, G. R., 1981: Auchenorrhyncha on roadside verges. A preliminary survey. – Acta Ent. Fenn., 38: 29-30.
Puton, A., 1886: Homoptera Am. Serv. (Gulaerostria Zett. Fieb.) Sect. 1. Auchenorrhyncha Dumér. Cicadina Burm. – Catalogue des Hémiptères (Héteroptères, Cicadines et Psyllides) de la fauna Paléarctique. 3e Ed., 3-100. Caen.
Quayum, M. A., 1968: Some studies on the host plant choice and wing dimorphism of *Javesella pellucida* (Fabr.) (Hom.: Araeopidae). – Acta Agric. Scand., 18: 207-221.
Raatikainen, M., 1960: The biology of *Calligypona sordidula* (Stål) (Hom., Auchenorrhyncha). – Annls Ent. Fenn., 26: 229-242.
- 1967: Bionomics, enemies and population dynamics of *Javesella pellucida* (F.) (Hom., Delphacidae). – Annls Agric. Fenn., 6, Suppl. 2, 149 pp.
- 1970: Ecology and fluctuations in abundance of *Megadelphax sordidula* (Stål) (Hom., Delphacidae). – Ibid., 9: 315-324.
- 1971: Seasonal aspects of leafhoppers (Hom., Auchenorrhyncha) fauna in oats. – Ibid., 10: 1-8.
- 1972: Dispersal of leafhoppers and their enemies to oatfields. – Ibid., 11: 146-153.
- & Tinnilä, A., 1959: The feeding and oviposition plants of the leaf-hopper *Calligypona pellucida* (F.) (Hom., Auchenorrhyncha) and the resistance of different oat varieties to the damage. – Publ. Finnish State Agric. Res. Board, 178: 101-109.
- & Vasarainen, A., 1964: Biology of *Dicranotropis hamata* (Boh.) (Hom., Araeopidae). – Annls Agric. Fenn., 3: 311-323.
- 1971: Comparison of leafhopper faunae in cereals. – Ibid., 10: 119-124.
- 1973: Early- and high-summer flight periods of leafhoppers. – Ibid., 12: 77-94.
- 1976: Composition, zonation and origin of the leafhopper fauna of oatfields in Finland. – Annls Zool. Fenn., 13: 1-24.
Razvyazkina, G. M., 1957: New and little known species of the genus *Macrosteles* (Homoptera-Cicadoidea). – Zool. Zh., 36: 521-528. (In Russian).
- & Pridantzeva, E. A., 1968: Leafhoppers of the group *Psammotettix striatus* L. (Homoptera, Cicadellidae) – vectors of virus diseases of cereals, their systematics and distribution. – Ibid., 47: 690-696. (In Russian).
Remane, R., 1958: Die Besiedlung von Grünlandflächen verschiedener Herkunft durch Wanzen und Zikaden im Weser-Ems-Gebiet. – Z. angew. Ent., 42: 353-400.
- 1959: *Lebradea calamagrostidis* gen. et spec. nov., eine neue Zikade aus Norddeutschland (Hom.

Cicadina Cicadellidae). – Zool. Anz., 163: 385-391.
- 1960: Zur Kenntnis der Gattung *Arthaldeus* Ribaut. – Mitt. münch. ent. Ges., 50: 72-82.
- 1961a: Revision der Gattung *Mocydiopsis* Ribaut (Hom. Cicadellidae). – Abh. math.-naturw. Kl. Akad. Wiss. Mainz, 1961: 101-149.
- 1961b: *Endria nebulosa* (Ball), comb. nov., eine nearktische Zikade in Deutschland (Hom. Cicadina, Jassidae). – NachrBl. bayer. Ent., 10: 73-76, 90-98.
- 1965: Beiträge zur Kenntnis der Gattung *Psammotettix* Hpt. – Zool. Beitr. (N. F.), 11: 221-245.
Reuter, O. M., 1880: Nya bidrag till Åbo och Ålands skärgårds Hemipter-fauna. – Meddn Soc. Fauna Flora fenn., 5: 160-236.
- 1886: (Nya former af Hemiptera). – Ibid., 13: 211.
Rey, C., 1891: Observations sur quelques Hémiptères-Homoptères et descriptions d'espèces nouvelles ou peu connues. – Revue fr. Ent., 10: 240-256.
- 1894: Remarques en passant. – Échange, 10: 45-46.
Ribaut, H., 1925: Sur quelques Deltocéphales du groupe *D. striatus* (L.) Then. – Bull. Soc. Hist. nat. Toulouse, 53-5-22.
- 1927: Trois espèces nouvelles du genre *Cicadula* (Homopt.). – Ibid., 56: 162-169.
- 1931a: Espèces nouvelles du groupe *Typhlocyba rosae*. – Ibid., 61: 333-342.
- 1931b: Les espèces francaises du groupe *Typhlocyba ulmi*. – Ibid., 61: 280-291.
- 1931c: Les espèces francaises des groupes *Erythroneura parvula* et *Erythroneura fasciaticollis*. – Ibid., 62: 499-516.
- 1933: Sur quelques espèces du genre *Empoasca*. – Ibid., 65: 150-161.
- 1934: Nouveaux Delphacides (Homoptera-Fulgoroidea). – Ibid., 66: 281-301.
- 1935a: Espèces nouvelles du genre *Agallia*. – Ibid., 67: 29-36.
- 1936a: Nouveaux *Deltocéphales* des groupes *abdominalis* et *sursumflexus* (Homoptera-Jassidae). – Ibid., 70: 259-266.
- 1936b: Homoptères auchénorhynques. I. (Typhlocybidae). – Faune de France, 31, 231 pp. Paris.
- 1938a: Un genre nouveau de la famille des Jassidae (Homoptera). – Bull. Soc. Hist. nat. Toulouse, 72: 97-98.
- 1938b: Le genre *Psammotettix* Hpt. (Homoptera-Jassidae). – Ibid., 72: 166-170.
- 1939: Nouveaux genres et nouvelles espèces de la famille des Jassidae (Homoptera). – Ibid., 73: 267-279.
- 1942: Démembrement des genres *Athysanus* Burm. et *Thamnotettix* Zett. (Homoptera-Jassidae). – Ibid., 77: 259-270.
- 1947: Démembrement du genre *Deltocephalus* Burm. (Homoptera-Jassidae). – Ibid., 81 [1946]: 81-86.
- 1948: Démembrement de quelques genres de *Jassidae*. – Ibid., 83: 57-59.
- 1952: Homoptères auchénorhynques. II (Jassidae). – Faune de France, 57, 474 pp. Paris.
- 1953: Trois espèces nouvelles du genre *Calligypona* (Homoptera-Araeopidae). – Bull. Soc. Hist. nat. Toulouse, 88: 245-248.
Rosen, H. von, 1956: Untersuchungen über drei auf Getreide vorkommende Erzwespen und über die Bedeutung, die zwei von ihnen als Vertilger von Wiesenzirpeneiern haben. – K. LautbrHögsk. Annlr, 23: 1-72.
Rossi, P., 1792: Fauna etrusca sistens Insecta quae in provinciis praesertim collegit —. 2, 348 pp. Liburni.
Sahlberg, C. (R.), 1842: Cicadae tres novae fennicae. – Acta Soc. Sci. Fenn., 1: 85-92.
Sahlberg, J., 1867: Hemiptera, samlade i Torneå Lappmark år 1867. – Notis. Sällsk. Faun. Fl. fenn. Förh. (n.s.) 8: 222-233.
- 1868: Entomologiska anteckningar från en resa i sydöstra Karelen sommaren 1866. I.

Orthoptera och Hemiptera. Ibid., 9: 159-197.
- 1871: Öfversigt af Finlands och den Skandinaviska halföns Cicadariae. - Ibid., 9 (12): 1-506.
- 1876: Nya finska Cicadarier. - Meddn Soc. Fauna Flora fenn., 1: 138.
- 1881: Bidrag till det Nordenfjeldske Norges insektfauna. - Forh. VidenskSelsk. Krist., 1880: 1-13.
Sanders, J. G., & De Long, D. M., 1917: The Cicadellidae (Jassoidea-Fam. Homoptera) of Wisconsin, with descriptions of new species. - Ann. ent. Soc. Am., 10: 79-95.
Schaefer, M., 1973: Untersuchungen über Habitatbindung und ökologische Isolation der Zikaden einer Küstenlandschaft (Homoptera: Auchenorrhyncha). - Arch. natursch. Landschaftsforsch., 13: 329-352.
Schiemenz, H., 1964: Beitrag zur Kenntnis der Zikadenfauna (Homoptera Auchenorrhyncha) und ihrer Ökologie in Feldhecken, Restwäldern und den angrenzenden Fluren. - Arch. Naturschutz, 4: 163-189.
- 1965: Zur Zikadenfauna des Geisings und Pöhlberges im Erzgebirge (Hom. Auchenorrhyncha). Eine faunistisch-ökologische Studie. - Zool. Beitr. (N. F.), 11: 271-288.
- 1969a: Die Zikadenfauna (Homoptera Auchenorrhyncha) mitteleuropäischer Trockenrasen - Untersuchungen zu ihrer Phänologie, Ökologie, Bionomie und Chorologie. - Abh. Ber. Naturk.Mus. - ForschStelle, Görlitz, 44: 195-205.
- 1969b: Die Zikadenfauna mitteleuropäischer Trockenrasen (Homoptera, Auchenorrhyncha). Untersuchungen zu ihrer Phänologie, Ökologie, Bionomie und Chorologie. - Ent.-Abh.Mus.Tierk.Dresden, 36: 201-280.
- 1971: Die Zikadenfauna (Homoptera Auchenorrhyncha) der Erzgebirgshochmoore. - Zool. Jb. (Syst.), 98: 397-417.
- 1975: Die Zikadenfauna der Hochmoore im Thüringer Wald und im Harz (Homoptera, Auchenorrhyncha). - Faun. Abh.st.Mus.Tierk. Dresden, 5: 215-233.
- 1976: Die Zikadenfauna von Heide- und Hochmooren des Flachlandes der DDR (Homoptera, Auchenorrhyncha). - Ibid., 6: 39-54.
- 1977: Die Zikadenfauna der Waldwiesen, Moore und Verlandungssümpfe im Naturschutzgebiet Serrahn (Homoptera, Auchenorrhyncha). - Ibid., 6: 297-304.
Schmutterer, H., 1953: Die Zikade Cicadella viridis (L.) als Roterlenschädling. - Forstw. Cbl., 72: 247-254.
Schrank, F. v. P., 1776: Beyträge zur Naturgeschichte. 137 pp. Augsburg.
- 1781: Enumeratio Insectorum Austriae indigenorum. 548 pp. Augustae Vindel.
- 1796: Sammlung naturhistorischer und physikalischer Aufsätze. 485 pp. Nürnberg.
- 1801: Fauna Boica, 2: 1-374. Ingolstadt.
Schulz, K., 1976: Zur Kenntnis der Gattung Jassargus Zachvatkin (Homoptera Auchenorrhyncha). (Diss.). 5+213 pp., plates. Marburg/Lahn.
Schwoerbel, W., 1957: Die Wanzen und Zikaden des Spitzberges bei Tübingen, eine faunistisch-ökologische Untersuchung (Hemipteroidea: Heteroptera und Cicadina = Homoptera auchenorrhyncha). - Z. Morph. Ökol. Tiere, 45: 462-560.
Schøyen, W. M., 1879: Supplement til H. Siebke's Enumeratio insectorum Norvegicorum Fasciculus I-II (Hemiptera, Orthoptera & Coleoptera). - Forh. VidenskSelsk. Krist., 3, 75 pp.
- 1889: Bidrag til Kundskaben om Norges Hemipter- og Orthopter-Fauna. - Ibid., 5: 1-13.
Scopoli, J. A., 1763: Entomologia Carniolica. 421 pp. Wien.
- 1772: Observationes Zoologicae. In: Annus historico-naturalis, 5: 75-125. Lipsiae.
Scott, J., 1870a: On certain British Hemiptera-Homoptera. Revision of the Family Delphacidae and descriptions of several new species of the genus Delphax of authors. - Entomologist's mon. Mag., 7: 22-29.
- 1870b: On certain British Hemiptera-Homoptera. Revision of the Family Cixiidae. - Ibid., 7:

118-123, 146-148.
- 1871: On certain British Hemiptera-Homoptera (Revision of the Family Delphacidae and descriptions of several new species of the genus *Delphax* of authors). - Ibid., 7: 193-196.
- 1874a: On certain British Hemiptera-Homoptera. Revision of the Bythoscopidae, and descriptions of some species not hitherto recorded as British. - Ibid., 10: 189-195, 235-242.
- 1874:b: On certain British Hemiptera-Homoptera. (Revision of the genus *Strongylocephalus*, and description of a new species.). - Ibid., 11: 120-123.
- 1875a: On certain British Hemiptera-Homoptera. - Ibid., 11: 229-232.
- 1876: On certain British Hemiptera-Homoptera (*Athysanus*). - Ibid., 12: 169-172.
Siebke, H., 1870: Om en i Sommeren 1869 foretagen entomologisk Reise gjennem Ringerike, Hallingdal og Valders. 71 pp. Christiania.
- 1874: Enumeratio insectorum norvegicorum. Fasciculus I, catalogum Hemipterorum et Orthopterorum continens. 60 pp. Christiania.
Signoret, V., 1865: Descriptions de quelques Hémiptères nouveaux. - Annls Soc. ent. Fr. (4) 6: 139-160.
Spinola, M., 1839: Essai sur les Fulgorelles, sous-tribu de la tribu des Cicadaires, ordre des Rhynchotes. - Ibid., 8: 133-337, 339-454.
Strand, E., 1899: Et lidet bidrag til Norges entomologiske fauna. - Ent. Tidskr., 20: 287-292.
- 1902: Norske fund av Hemiptera. - Ibid., 23: 257-270.
- 1906: Bidrag til det sydlige Norges Hemipterfauna. - Arch. Math. naturv., 27: 1-9.
- 1913: Neue Beiträge zur Arthropodenfauna Norwegens nebst gelegentlichen Bemerkungen über deutsche Arten. XV. Homoptera. - Nytt Mag. Naturvid., 51: 270-274.
Strübing, H., 1955: Beiträge zur Ökologie einiger Hochmoorzikaden (Homoptera-Auchenorrhyncha). - Öst. zool. Z., 6: 566-596.
- 1956: Über Beziehungen zwischen Ovidukt, Eiablage und natürlicher Verwandtschaft einheimisher Delphaciden. - Zool. Beitr. (N. F.), 2: 331-357.
- 1958: Lautäusserung - der entscheidende Faktor für das Zusammenfinden der Geschlechter bei Kleinzikaden (Homoptera-Auchenorrhyncha). - Ibid., 4: 15-21.
- 1959: Lautgebung und Paarungsverhalten von Kleinzikaden. - Verh. dt. zool. Ges. Münster/ Westf., 1958: 118-120.
- 1960: Eiablage und photoperiodish bedingte Generationsfolge von *Chloriona smaragdula* Stål und *Euidella speciosa* Boh. (Homoptera-Auchenorrhyncha). - Zool. Beitr. (N. F.), 5: 301-332.
- 1965: Das Lautverhalten von *Euscelis plebejus* Fall. und *Euscelis ohausi* Wagn. (Homoptera-Cicadina). - Ibid., 11: 289-341.
- 1970: Zur Artberechtigung von *Euscelis alsius* Ribaut gegenüber *Euscelis plebeius* Fall. (Homoptera-Cicadina). Ein Beitrag zur Neuen Systematik. - Ibid., 16: 441-478.
- 1977: Lauterzeugung oder Substratvibration als Kommunikationsmittel bei Kleinzikaden? (diskutiert am Beispiel von *Dictyophara europaea* - Homoptera-Cicadina: Fulgoroidea). - Ibid., 23: 323-332.
Stål, C., 1853: Nya svenska Homoptera. - Öfvers. K. VetenskAkad. Förh., 10: 174-177.
- 1854: Kort öfversigt af Sveriges *Delphax*-arter. - Ibid., 11: 189-197.
- 1858: Nya svenska Hemiptera. - Ibid., 15: 355-358.
- 1859a: Beitrag zur Hemipteren-Fauna Sibiriens und des Russischen Nord-Amerika. - Stettin. ent. Ztg, 19: 175-198.
Šulc, K., 1911: Über Respiration, Tracheensystem und Schaumproduktion der Schaumcikadenlarven (Aphrophorinae-Homoptera). - Z. wiss. Zool., 99: 147-188.
Taksdal, G., 1977: Auchenorrhyncha and Psylloidea collected in strawberry fields. - Norw. J. Ent., 24: 107-110.
Then, F., 1886: Katalog der österreichischen Cicadinen. - Jahres-Ber. Gymn. Theresianischen

Akad. Wien, 1886: 1-59.

- 1896: Neue Arten der Cicadinen-Gattungen *Deltocephalus* und *Thamnotetttix*. - Mitt. naturw. Ver. Steierm., 32: 165-197.
- 1897: Fünf Cicadinen-Species aus Österreich. - Ibid., 33: 102-116.
- 1898: Über einige Merkmale der Cicadinen. *Deltocephalus rhombifer* und *Deltocephalus putoni*. - Ibid., 34: 40-52.
- 1902: Zwei Species der Cicadinen-Gattung *Deltocephalus*. - Ibid., 38: 186-192.

Thomson, C. G., 1869: Genus *Jassus*. Conspectus specierum Sueciae.-Opuscula Entomologica, 1: 44-77. Lund.

- 1870: Öfversigt af de i Sverige funna arter af slägtet *Pediopsis* Burm. - Ibid., 3: 316-321.

Thunberg, C. P., 1784: Nova insectorum species descripta. - Nova Acta R. Soc. Scient. upsal., 4: 1-28.

Tjeder, B., 1937: Holidays in southern Scania. - Ent. Tidskr., 58: 161-165.

- 1951: Med Kiviks sand i skorna. - Sveriges natur, 1951: 58-75.

Tollin, C., 1851: Über Kleinzirpen, besonders über die Gattung *Typhlocyba* nebst Beschreibung einiger neuen Arten. - Stettin. ent. Ztg, 12: 67-74.

Trolle, L., 1966: Nye danske cikader (Hemiptera, Cicadellidae). - Flora og Fauna, 72: 93-100.

- 1968: Nye danske cikader 2 (Hemiptera, Cicadellidae). - Ibid., 74: 113-121.
- 1973: Cicadeslægten *Stenocranus* Fieber, 1866, i Danmark. - Ibid., 79: 12-14.
- 1974: Danske Typhlocybiner (Auchenorrhyncha, Cicadellidae, Typhlocybinae). - Ent. Meddr, 42: 53-62.
- 1982: Status over Bornholms cikader. - Fjælstaunijn, 2/82: 76-82.

Tullgren, A., 1916a: En ny strit - *Typhlocyba bergmani* n.sp. - från Norge. - Ent. Tidskr., 37: 65-69.

- 1916b: Rosenstriten (*Typhlocyba rosae* L.) och en ny äggparasit på densamma. - Meddn Cent-Anst.Förs Väs. JordbrOmråd. Stockh., 132: 3-13.
- 1925: Om dvärgstriten (*Cicadula sexnotata* Fall.) och några andra ekonomiskt viktiga stritar. - Ibid., 287: 1-71.

Turton, W., 1802: A general system of Nature, 2. 719 pp. London.

Törmälä, T., & Raatikainen, M., 1976: Primary production and seasonal dynamics of the flora and fauna of the field stratum in a reserved field in Middle Finland. - J. Sci. Agric. Soc. Finland, 48: 363-385.

Uhler, Ph. R., 1877: Report upon the insects collected by P. R. Uhler during the explorations of 1875. - Bull. U. S. geol. geogr. Territ., 3 (2): 355-475.

Vacke, J., 1962: Contribution to study of vectors of oat sterile-dwarf virus. - Plant Virology, Proc. 5th Conf. Czechoslov. Plant Virologists, 335-338. Prague.

Van Duzee, E. P., 1916: Check list of Hemiptera (excepting the Aphididae, Aleurodidae, and Coccidae) of America north of Mexico, i-xi, 1-111. New York.

- 1917: Catalogue of the Hemiptera of America north of Mexico excepting the Aphididae, Coccidae, and Aleurodidae. - Univ. Calif. Publs Ent., 2: i-xiv, 1-902.

Vidano, C., 1958: Le Cicaline italiane della vite (Hemiptera Typhlocybidae). - Boll. Zool. agr.Bachic., (II) 1 [1957-58]: 61-115.

- 1959: Possibili rapporti tra Emitteri Tiflocibidi e degenerazione infettiva della Vite. - Not. sulle malattie delle piante, 47-48: 219-226. Milano.
- 1965: A contribution to the chorological and oecological knowledge of the European Dikraneurini (Homoptera Auchenorhyncha). - Zool. Beitr. (N. F.), 11: 343-367.
- & Arzone, A., 1978: Typhlocybinae on officinal plants. - Auchenorrhyncha newsletter I: 27-28. - Meded.Lab.Ent. Wageningen no. 306.

Vilbaste, J., 1961: Neue Zikaden (Homoptera: Cicadina) aus der Umgebung von Astrachan. - Eesti NSV Tead. Akad. Toim., 10: 315-331.

- 1962: Über die Arten *Rhopalopyx preyssleri* (H.-S.) und *Rh. adumbrata* (C. R. Sahlberg) (Homoptera, Iassidae). – Notul. ent., 42: 62-66.
- 1965: Über die Zikadenfauna Altais. 144 pp. Tartu. (In Russian).
- 1968: Preliminary key for the identification of the nymphs of North European Homoptera Cicadina. – Annls Ent. Fenn., 34: 65-74.
- 1971: Eesti tirdid I. 284 pp. Tallinn.
- 1973: On North-European species of the genus *Limotettix* J. Sb., with notes on North-American species (Homoptera, Cicadellidae). – Eesti NSV Tead.Akad. Toim., 22: 197-209.
- 1974: Preliminary list of Homoptera-Cicadinea of Latvia and Lithuania. – Ibid., 23: 131-163.
- 1980: On the Homoptera-Cicadinea of Kamchatka. – Annls zool., 35: 367-418.
- 1982: Preliminary key for the identification of the nymphs of North European Homoptera Cicadinea. II: Cicadelloidea. – Annls zool. Fenn., 19: 1-20.
Vondraček, K., 1949: Contribution to the knowledge of the sound-producing apparatus in the males of leafhoppers. – Acta acad.Sci.nat. moravo-siles., 21: 1-36.
Wagner, W., 1935a: Die Zikaden der Nordmark und Nordwest-Deutschlands. – Verh. Ver. naturw. Heimatsforsch., 24: 1-44.
- 1935b: Beitrag zur Homopteren-Fauna Dänemarks und Beschreibung von drei neuen Varietäten aus der Gattung *Philaenus* Stål. – Ent. Meddr, 29: 162-171.
- 1937a: Neue Homoptera-Cicadina aus Norddeutschland. – Verh. Ver. naturw. Heimatsforsch., 25: 69-73.
- 1937b: Zur Synonymie der deutschen *Aphrodes*-Arten (Hem., Hom.). – Ibid., 26: 65-70.
- 1938: (Homopt. Delphacidae). Zur Synonymie der *Kelisia guttula* Germ. – Bombus, 1, Nr. 4: 12.
- 1939: Die Zikaden des Mainzer Beckens. Zugleich eine Revision der Kirschbaumschen Arten aus der Umgebung von Wiesbaden. – Jb. nassau. Ver. Naturk., 86: 77-212.
- 1940: Zwei neue Zikaden-Arten aus der Umgebung von Hamburg. – Verh. Ver. naturw. Heimatsforsch., 28: 110-113.
- 1941: Die Zikaden der Provinz Pommern. – Dohrniana, 20: 95-184.
- 1944: (Hom. Jassidae). Zwei neue *Oncopsis*. – Bombus, 26/29: 128-131.
- 1950: Die salicicolen *Macropsis*-Arten Nord- und Mitteleuropas. – Notul. ent., 30: 81-114.
- 1950b: *Balclutha boica* n.sp., eine neue Jasside aus Bayern. – Ber. naturf. Ges. Augsburg, 3: 97-100.
- 1951: Verzeichnis der bisher in Unterfranken gefundenen Zikaden. – Nachr. naturw.Mus. Aschaffenb., 33: 1-54.
- 1952: Bemerkungen zur Zikadenfauna des nördlichen Westdeutschlands. – Faun. Mitt. NDtl., 2: 2-4.
- 1955a: Neue mitteleuropäische Zikaden und Blattflöhe. – Ent. Mitt. Zool. StInst. zool.Mus. Hamburg, 1955: 163-194.
- 1955b: Die Bewertung morphologischer Merkmale in den unteren taxonomischen Kategorien, aufgezeigt an Beispielen aus der Taxonomie der Zikaden. – Mitt. hamb.zool.Mus.Inst., 53: 75-108.
- 1958: Über eine Zikaden-Ausbeute vom Grossen Belchen im Schwarzwald (Homoptera Auchenorrhyncha). – Ent. Mitt. zool. StInst. zool. Mus. Hamburg, 1958: 435-443.
- 1963: Dynamische Taxionomie, angewandt auf die Delphaciden Mitteleuropas. – Mitt. hamb. zool. Mus. Inst., 60: 111-180.
- 1964: Die auf Rosaceen lebenden *Macropsis*-Arten der Niederlande. – Ent. Ber., Amst., 24: 123-136.
- 1968: Die Grundlagen der dynamischen Taxonomie, zugleich ein Beitrag zur Structur der Phylogenese. – Abh. Verh. naturw.Ver.Hamburg (N. F.), 12 [1967]: 27-66.
- & Franz, 1961: Unterordnung Homoptera Überfamilie Auchenorrhyncha (Zikaden). – Die

Nordost-Alpen im Spiegel ihrer Landtierwelt, 2: 74-158. Innsbruck.

Wahlgren, E., 1915: Det öländska alvarets djurvärld I. – Ark. Zool., 9, 135 pp.

Wallengren, H. D. J., 1851: Hemiptera och Lepidoptera funna i nord-östra Skåne. – Öfvers.K. VetenskAkad. Förh., 1850: 252-254.

– 1870: Anteckningar i entomologi. – Ibid., 1870: 145-182.

Walsh, B. D., 1862: Fire blight. Two new foes of the apple and pear. – Prairie Farmer (n. s.), 10: 147-149.

Walter, S., 1975: Larvenformen mitteleuropäischer Euscelinen (Homoptera, Auchenorrhyncha). – Zool. Jb. (Syst.), 102: 241-302.

– 1978: Larvenformen mitteleuropäischer Euscelinen (Homoptera, Auchenorrhyncha) Teil II. – Ibid., 105: 102-130.

Weaver, C. R., & King, D. R., 1954: Meadow spittlebug. – Res. Bull. Ohio agric. exp.Sta., 741, 99 pp.

Wilson, M. R., 1978: Descriptions and key to the genera of the nymphs of British woodland Typhlocybinae (Homoptera). – Systematic Entomology, 3: 75-90.

Wüstnei, W., 1895: Beiträge zur Insektenfauna Schleswig-Holstein. Sechstes Stück. – Schr.naturw.Ver. Schlesw.-Holst., 10: 263-279.

Young, D. A., 1952 (including Young & Christian, 1952): A reclassification of Western Hemisphere Typhlocybinae (Homoptera, Cicadellidae). – Kans. Univ. Sci. Bull., 35 (1): 217 pp.

Zachvatkin, A. A., 1929: Description d'un nouvelle espèce du genre *Edwardsiana* Jaz. 1929 (Homoptera, Eupterygidae) des environs de Moskou. – Ént. Obozr., 23: 262-265.

– 1933: Sur quelques Homoptères intéressants de la faune Italienne. – Memorie Soc. ent. ital., 12: 262-272.

– 1935: Notes on the Homoptera-Cicadina of Jemen. – Uchen. Zap. mosk. gos. Univ., 4: 106-115.

– 1938: Un *Limotettix* nouveau des environs de Moscou. – Bull. Soc. Nat. Moscou (Sect.Biol.) (n.s.), 47: 285-287.

– 1946: Studies on the Homoptera of Turkey I-VII. – Trans. R. ent. Soc. Lond., 97: 148-176.

– 1947: Homoptera-Cicadina from north-western Persia. I. – Ént. Obozr., 28: 106-115.

– 1948a: Novye cikady (Homoptera-Cicadina) srednerusskoy fauny. – Nauch.-met. Zap. gl. Upr. Zap., Moskva, 11: 177-185. (In Russian).

– 1948b: Novye i maloizvestnye cicadiny iz okskogo zapovednika. – Ibid., 11: 186-197. (In Russian).

– 1953a: Biologo-sistematicheskie zametki o cikadinakh sredne-russkoy fauny. – Sbornik nauchnykh rabot. Moskva: 205-209. (In Russian).

– 1953b: Cicadiny okskogo gosudarstvennogo zapovednika. – Ibid., 211-223. (In Russian).

– 1953c: Cicadiny peskov astrakhanskogo zapovednika. – Ibid., 225-236. (In Russian).

– 1953d: Cicadiny, sobrannye V. S. Elpatevskim na beregakh ozera Knubsugul (B. Kosogol). – Ibid., 247-250. (In Russian).

Zetterstedt, J. W., 1828: Ordo III. Hemiptera. Fauna Insectorum Lapponica 1 (563 pp.): 513-550. Hammone.

– 1838-40: Insecta lapponica descripta. Sectio prima (1838): 1-314. Sectio quinta (1840): 1018-1140.

Ziegler, H., & Ziegler, I., 1958. – Z. vergl. Physiologie, 40: 549-555.

Index

Synonyms are given in italics. The number in bold refers to the main treatment of the taxon.

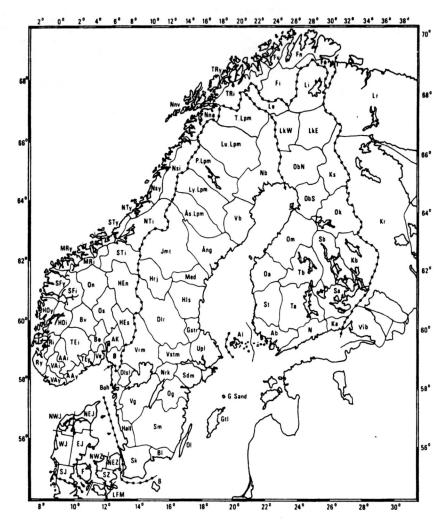

List of abbreviations for the provinces used throughout the text, on the map and in the following tables.

DENMARK

SJ	South Jutland	LFM	Lolland, Falster, Møn
EJ	East Jutland	SZ	South Zealand
WJ	West Jutland	NWZ	North West Zealand
NWJ	North West Jutland	NEZ	North East Zealand
NEJ	North East Jutland	B	Bornholm
F	Funen		

Printed in the United States
By Bookmasters